Lecture Notes in Chemistry

44

Supercomputer Simulations in Chemistry

Proceedings of the Symposium on Supercomputer Simulations in Chemistry, held in Montreal August 25−27, 1985, sponsored by IBM-Kingston and IBM-Canada

Edited by M. Dupuis

Springer-Verlag

Berlin Heidelberg New York London Paris Tokyo

Editor

M. Dupuis
IBM Corporation, Dept. 48B MS 428, Data Systems Division
Neighborhood Road, Kingston, New York 12401, USA

ISBN 3-540-17178-9 Springer-Verlag Berlin Heidelberg New York
ISBN 0-387-17178-9 Springer-Verlag New York Berlin Heidelberg

Printing and binding: Druckhaus Beltz, Hemsbach/Bergstr.
2152/3140-543210

Preface

Awareness of the need and potential of supercomputers for scientific and engineering research has grown tremendously in the past few years. It has culminated in the Supercomputer Initiative undertaken two years ago by the National Science Foundation and presently under full development in the United States. Similar initiatives are under way in several European countries and in Japan too. Thus the organization of a symposium on 'Supercomputer Simulations in Chemistry' appeared timely, and such a meeting was held in Montreal (Canada) in August 1985, sponsored by IBM-Kingston and IBM-Canada, and organized by Dr. Enrico Clementi and Dr. Michel Dupuis. In connection with this, IBM's support of the Cornell University Supercomputer Center, several projects in the IBM Research Division, the experimental parallel engine (lCAP) assembled at IBM-Kingston, and the announcement (Fall 1985) of an add-on vector feature to the 3090 IBM mainframe underscore IBM's commitment to high-end scientific/engineering computing.

The papers presented in this volume discuss topics in quantum mechanical and statistical mechanical simulations, both of which test the limits of computer hardware and software. Already a great deal of effort has been put into using vector supercomputers in these two areas. Much more is needed and, without doubt, is bound to happen. To start, an historical perspective of computational quantum chemistry is provided by Professor Löwdin. The contribution by Ohno and co-workers gives an indication of the present status of Japanese supercomputers. Kutzelnigg et al., Bauschlicher et al., and Guest et al. describe three significant efforts dealing with various aspects of programming quantum mechanical methods on vector supercomputers, and advances already observed in applying these methods to important problems of chemical physics, and surface science. Finally Truhlar et al. discuss the state-of-the-art in the field of chemical dynamics where today's studies contribute significantly to the detailed understanding of gas phase collision dynamics.

An alternative to vector supercomputers is discussed by Clementi et al., in their presentation of the loosely coupled array of processors (lCAP). This system is ideally suited to quantum and statistical simulations. Lie and et al., and Scheraga et al. present two studies carried out on the lCAP system. Indeed the approach to the determination of the three-dimensional structure of proteins is CPU intensive and inherently 'parallelizable.'

Statistical simulations were born with the advent of electronic computers. The state-of-the-art is reviewed by Alder and various examples are presented by Catlow et al., Lie et al., and Heinzinger et al. in the area of solid state chemistry and solution chemistry. The contribution by Lester et al. contains Quantum Monte-Carlo studies of molecular systems, some of which were carried out on a parallel computer also.

One of the ever present difficulties in quantum mechanical applications is the Input/-Output of data. The same hurdle, maybe even higher, exists in the field of High-Energy Physics. The problem is exacerbated by the ever increasing speed of CPU's. Freitag reports on the present and future of data storage. Finally Butscher gives an overview of supercomputing in the seismic industry. Seismic applications benefit from vector hardware, and also from parallel processing as shown by Butscher.

Without doubt the combination of fast scalar, vector and parallel processing already existing and more and more available will lead to significant scientific discoveries. This is an exciting time for computational chemists. Says Prof. Kenneth G. Wilson, Nobel laureate: "If you look at the opportunities being brought on by the computer revolution... nothing compares with this period of history but The Renaissance." We can only wonder on the foresight of Prof. Robert S. Mulliken, who in his 1966 Nobel Lecture stated: "... I would like to emphasize strongly my belief that the era of computing chemists, when hundreds if not thousands of chemists will go to the computing machine instead of the laboratory for increasingly many facets of chemical information, is already at hand."

Acknowledgement: It is a pleasure to acknowledge the help of Ms. Alberta Martino and Mr. Ray Thiede in the organization of the symposium, and Ms. Maria Van Duyne for her careful preparation of this volume.

Michel Dupuis
IBM Corporation
Kingston, N.Y. USA

Contents

SOME ASPECTS ON THE HISTORY OF COMPUTATIONAL QUANTUM CHEMISTRY IN VIEW OF THE DEVELOPMENT OF THE SUPERCOMPUTERS AND LARGE-SCALE PARALLEL COMPUTERS

P.O. Löwdin

Quantum Theory Project
Departments of Chemistry and Physics
university of Florida
Gainesville, Florida 32611

and

Professor Emeritus of Quantum Chemistry
Uppsala University
Uppsala, Sweden
NFCR Senior Investigator

1. Classical Quantum Mechanics

When Max Planck[1] introduced the quantum postulate $E = h\nu$ in 1900, the main purpose was to give a theoretical description of the black-body radiation in agreement with the experimental experience. When Albert Einstein[2] in 1905 introduced the quantization of the electromagnetic waves according to the law $E = h\nu$, the idea was to give a theoretical explanation of the photo-electric effect. When Niels Bohr[3] in 1913 developed the first model of the hydrogen atom based on the quantization of angular momenta, $pa = nh/2\pi$, the success was guaranteed by the fact that his theoretical results were in agreement with the experimental spectra. Somewhat later Sommerfeld[4] introduced the three quantum numbers (n, l, m) describing the elliptical orbits, and in 1918 he could successfully explain the fine-structure of the hydrogen atom spectra by using the special theory of relativity and the fine-structure constant $\alpha = 2\pi e^2/hc = 1/137$.

In chemistry, Lewis[5] had used the new electronic atomic model to introduce the concepts of the inert rare-gas shells and the valence electrons and the idea that the covalent chemical bond is represented by an electron-pair shared between two atoms. Using the concept of the electron lone-pair, he formulated a new basic-acid theory, whereas Huggins[6] in 1918 used it to develop the first theory of the hydrogen bond: a proton shared between two electron lone-pairs.

Using the properties of the rare gas shells and the magic numbers $Z = 2, 8, 8, 18, 18, 32, \dots$, which are all double squares $2n^2$, Pauli[7] formulated the famous "exclusion principle." By means of the three quantum numbers (n, l, m), the "Aufbau Principle," Pauli's rule, and the rule of thumb that the electronic orbitals have energies arranged after increasing $(n + l; n)$, Bohr[8] could finally explain the repetition of the numbers in the "magic series" and the overall structure of the period system of the chemical elements, which was a great triumph, of course.

It should be observed, however, that, even if the classical quantum theory developed in the years 1910-1923 was a great conceptual success, it could not quantitatively describe the properties of a system having more than one electron--something was definitely missing.

2. Development of Modern Quantum Theory

In classical quantum mechanics, the electron was considered as a particle with definite mass m and charge e. In 1924, Louis de Broglie introduced the idea that if electromagnetic waves could be quantized, perhaps all "quanta" would also be waves. Combining Planck's Law $E = h\nu$ with Einstein's energy-mass relation $E = mc^2$ and the standard relation $c = \nu\lambda$, he obtained for the momentum p of a light-quantum or photon: $p = mc = mc^2/c = E/c = h\nu/\lambda\nu = h/\lambda$. He could show that the two relations

$$E = h\nu \ , \qquad p = h/\lambda \tag{2.1}$$

were relativistically invariant and further that, if this idea was applied to an electron in the hydrogen atom moving in a Bohr circle with radius a, Bohr's law for the quantization of the angular momentum $pa = nh/2\pi$ would take the simple form $2\pi a = n\lambda$. This means that Bohr's stationary states would correspond to "standing waves" with an integer number of waves around the circle. In this way, de Broglie introduced the idea of "wave-corpuscle parallelism" into modern physics.

In classical physics, the field of "wave dynamics" had been rather well developed since the early 1800's with important applications to ocean waves, thermal waves, elastic waves, electromagnetic waves, etc., and --e.g. from the theory of music instruments--it was well-known that standing waves associated with the so-called eigen-frequencies of the instruments would mathematically correspond to eigenvalue problems.[9] It was still somewhat surprising when Erwin Schrödinger[10] in 1925 could apply de Broglie's ideas to the interior of the hydrogen atom and develop a "wave mechanics," in which Bohr's classical circle was replaced by a spherical charge distribution and Sommerfelds' ellipses by more complicated charge distributions corresponding to dumbbells, etc. The new theory gave not only the three quantum numbers (n, l, m) in a correct form with new interpretations of the quantum number $l = 0, 1, 2, \ldots$ n-1, but it also permitted the evaluation of spectral intensities. There was hence no question that the new theory was superior to the classical quantum mechanics.

It was perhaps even more surprising that, in 1925, there were three different and independent formulations of a new quantum theory. In addition to Schrödinger's *wave mechanics*, in which quantities like the position x and momentum p of a "particle" were considered as *operators*, Heisenberg, Born and Jordan[11] formulated a matrix mechanics, in which these quantities were interpreted as *matrices*, whereas finally Dirac[12] formulated a *q-number theory*, in which the non-commutative quantities x and p were left undefined and more or less *contentless*. The only thing the three approaches seemed to have in common was the validity of the commutation relation

$$px - xp = \frac{h}{2\pi i} \cdot 1 \ , \tag{2.2}$$

but it was soon proven[13] that, even if the three approaches would have completely different interpretations, they would give essentially the same physical results as expressed in numbers.

In Niels Bohr's Copenhagen school, it was emphasized that physics should primarily deal with measurable quantities, and the measurement itself--including the object, the apparatus, and the observer--became of fundamental importance. Heisenberg analyzed a thought-experiment, in which one measured the position x of a particle by means of a microscope having the open angle 2α. He stressed that, in order for the observer to *see* the particle, one would need at least one photon of light having the momentum $p = h/\lambda$. The photon would then be scattered into the microscope inside a cone with opening 2α with all ray contributing the same image point, and the uncertainties in the direction of the vector $\bar{p}$ having the length $p = h/\lambda$ would then leave the electron with an uncertainty Δp in its momentum given by the formula $\Delta p = 2 p \sin\alpha$. This uncertainty could be diminished by diminishing the angle α but for very small α one would then have diffraction phenomena, which would influence the resolution power of the microscope leading to an uncertainty Δx in the position corresponding to the formula $\Delta x \sin \alpha = \lambda/2$. Hence one has $\Delta p \cdot \Delta x \sin \alpha = 2p \sin \alpha$. $\lambda/2 = h \sin \alpha$, which gives Heisenberg's *uncertainty relation*[14]

$$\Delta p \cdot \Delta x \sim h \ . \tag{2.3}$$

In classical physics, one could simultaneously measure the position x and the momentum $p = m\, dx/dt$ of a particle at the time t by following it in its orbit. Since this would be impossible in the new quantum theory, on would have to abandon the concept of the *classical orbit*-- something which was strongly opposed by Einstein. Heisenberg[15] realized that this would have fundamental implications for two- and many-electron systems. If one considers a two-electron system, as e.g. the helium atom, one could in classical physics follow the two electrons 1 and 2 in their orbits and identify them at all times, whereas--in the new quantum theory--this would be impossible. According to Heisenberg, this implies that if one tries to label the particles, all physical results must be independent of the labeling, so that

$$F(1,2) = F(2,1) \ , \tag{2.4}$$

i.e. they are invariant under permutations of the indices. This introduced a new symmetry principle into physics which became the basis for the *exchange phenomenon*.

Heisenberg showed that there exists two types of states of the helium atom, which he called ortho- and para-helium. Suddenly there existed a completely new possibility to treat physical systems with more than one electron, and the question was whether one had really broken through the barrier which had limited the classical quantum theory.

3. Many-Electron Schrödinger Equation in Configuration Space

Even if Lewis[5] had shown that the covalent chemical bond corresponded to an electron pair shared between two atoms and obtained many important qualitative chemical results in this way, no one--not even Bohr[16] had obtained any meaningful quantitative results as to the properties of the bond. In 1927, Heitler and London[17] introduced the new exchange phenomenon (2.4) into a study of the ground state of a hydrogen molecule H_2 consisting of two hydrogen atoms, a and b. Denoting the electronic coordinates by r_1 and r_2, and plotting the energy as a function of the internuclear coordinate R, they could show that a wave function of the type a(1)b(2) would give only a shallow minimum in the wrong place, whereas a symmetrized wave function

$$\frac{1}{2}\left[a(1)b(2) + b(1)a(2)\right] \tag{3.1}$$

has a deep minimum corresponding approximately to the bond energy at the correct distance $R = R_0$. Even if they did not carry out the calculation in great deal, the Heitler-London paper represents nevertheless the birth of the field of *quantum chemistry*.

In order to treat a system of particles k with mass m_k and charge e_l in Schrödinger's wave mechanics, one used a *configuration space* in which each point is represented by all particle coordinates and a Hamiltonian H of the form:

$$H = \sum_{k} \frac{\vec{p}_k^2}{2m_k} + \sum_{k<l} \frac{e_k e_l}{r_{kl}} , \tag{3.2}$$

where $\vec{p}_k = (h/2\pi i)\vec{\nabla}_k$, which is simply the sum of the kinetic energy and the Coulomb energy. The time-dependent Schrödinger equation for the wavefunction Ψ has then the form

$$-\frac{h}{2\pi i}\frac{\partial\Psi}{\partial t} = H\Psi \tag{3.3}$$

whereas the stationary states are determined by the solutions to the eigenvalue problem:

$$H\Psi = E\Psi , \tag{3.4}$$

subject to the boundary conditions that--for closed states--the wave function is absolutely quadratically integrable (L^2), and--for scattering states-- the absolute value should stay finite almost everywhere.

If all the particles are identical ($m_k = m$, $e_k = e$), the Hamiltonian is symmetric in all particle indices, and if P is an arbitrary permutation of these indices, on has

$$PH = HP , \tag{3.5}$$

i.e. the permutations P are constants of motion. This is a more precise mathematical formulation of the exchange phenomenon (2.4).

The big question was now whether this new idea of using a multi-dimensional configuration was really valid, i.e. if the theoretical results obtained by solving (3.4) would agree with experimental experience. The first test was carried by the Norwegian theoretical physicist, Egil Hylleraas, in 1928 when working in Göttingen, Germany. Hylleraas[18] calculated the ground state energy of the helium atom by using a correlated wave function Ψ, which contained also the interelectronic distance r_{12}, and his theoretical result agreed very well with the experimental one--of course, he had to make careful estimates of the relativistic corrections, since relativity is automatically built into nature. It was a great triumph for the new theory, and no one doubted that one could now go from N = 2 to any arbitrary N-value, including N = 6.10^{23} --even if this is somewhat drastic extrapolation.

Egil Hylleraas may be considered the father of *computational* quantum physics and quantum chemistry, and it was hence natural that the 1963 Sanibel Symposium on Quantum Chemistry and Solid-State Theory was organized in his honor; the impact of Hylleraas' work[19] on our field is rather well-reflected in the proceedings[20] from this conference.

In their study of the H_2-molecule, Heitler and London[17] had only estimated the value of a quantity now called the exchange integral:

$$(ab \mid ba) = \int \frac{a(1)b(1)a(2)b(2)}{r_{12}} \, dv_1 \, dv_2 \ , \tag{3.6}$$

and the first calculation was carried out in 1928 by Sugiura.[21] The result showed that the Heitler-London theory was good enough to give an explanation of the nature of the covalent chemical bond, but that the wave (3.1) was not good enough to give any details. A more accurate ground state energy in excellent agreement with the experimental result was rendered in 1933 by James and Coolidge,[23] who used a symmetric correlated wave function of the Hylleraas type. However, during the period 1928-1933, there were also many interesting developments in the fundamental theory.

First of all, the *electronic spin* had been established by Goudsmit and Uhlenbeck.[23] Let us introduce the spin-coordinate $\zeta = \pm 1/2$ and the two spin-functions $\alpha = \alpha(\zeta)$ = $\binom{1}{0}$ and $\beta = \beta(\zeta) = \binom{0}{1}$. It was realized that, in order to fulfill the Pauli exclusion principle, the total space-spin wave function ought to be *antisymmetric*. For two electrons, there is obviously only one antisymmetric spin function $(\alpha_1\beta_2 - \beta_1\alpha_2)$ but three symmetric ones: $\alpha_1\alpha_2$, $(\alpha_1\beta_2 + \beta_1\alpha_2)$, $\beta_1\beta_2$. Since the space function (3.1) is symmetric, the antisymmetric product

$$^1\Psi = \frac{1}{2}[a(1)b(2) + b(1)a(2)](\alpha_1\beta_2 - \beta_1\alpha_2) \tag{3.7}$$

is apparently a spin singlet (S = 0), whereas the product functions

$$^3\Psi = \frac{1}{2} \left[a(1)b(2) - b(1)a(2) \right] \begin{bmatrix} \alpha_1\alpha_2 \\ \frac{1}{2}(\alpha_1\beta_2 + \beta_1\alpha_2) \\ \beta_1\beta_2 \end{bmatrix} \tag{3.8}$$

represent a spin triplet (S = 1).

As a result of considerations of this type, Heisenberg[26] could in 1928 construct the first model of *ferromagnetism* based on the sign of the so-called "exchange integral":

$$J = \frac{1}{2} \left({}^1E - {}^3E \right) \ , \tag{3.9}$$

which is essentially half the difference between the singlet and triplet energies. if J > 0, then the triplet energy is lower than the singlet energy, the spins align in a parallel fashion, and one has a "ferromagnetic" ground state. On the other hand, if J < 0, the singlet energy is lowest, and the spins align in an antiparallel way; one then has a model of what today is called "antiferromagnetism". Even if it is easy to evaluate the exchange integral in terms of the approximate wave functions (3.7) and (3.8), it took a remarkably long time before all the relevant terms were included.[25]

Using the exchange integral (3.9), one can now write the energy of the singlet and triplet states in the form

$$E = \frac{1}{2} \left({}^1E + {}^3E \right) \pm \frac{1}{2} \left({}^1E - {}^3E \right) = \frac{1}{2} \left({}^1E + {}^3E \right) + \eta J \ , \tag{3.10}$$

where $\eta = +1$ for the singlet and $\eta = -1$ for the triplet. If $\vec{S} = \vec{s}_1 + \vec{s}_2$ is the total spin, its square $(\vec{S})^2 = 3/2 + 2\vec{s}_1\vec{s}_2$ has the eigenvalues $v = S(S + 1)$, i.e. one has $v = 0$ for the singlet (S = 0) and $v = 2$ for the triplet (S = 1). This gives the relation $\eta = 1 - v = -(1/2 + 2\vec{s}_1\vec{s}_2)$ and further

$$E = \frac{{}^1E + 3\,{}^3E}{4} - 2J\vec{s}_1\vec{s}_2 \ . \tag{3.11}$$

This formula, which is exact for N = 2, forms the basis for the vector-coupling model, which still plays a fundamental role in spin lattice theory.

In 1929, John C. Slater[26] introduced the idea that by using a combined space-spin coordinate $x_k = (r_k, \zeta_k)$, one could write the total wave function in the form

$$\Psi = \Psi(x_1, x_2, \dots x_N) \ , \tag{3.12}$$

and the Pauli exclusion principle would then correspond to the antisymmetry condition:

$$P\Psi = (-1)^P \Psi \ , \tag{3.13}$$

where P is any permutation of the coordinates $X = (x_1, x_2, ...x_N)$ and p its parity. In this connection, the binary product

$$< \Psi_1 \mid \Psi_2 > = \int \Psi_1^*(X)\Psi_2(X)dX \tag{3.14}$$

had to be extended to include not only integrations over the space coordinates r_1, r_2, ...r_N but also summations over the spins ζ_1, ζ_2, ...ζ_N If $\phi = \phi (x_1, x_2, ...x_N)$ is an arbitrary trial wave function, the expectation value $<H>$ of the energy is calculated by using the expression:

$$< H > = \frac{< \Phi \mid H \mid \Phi >}{< \Phi \mid \Phi >} \tag{3.15}$$

Today we take all these "standard expressions" for granted, but they are really the results of a fairly long historical development.

4. Conceptual Development of Quantum Chemistry

The Independent Particle Model and the Hartree-Fock Scheme

In the independent-particle-model (IPM) for treating a many-particle system suggested by Bohr[27] in 1923, each particle moves in the external field and the average field of all the other particles in the system. The model was taken into modern atomic theory in 1928 by Hartree,[28] who intuitively constructed a one-particle Schrödinger equation to be solved iteratively by a "self- consistent-field" (SCF) procedure and succeeded in describing the structure of many-electron atoms in this way.

In the treatment of molecular systems, the problem of the motion of the atomic nuclei turned out to be very difficult. However, since the mass of each nucleus is so much larger than the mass of the electrons, one can in a first approximation,[29] assume that they are all infinitely heavy and situated in fixed positions. Denoting the atomic number of the nucleus g by Z_g, the Hammiltonian (3.2) takes the simplified form:

$$H = H_{(0)} + \sum_i H_i + \sum_{i<j} H_{ij} \ , \tag{4.1}$$

where

$$H_{(0)} = e^2 \sum_{g<h} \frac{Z_g Z_h}{r_{gh}} \tag{4.2}$$

is the Coulomb repulsion of the nuclei,

$$H_i = \frac{\vec{p_i}^2}{2m} - e^2 \sum_g \frac{Z_g}{r_{ig}} \qquad (4.3)$$

is the sum of the kinetic energy of electron i and its potential energy in the nuclear framework, and

$$H_{ij} = \frac{e^2}{r_{ij}} \qquad (4.4)$$

represents the interelectronic Coulomb repulsion. We note that the indices i and j go over the electrons, whereas the indices g and h go over the atomic nuclei.

In 1930, Slater[30] showed that the Hartree equation could be derived from the Hamiltonian (4.1) and the variation principle

$$\delta <H> = 0 \ , \qquad (4.5)$$

provided that the total wave function Φ is approximated by a product of one-electron functions. Slater and Fock[31] independently showed further that, if the total wave function Φ is approximated by a single determinant

$$\Phi = (N!)^{-1/2} \ | \ \psi_1, \psi_2, \dots \psi_N \ | \ , \qquad (4.6)$$

formed by N one-electron functions $\psi_k(x)$ for k = 1,2, ...N, then the determinant is essentially invariant under linear transformations of these functions, which may hence be chosen orthonormal. It is easily shown that the determinant is determined by a projector ρ having the kernel:

$$\rho(x_1, x_2) = \sum_{k=1} \psi_k(x_1)\psi_k^*(x_2) \qquad (4.7)$$

and the properties

$$\rho^2 = \rho, \ \ \rho^\dagger = \rho, \ \ Tr \ \rho = N \ . \qquad (4.8)$$

Using the variation principle (4.5), one is then led to a system of N non-linear coupled integro-differential equations of the form:

$$H_{eff}(1) \ \psi_k(x_1) = \sum_{l=1}^{N} \psi_l \ \lambda_{lk} \ , \qquad (4.9)$$

where the coefficients $\lambda = \{\lambda_{lk}\}$ form a matrix of "Lagrangian multipliers" and the effective Hamiltonian H_{eff} has the form:

$$H_{eff}(1) = H_1 + e^2 \int dx_2 \; \frac{\rho(x_2, x_2) - \rho(x_1, x_2)P_{12}}{r_{12}} \; , \tag{4.10}$$

here P_{12} is a permutation operator having the property $P_{12} f(x_1) = f(x_2)$ for an arbitrary function f. We note that, according to Dirac,[32] this effective Hamiltonian is self-adjoint and may be interpreted in terms of the independent-particle-model, and that the very last term containing the operator P_{12} is the so-called "exchange term," which subtracts the self-interaction and takes care of the effect of the Pauli exclusion principle.

Introducing the unitary transformation U, which brings the self-adjoint matrix λ to diagonal form ε, so that $U^\dagger \lambda U = \varepsilon$, and the transformed functions $\psi = \psi U$, one may initially bring the equations (4.9) to the special form:

$$H_{eff}(1)\psi_k(x_1) = \varepsilon_k \psi_k(x_1) \tag{4.11}$$

which is analogous to a one-particle Schrödinger equation; here the eigenvalues ε_k are interpreted as the one-electron energies.

We note that the relation (4.6) defines a *Slater determinant,* that the kernel $\rho(x_1, x_2)$ in (4.7) is referred as the *Fock-Dirac density matrix,* that the relations (4.9) and (4.11) are known as the *Hartree-Fock equations,* and that the solutions $\psi_k(x_1)$ to the eigenvalue problems (4.11) are sometimes called the *canonical* Hartree-Fock functions; it should be observed that there are no non-diagonal Lagrangian multipliers $\lambda_{kl} \neq 0$ for $k \neq l$ in the relations (4.11).

In order to solve the SCF-equations without exchange in the atomic problem, Hartree had worked out a numerical procedure for solving the one-dimensional radial Schrödinger equation. The first example of a numerical solution of the Hartree-Fock equations (4.9) was given by Fock and Petrashen[34] for N = 3 in an application to the ground state of the Li-atom in 1932. Since they wanted to keep the ^{2}S symmetry of the configuration $(1s)^2 2s$, they found somewhat surprising that there was a non-diagonal Lagragian multiplier $\lambda_{12} \neq 0$ which could not be transformed away. We will comment about the effect of such symmetry restrictions in greater detail at a later stage.

In 1935 Koopmans[35] showed that the one-electron energies ε_k could be interpreted as approximate *ionization energies.* It should perhaps be mentioned that Tjalling Koopmans later went into economy and shared the 1975 Prize in Economic Sciences in Memory of Alfred Nobel together with Leonid Katorovich "for their contributions to the theory of optimum allocation of resources." It is evident that the variation principle has rather general fields of applications.

For a more complete review of the early developments in the Hartree-Fock method, the reader is referred to Hartree's book[36] and for a review of the numerical integration methods used elsewhere.[37]

Molecular Orbitals, Hybridization and the Hückel Scheme for Conjugated Systems.

In the original independent-particle-model,[25] the electrons were moving in classical orbits, which--in the transfer to the new quantum theory--had to be replaced by one-electron functions describing three-dimensional charge or probability amplitudes $\psi = \psi(r)$, for which Mulliken[38] suggested the new name *orbital*. In 1928, Mulliken and Hund[39] had introduced the idea that, in many molecules, the valence electrons were not localized in the covalent bonds but distributed all over the molecule in what was called *molecular orbitals* (MO's). These could conveniently be formed by linear combinations of atomic orbitals (LCAO's), and the MO-LCAO method became soon one of the cornerstones of the new quantum chemistry.

The simplest example is provided by the H_2-molecule with the two atomic orbitals, a and b, in which the combinations $\phi = a + b$ and $\chi = a - b$ represent *molecular orbitals* of a bonding respective non-bonding nature. It is evident that a wave function of the type

$$\Phi(1, 2) = \phi(1)\phi(2) \tag{4.12}$$

automatically satisfies the symmetry requirement, and that i represents a covalent bond, since the highest electron density is *between* the two atomic nuclei, whereas the wave function $\Xi\,(1,2) = \chi(1)\chi(2)$ is symmetric but has a node between the nuclei.

It should be observed that, in 1928, Bloch[40] had introduced similar ideas into solid-state theory, in which the electrons would move all over the crystal in what we may now call *Block orbitals;* this paper forms the basis for the modern *band theory of solids.*

In certain molecular systems with high symmetry or in crystals with translational symmetry, the molecular orbitals may often be constructed by symmetry considerations by using the projection operators P_{kk}^z associated with various irreducible representations. For molecular systems in general, however, the molecular orbitals would have to be calculated by using some form of the independent-particle-model. It would take until the late 1940's until the MO-LCAO method was combined properly with the Hartree-Fock approximation and computational schemes could be worked out in detail, and we will return to this questions in a later section.

In 1930, Slater[41] compared the energies of the H_2-molecule obtained in the Heitler-London approximation based on the so-called *valence bond* (VB) function (3.7) or, in abbreviated form:

$$\Psi(1, 2) = \frac{1}{2}\,(ab + ba)\,(\alpha\beta - \beta\alpha) \tag{4.13}$$

and the results obtained from the MO-function

$$\Phi(1, 2) = \frac{1}{2}\,(aa + ab + ba + bb)\,(\alpha\beta - \beta\alpha)\ . \tag{4.14}$$

He pointed out that the latter wave function, because of the large amount of "ionic character" represented by the terms $a_1 a_2$ and $b_1 b_2$, would have a completely wrong asymptotic behavior for separated atoms ($R = \infty$), and that this defect would influence the entire Hartree-Fock approximation, which would never be a good basis for any theory of magnetic properties. He also emphasized the fundamental role of the *overlap integrals*

$$S_{ab} = \int a^*(1)b(1) \; dv_1 \; , \qquad\qquad (4.15)$$

associated with atomic orbitals, a and b, on different atoms, and the need to take them into full account. He pointed out that, for a solid, the energy (3.15) may take the form ∞ / ∞ as the volume increases and the question of whether there would be a "non-orthogonality catastrophe" in solid-state theory was intensely discussed in the literature.[42] We will return to this problem in a later section.

In using atomic orbitals (AO's) of s-, p-, d-type, etc., it was early realized that the s-orbital corresponds to a spherically symmetric electronic distribution, whereas the p-orbital would correspond to a dumb-bell consisting of two spheres having different signs. In 1932, it was discovered by Slater[43] and by Pauling[44] that, if one formed a superposition of "hybrid orbital" of the form h = as + bp, the two parts with the same sign would enhance each other, and one would get an orbital with a specific extension in space corresponding to a "directed valency;" here the coefficients a and b are assumed to be positive and satisfying the relation $a^2 + b^2 = 1$. The hybrid, which has its extension in the direction of the unit vector $\vec{n} = (\cos \alpha, \; \cos \beta, \; \cos \gamma)$ has the specific form:

$$H = as + b(p_x \cos \alpha + p_y \cos \beta + p_z \cos \gamma) \; . \qquad\qquad (4.16)$$

From the four orthonormal atomic orbitals (s, p_x, p_y, p_z), one can by linear combination form four orthonormal hybrids (h_1, h_2, h_3, h_4). The orthogonality condition gives directly:

$$< h_1 \mid h_2 > = a_1 a_2 + b_1 b_2 \; (\cos \alpha_1 \cos \alpha_2 + \; \cos \beta_1 \cos \beta_2 + \; \cos \gamma_1 \cos \gamma_2)$$

$$= a_1 a_2 + b_1 b_2 \cos \gamma_{12} = 0 \; , \qquad\qquad (4.17)$$

where γ_{12} is the geometrical angle between the two hybrids. Introducing the quantity $n = b^2/a^2$ and the symbol $h = sp^n$, one gets then:

$$\cos \gamma_{12} = - \frac{1}{\sqrt{n_1} \sqrt{n_2}} \; . \qquad\qquad (4.18)$$

Since the total amount of s-character has to be preserved, one has further the auxiliary condition[45]

$$\sum_{k=1}^{4} a_k^2 = \sum_{k=1}^{4} \frac{1}{n_k + 1} = 1 \qquad (4.19)$$

It is easily seen that it is possible to construct four orthonormal hybrids sp^{n_1}, sp^{n_2}, sp^{n_3} and sp^{n_4}, as long as the four numbers n_1, n_2, n_3 and n_4 satisfy the relation (4.19). The most popular hybrids are, of course, the four sp³ hybrids corresponding to the tetrahedral arrangement and the three sp² hybrids forming 120° angles with each other in a plane supplemented by a fourth hybrid $sp^{\infty} = \pi$, which is a p-orbital perpendicular to the plane.

In his studies of the *conjugated systems*, which are planar organic molecules having alternating single and double bonds, as e.g., the aromatic hydrocarbons, Hückel[46] in 1933 introduced the idea that the single bonds in the molecular plane are associated with carbon sp² hybrids, whereas the double bonds are associated with electrons in molecular orbitals formed by linear combinations of the atomic π-orbitals of the carbon atoms involved. Taking the benzene molecule an example, one has six π-orbitals ϕ_μ with $\mu = 1,2,...6$, and one can then construct six linearly independent molecular orbitals:

$$\psi_k = \sum_{\mu=1}^{6} \phi_\mu C_{\mu k} \quad . \qquad (4.20)$$

The coefficients $\mathbf{C} = \{C_{\mu k}\}$ are often determined by pure symmetry considerations. In principle, they should be derived by solving a one-electron Schrödinger equation:

$$H_{eff}\,\psi_k = \varepsilon_k \psi_k \quad , \qquad (4.21)$$

where H_{eff} is an intuitively constructed "effective one-electron Hamiltonian." Using the variation principle, one obtains instead of (4.21) a system of linear equations

$$\mathbf{HC} = \varepsilon \Delta C \quad , \qquad (4.22)$$

where the hamiltonian matrix $\mathbf{H} = \{H_{\mu\nu}\}$ and the metric matrix $\Delta = \{\Delta_{\mu\nu}\}$ have the following elements, respectively:

$$H_{\mu\nu} = <\phi_\mu \mid H_{eff} \mid \phi_\nu> ; \ \Delta_{\mu\nu} = <\phi_\mu \mid \phi_\nu> \quad . \qquad (4.23)$$

For the matrix elements, Hückel then introduced the following simplifying assumptions:

$$H_{\mu\mu} = \alpha, \ H_{\mu\nu} = \begin{cases} \beta, \text{ for nearest neighbors} \\ 0, \text{ otherwise} \end{cases} \qquad (4.24)$$

$$\Delta_{\mu\mu} = 1, \ \Delta_{\mu\nu} = \begin{cases} S, \text{ for nearest neighbors} \\ 0, \text{ otherwise} \end{cases} \qquad (4.25)$$

where α and β are semi-empirical parameters to be determined from known experimental data. In the original Hückel scheme, the overlap integral S was often neglected, i.e., one put S = 0.

In his treatment of the benzene molecule, Hückel found that he had a total of six electrons available for the six molecular orbitals ψ_k (k = 1, 2, ...6). He found further that, if he put three pairs of electrons with antiparallel spins in the molecular orbitals having the lowest energies, then there would be three molecular orbitals left unoccupied. Hence one distinguishes between occupied orbitals denoted by the symbol π and the unoccupied orbitals denoted by the symbol π^*; it was evident that the transitions $\pi \rightarrow \pi^*$ would be essential for the description of the electronic spectra of the conjugated systems. In this description, the six electrons in the π-orbitals would be mobile electrons travelling around the benzene ring, and it was clear that these "delocalized" electrons in some way corresponded to the various arrangements of the "double bonds" in the classical Kekule and Dewar structures.

In 1932, Hückel[47] and somewhat later Pauling and Wheland[48] used the new approach to study the "directing power" of various substituents in conjugated systems and particularly to explain the so-called "alternating effects."

It was also shown that the results of the original Hückel scheme with S = 0 were not qualitatively changed if the effect of a non-vanishing overlap integral S was incorporated into the theory.[49] In this connection the important concept of the "topological matrix," having 1's for the elements associated with nearest neighboring carbon atoms and 0's everywhere else, was introduced into the theory. It is still of importance in the so-called graph-theoretical considerations.

The Hückel method became one of the most powerful theoretical tools in the study of the conjugated systems in inorganic chemistry--at least qualitatively. The development showed, however, that the semi-empirical parameters α and β which were good for describing the ground state of a specific molecule were not particularly good for describing the excited electronic states of the same molecule, and the quantitative theory of electronic spectra based on this approach came for some time to a standstill. It is hence remarkable that, even if the Hückel method was only partly quantitatively successful, it turned out to be extremely useful qualitatively as a basis for new concepts. To illustrate this point, I will here temporarily break the historical description and briefly review some of the later developments leading up to the situation of today.

In the 1950's, Kenichi Fukui[50] and his collaborators introduced the idea that the electrons, which are most likely to be involved in various chemical reactions, are the so-called "frontier electrons" which have the highest molecular orbital energies. This means that the frontier electrons are situated in the highest occupied molecular orbital (HOMO), whereas there are no electrons in the lowest empty molecular orbital (LEMO). The concepts of the frontier electrons, the HOMO- and LEMO-energies, and the HOMO-LEMO energy gap was going to stimulate the publication of hundreds of papers in theoretical organic chemistry.

As an example, we will consider two molecules, A and B, where the HOMO of A is energetically higher than the LEMO of B. If these two molecules are brought into contact, one or two electrons from A will automatically go into B, and one has a *charge-transfer reaction* with A as a donor and B as an acceptor. This idea has been utilized by, e.g., the Pullmans[51] in their studies of quantum biochemistry.

During the last two decades, the idea of the frontier electrons has been used in a particularly fruitful way in organic chemistry in the formulation of the so-called Woodward-Hoffman rules.[52] It is evident that even if one has great difficulties in understanding and interpreting the "phases" of the wave functions in quantum theory, the *signs* of the atomic orbitals used in hybridization or in the formation of molecular orbitals play a fundamental role in the study of the formation and breaking of chemical bonds. The simplest example is provided by the hydrogen molecule with two positive atomic orbitals, a and b, where the combination $\phi = a + b$ is a bonding orbital, whereas the combination $\chi = a - b$ is an anti-bonding orbital. A characteristic feature of the π-orbitals of all conjugated systems is that they are antisymmetric with respect to the molecular plane, in which they have a node, and that they change signs under reflection in this plane. In studying the possible reactions between two conjugated systems or within a single system undergoing rotations, Woodward and Hoffman examined the "sign pattern" of the frontier electrons in the HOMO's involved and formulated the rule that, if the frontier orbitals may be combined into a "bonding orbital" in the product, the reaction is usually permitted, whereas, if they are combined into an anti-bonding orbital, the reaction is forbidden.

These simple conceptual considerations have been of such fundamental importance for the experimental organic chemists that together Fukui and Hoffman were awarded the 1981 Nobel Prize in Chemistry. Today the Woodward-Hoffman rules are an important "rule of thumb" in organic chemistry, but it still remains for the computational quantum chemists to work out the finer quantitative details connected with these rules--in fact, this problem should be a great challenge for the computational audience gathered here.

The Hückel scheme is, of course, a special case of the general molecular-orbital method, and the computational aspects of the MO-LCAO approach will be reviewed in greater detail in a later section.

Valence Bond Method and the Idea of Chemical Resonance.

In the Heitler-London description of the hydrogen molecule, the total singlet wave function for the covalent bond a - b is given by the expressions (3.7), (4.13) or

$$(ab + ba)\,(\alpha\beta - \beta\alpha) \quad . \tag{4.26}$$

In the years 1928-32, the question was how one could generalize this valence bond approach to a many-electron molecule with several covalent chemical bonds a - b, c - d, e - f, ... , and important contributions were made by several authors.[53] It finally became clear that the singlet valence bond (VB) wave function Ψ should be written in the form:

$$\Psi = O_{AS}(ab + ba)\,(cd + dc)\,(ef + fe) \ \dots \ (\alpha\beta - \beta\alpha)\,(\alpha\beta - \beta\alpha)\,(\alpha\beta - \beta\alpha) \ \dots \tag{4.27}$$

where it is understood that one should put the indices 1, 2, 3, 4, 5, 6, ... in order in the space and spin functions, and where 0_{AS} is the antisymmetric projection operator

$$O_{AS} = (N!)^{-1} \sum_{P} (-1)^{P} P \; , \tag{4.28}$$

where p is the parity of the permutation P of the space-spin coordinates x_1, x_2 ... x_N.

The treatment of the double bonds in the conjugated systems was still somewhat ambiguous. In studying the two Kekule structures of benzene, both Slater[54] and Pauling[55] had realized that a linear combination of the corresponding wave functions in the variation principle could lead to a lower energy than the individual terms and that this would correspond to some form of "chemical resonance."

Apparently one could use the wave functions corresponding to the various chemical structures as a *basis* for calculating the total wave function, and the main problem is then to avoid any linear dependencies. In this connection, one could use some rules developed by Rumer[56] for constructing linearly independent wave functions leading to the so-called "canonical valence-bond structures." Taking the benzene molecule as an example, the canonical forms are represented by the two Kekule-structures and the three Dewar structures as shown in equation (4.29).

The first explicit recipes for calculating the matrix elements with respect to the Hamiltonian H involving such canonical structures as basis elements were given by Pauling[57] in 1932 without any form of proof, and it was hence essential to show that they were correct. It turned out to be much more difficult and cumbersome to prove that Pauling's recipes were correct than one ever anticipated,[58] until finally Shull[59] in 1969 gave a very simple and elegant proof.

During many years, the valence bond method developed parallel with the molecular-orbital method, and it turned out that both had their advantages and disadvantages.[60]

It should perhaps be observed that the valence-bond method has never lost its popularity among the quantum chemists, and that it has been further developed by Kotani, McWeeny, Simonetta, and others.[61] During the last fifteen years, considerable computational efforts based on the VB-method and its generalizations have been carried out by Goddard.[62] Even the mathematical structure of the valence-bond functions have been studied in greater detail.[63]

The mathematical formulation of the valence bond method is to a large extent due to Pauling.[57] It is hence interesting to observe that during the decade to follow, Pauling[64] developed the idea of "chemical resonance" into an almost independent semi-empirical branch of quantum chemistry, which turned out to be both qualitatively and quantitatively useful in many connections, even when the link to the original VB-method was lost.

Kekule and Dewar structures of benzene. (4.29)

5. Calculation of Molecular Integrals; the Early Development; Some Remarks about the Time-Period 1938-48

During the 1930's, scientists studying the electronic structure of molecules calculated the integrals occurring in the expectation value (3.15) of the Hamiltonian as they came along, and no systematic attempts were made to solve this problem on a large-scale basis. It very soon became clear that the most difficult integrals were connected with the interelectronic Coulomb repulsion, and that they--in both the MO- and VB-methods--could be reduced to integrals of the form:

$$(ab \mid cd) = \int \frac{a^*(1)b(1) \ c^*(2)d(2)}{r_{12}} \ dv_1 dv_2 \ , \tag{5.1}$$

where a, b, c and d are atomic orbitals associated with the atomic nuclei of the molecule. Depending on the number of nuclei involved, one distinguishes between one-, two-, three-, or four-center integrals. The atomic orbitals were usually of what one now calls Slater-type orbitals (STO's) of the form

$$r^n e^{-\zeta r} Y_{lm}(\theta, \phi) \tag{5.2}$$

where the angular functions Y_{lm} are the standard spherical harmonics.

The first attempt to calculate such molecular integrals on a systematic basis was probably started in 1938 by Masao Kotani[65] and his co-workers in Tokyo, Japan.

Because of the Second World War, little was known in the West about this work. Kotani apparently used the idea of *parallel computing*, but the "computers" were graduate students or post-doctoral fellows working on primitive desk machines. The results were anyway remarkable. When the "Tables of Molecular Integrals for Hydrogen-Like Functions" became available in the early 1950's, they aroused a great deal of interest all over the world both because of the numerical content and, in particular, because of the skillful group-theoretical analysis of the mathematical structure of the many-electron wave function and the connection between its space and spin parts.

When I started my doctoral studies in theoretical physics at Uppsala University in 1938, I was mainly interested in the relation between quantum theory and the special theory of relativity and the nature of the elementary particles, but the progress of the work was slow. During the war, Sweden was scientifically very isolated, and I was hence very happy when, in 1946, I received a fellowship to go to Zurich to study with Wolfgang Pauli. The new atmosphere was exceedingly stimulating with interesting seminar discussions involving Pauli, Heitler, Wentzel, Stueckelberg and others. My simple "warming-up problem" was the strong- coupling meson theory for nuclear forces, but my progress was evidently slow. Over the weekends, I loved to climb in the Alps, and I remember that one Monday morning, when I returned to the laboratory, Pauli said: "Löwdin, I hope that you are a better climber than physicist, otherwise you will kill yourself." Pauli also told me that if I was getting tired of seeing the infinities in the field theory artificially removed in more or less ingenious ways, I should go into some other field--perhaps solid-state theory, which was more pedestrian but still "real physics." At the end of my stay in Zurich, I decided that--after eight years in field theory with little progress-I should try to make a change.

On the suggestion of Professor Ivar Waller in Uppsala, I started working on the problem of failure of the Cauchy relations for the elastic constants c_{ij} of the alkali halides. This problem had been unsolved since 1827, when Cauchy proved that, if the forces in a crystal are of two-body type, when the 21 elastic constants c_{ij} ($= c_{ji}$) for i,j = 1, 2, ... 6 would automatically fulfill six auxiliary conditions: the Cauchy relations. All experiments indicated, however, that in a real crystal, the Cauchy relations are *not* satisfied.

In a cubic crystal, there are--because of the symmetry--only three independent elastic constants c_{11}, c_{12} and c_{44}, and there is only one Cauchy relation $c_{12} = c_{44}$. My problem[66] was to explain by means of quantum theory why the last relation was not valid for the alkali halides of this type $LiCl$, $NaCl$, KCl, ... LiF, NaF, KF, ...

There had been many attempts before to treat this problem. Over the years, some leading theoreticians had more or less strictly proven that the Cauchy relations should be valid as soon as the forces in the crystal were invariant under translations and rotations of the crystal as a solid body. For some time, it was believed that the failure of the Cauchy relations must depend on the existence of forces lacking these invariance properties, or on some unknown errors in the infinitesimal theory of elasticity. Fortunately for the outcome of my research, these "conjectures" turned out to be wrong: the failure of the Cauchy relations could be shown to follow directly from the existence of translationally and rotationally invariant *three- and four-body forces* connected with three- and four- center integrals of the type (5.1), and these results would be valid even if one started from *finite* elastic deformations. In addition, the theoretical values for the elastic constants c_{11}, c_{12}, and c_{44} were in reasonable agreement with the experimental results-- only in one case (LiF) were they too good to be true.

The reason I report about this early work here is the computational aspects. The atomic orbitals used in these calculations were the *numerical* SCF-functions for the ions, which had been evaluated previously by other authors. The technique used in calculating the molecular integrals of type (5.1) was based on an idea from the 1930's[67] that if one could expand the atomic orbital situated on one center into spherical harmonics around another center, then one could reduce the two-, three-, and four-center integrals to one-center integrals which could be evaluated by numerical integration. For me it was necessary to extend this technique both to numerical atomic orbitals and to combinations of Slater-type orbitals approximating the given SCF-functions.

Intrigued by the idea of "parallel computations," I succeeded in hiring 10 graduate students who carried out numerical integrations by means of Simpson's rule with correction term by using simple algorithms and FACIT desk-computers. In six months, we calculated the values of about 10,000 two-center integrals, 400 three-center integrals and 300 four-center integrals, and, using these results, I could explain the cohesive and elastic properties of a large number of alkali halides as well as their behavior under very high pressure in both the body-centered and face-centered crystal forms.

If an atomic orbital situated on one center is expanded in spherical harmonics around another center, the expansion coefficients have become known as α-functions. It is perhaps remarkable that, during the decades to follow, this simple technique has undergone many improvements and developments,[68] and that some new results were also presented at the Montreal congress[69] in 1986.

Another computational problem of the alkali halides which had to be settled in my study was the *non-orthogonality problem* associate with the fact that the SCF-functions on neighboring ions were overlapping to a large extent, and that the overlap integrals hardly could be neglected. Denoting the normalized atomic orbitals by $\phi = \{\phi_\mu\}$, they have the metric matrix $\Delta = <\phi \mid \phi>$ with the elements $\Delta_{\mu\nu} = <\phi_\mu \mid \phi_\nu> = \delta_{\mu\nu} + S_{\mu\nu}$ where $S_{\mu\mu} = 0$ and the overlap integrals $S_{\mu\nu}$ for $\mu \neq \nu$ are assumed to be fairly small. In using the truncated basis ϕ, the main problem is now to construct an orthonormal set $\varphi = \phi A$ with the property $<\varphi \mid \varphi> = 1$. The conventional Schmidt procedure utilizing successive orthonormalization, which is of such a fundamental importance in mathematics, is useless in this connection since it destroys the symmetry of the crystal. It was early shown, however, that, if one puts $A = \Delta^{-1/2}$, the new set

$$\varphi = \phi \Delta^{-1/2} \tag{5.3}$$

would be orthonormal and keep the symmetry of the crystal; this procedure has hence become known as *symmetric orthonormalization*. In fact, if the basis set ϕ, undergoes a unitary transformation R, so that $\phi' = \phi R$, then the set φ undergoes the same transformation.[71]

It should be observed that, since the metric matrix Δ is self-adjoint and positive definite, it may be brought to diagonal form μ with positive eigenvalues μ_k by means of a unitary transformation U, and the positive inverse square root $\Delta^{-1/2}$ is then strictly defined by the relations

$$U^\dagger \Delta U = \mu, \quad \Delta^{-1/2} = U\mu^{-1/2}U^\dagger \quad , \tag{5.4}$$

which are also useful from the computational point of view.[72] In the case of small overlap, as, e.g., in the alkali halides, one can also use the expansion

$$\Delta^{-1/2} = (1 + S)^{-1/2} = 1 - \frac{1}{2}S + \frac{3}{8}S^2 + \cdots \tag{5.5}$$

which is convergent when all the eigenvalues s_k of the overlap matrix S are situated between -1 and +1. Putting his expansion into (5.3), one obtains the relation

$$\phi_\mu = \phi_\mu - \frac{1}{2}\sum_\alpha \phi_\alpha S_{\alpha\mu} + \frac{3}{8}\sum_{\alpha\beta} \phi_\alpha S_{\alpha\beta}S_{\beta\mu} \tag{5.6}$$

which, in a first approximation, describes the "overlap deformation" of the atomic orbitals when the ions are put together in the crystal.

It is generally assumed that the basis functions $\phi = \{\phi_\mu\}$ are *linearly independent*, which is true if and only if the associated Gramm determinant is non-vanishing: $|\Delta| \neq 0$. The main reason for this assumption is, of course, that--if H is the Hamiltonian under consideration and the set ϕ is linearly dependent--then the secular equation:

$$|<\phi \mid H - z \cdot 1 \mid \phi>| = 0 \quad , \tag{5.7}$$

becomes *identically vanishing* for all values of the complex variable z, which means that, either one has to change the basis set and remove the linear dependencies, or the eigenvalue problem has to be reformulated.[73] The smallest eigenvalue μ_1 of the metric matrix Δ is called the "measure of linear independence" of the set $\phi = \{\phi_\mu\}$. If one considers all functions $\theta = \phi d$ formed from the basis set ϕ by means of all vectors d satisfying the condition $d^\dagger d = 1$, one has

$$|| \theta ||^2 = <\theta \mid \theta> = d^\dagger \Delta d > \mu_1 \quad , \tag{5.8}$$

i.e., the set ϕ can be linearly independent, if and only if $\mu_1 = 0$. The real nightmare in the calculations is the existence of "approximate linear dependencies," which are indicated by the fact that the smallest eigenvalues μ_1, μ_2, ... are so small that, considering the number of significant figures in the computations, they are practically vanishing.[74] It should be observed that all the basis sets conventionally used in quantum chemistry very quickly become approximately linearly dependent.[75] This means that, if one would like to improve the accuracy of an energy calculation, one has to increase not only the size of the basis set but also the number of significant figures in the calculations.

In order to clarify this point in greater detail, one more type of orthonormalization was introduced in 1955[76] by considering the orthonormal orbitals $\chi = \psi U$ or

$$\chi = \phi U = \phi \Delta^{-1/2} U = \phi U \mu^{-1/2} \quad , \tag{5.9}$$

or

$$\chi_k = \frac{1}{\sqrt{\mu_k}} \sum_\alpha \phi_\alpha U_{\alpha k} \quad . \tag{5.10}$$

The relation (5.9) forms the basis for the *canonical orthonormalization procedure* If the symmetric orbitals ϕ given by (5.6) are still fairly localized, the canonical orbitals χ are usually completely delocalized and have the character of "topological orbitals." The canonical orbitals χ_k are characterized by the eigenvalues μ_k of the metric matrix Δ, and they have the advantage that, if there are any approximate linear dependencies, one can easily omit the orbitals χ_1, χ_2, ... associated with the smallest eigenvalues μ_1, μ_2 ...

For a more complete survey of the various types of orthonormalization procedures used in quantum chemistry and solid-state theory, the reader is referred elsewhere.[77]

Start of the Era of the Electronic Computers.

Except for a few analog machines of the type "differential analyzers," etc., practically all computations in atomic, molecular and solid-state physics based on modern quantum theory had, during the first twenty years, been carried out on various types of desk calculators, but a new development was underway. In his memoirs-Feynman has described how during the years 1942-45 a series of IBM *punch-card machines* were used for calculational purposes in, e.g., Los Alamos. My own first contact with this idea was in the Fall of 1948 in Uppsala, where some IBM representatives skillfully demonstrated the underlying principles. I remember that an Uppsala professor of mathematics finally remarked, "I believe that IBM could do something much better," but I do not know what type of interpretation of the representatives gave this comment--I only know that 10 years later they came back with something much better, the IBM 650.

The mathematical principles for large-scale calculations go back to Babbage, Turner, von Neumann and others and to the use of Boolean algebra, but there had never previously been enough mechanical and/or electrical equipment--what we now call "hardware"--available to put them into practical use. Here is not the place to review in detail the development of the electronic computers.

Let me only remind you that, at the end of the Second World War, the U.S. scientists had constructed the famous ENIAC, which was using *electronic tubes* as the main computational and memory elements. In the years 1948-52, universities and research centers all over the world started constructing their own electronic computers, and the MANIAC, the ILIAC, the WHIRLWIND, etc., became well-known as prototypes. This was a very cumbersome period, since the construction of the "hardware" for the computational units based on electronic tubes was rather difficult, and new types of memory units had to be invented and tested over and over again. All the routine calculations were carried out in "machine language," and the work with the coding was difficult and time-consuming and the results could usually not be transferred from one machine to another. This was the great pioneering time characterized by "individuality."

During the years 1949-50, R.S. Mulliken in Chicago, J.C. Slater at M.I.T., and D. Hartree in Cambridge (England) started realizing the potential of the electronic computers and became interested in large-scale computations in atomic, molecular and solid-state theory. Even if a great deal of pioneering work was carried out on the various university machines, the calculations were cumbersome and time-consuming --still the new approach seemed very promising.

One important change was the development of the *commercial electronic computers* manufactured by IBM, UNIVAC and other companies, since the scientists had no longer to worry about the hardware. In addition, the machines of the same brand were often compatible, so that one could transfer programs from one to another. The greatest break-through came perhaps with the construction of the "compilers," so that the programs could be written in some form of mathematical language and not in the "machine-code."

After careful analysis of the underlying quantum theory and the mathematical structure of expectation values of the type (3.15), the quantum chemists could now start writing specific programs for molecular calculations which--after proper testing in pilot calculations--could be used in large-scale production calculations. After a rather long trial period, computational quantum chemistry was finally on its way.

The first generations of the electronic computers were bulky and heavy, since they were based on the use of electronic vacuum tubes. In the late 1940's, Bardeen, Brattain and Shockley at Bell Telephone Laboratories had discovered the *transistor,* for which they were awarded in 1953 Nobel Prize in Physics. The transistor is a small amplifying device, which in many cases not only replaces the electronic tube but has much richer fields of application. It has completely revolutionized the entire electronic industry and, particularly, the computers. With the discovery of the integrated circuit and the silicon chip, the modern electronic computers have become smaller and faster--and even when they are equipped with gigantic memories--comparatively inexpensive. With a great deal of standard software for molecular calculations available, the electronic computers have finally become wonderful tools for the theoretical chemists. I will comment somewhat more about the modern computers at a later stage.

Some Developments of the Theoretical Methods during the Years 1948-58.

Even if the Hartree-Fock method had been successfully applied to atoms, it had only rarely been applied to molecules in a quantitative way.[79] In 1949, Mulliken[80] felt that time had come to develop the molecular-orbital method based on the MO-LCAO approach in greater detail and that perhaps calculations could be performed using all new large-scale computers. In the ASP-MO-LCAO-SCF method, the total wave function is approximated by an antisymmetrized product of molecular orbitals formed by linear combinations of atomic orbitals in a self-consistent-field way. Starting from the atomic orbitals $\phi = \{\phi_\mu\}$ having the metric matrix $\Delta = <\phi \mid \phi>$, one may now write the molecular orbitals $\psi = \{\psi_k\}$ in the form

$$\psi = \phi C \quad , \tag{5.11}$$

or

$$\psi_k = \sum_\alpha \phi_\alpha \, C_{\alpha k} \quad , \tag{5.12}$$

which is analogous to (4.20). The orthonormality condition now takes the form:

$$<\psi \mid \psi> = C^\dagger \Delta C = 1 \quad . \tag{5.13}$$

In the Hartree-Fock scheme, the entire physical is described by the Fock-Dirac operator $\rho = \mid \psi><\psi \mid$ having the kernel (4.7) and the properties (4.8). Substituting the expression (5.11) into ρ, one gets directly

$$\rho = \mid \psi><\psi \mid = \mid \phi> CC^\dagger <\phi \mid = \mid \phi> \, R \, <\phi \mid \quad , \tag{5.14}$$

where $R = CC^\dagger$ is the charge- and bond-order matrix introduced by Coulson and Longuet-Higgins.[81] It has the properties:

$$R\Delta R = R, \quad R^\dagger = R, \quad Tr(R\Delta) = N \tag{5.15}$$

which are analogous to (4.8). Using the variation principle (4.5), one obtains the Hartree-Fock equations (4.9) and (4.11), in which the effective Hamiltonian H-- has the form (4.10). Since (4.11) has the form of an eigenvalue problem for the self-adjoint operator H_{eff}, it is equivalent with the variation principle:

$$\delta <\psi_k \mid H_{eff} \mid \psi_k> = 0 \quad . \tag{5.16}$$

Using (5.12), one gets immediately the standard type of a system of linear equations:

$$\sum_\nu <\phi_\mu \mid H_{eff} - \varepsilon_k \cdot 1 \mid \phi_\nu> C_{\nu k} = 0 \quad , \tag{5.17}$$

with the secular equation:

$$\mid <\phi_\mu \mid H_{eff} - \varepsilon \cdot 1 \mid \phi_\nu> \mid = 0 \quad , \tag{5.18}$$

in which H_{eff} has the form (4.10) with the density matrix (5.14) expressed in the form:

$$\rho(x_1, x_2) = \sum_{\alpha\beta} \phi_\alpha(x_1) R_{\alpha\beta} \, \phi_\beta^*(x_2) \ . \tag{5.19}$$

The first self-consistent-field (SCF) scheme for calculating molecular orbitals based on the coefficients $C = \{C_{ak}\}$ was developed in 1949 by Roothaan.[82] It is based on the iterative cycle:

$$\underset{\longleftarrow}{\,C \ \rightarrow \ H_{eff} \ \rightarrow \ C\,|} \tag{5.20}$$

In this context, a calculation is said to have become *self-consistent*, when further iteration does not change the significant figures of the result; we note that this is not a strict proof for mathematical convergence, but only a practical rule to stop the iterations.

There are several modifications of this molecular SCF-scheme;[83] instead of starting from the coefficients $C = \{C_{\mu k}\}$, one can start from an estimate of the charge- and bond-order matrix $\mathbf{R}$ and put the emphasis of self-consistency on this matrix.

It was evident that, in the MO-LCAO scheme, it was necessary to evaluate a large number of molecular integrals, and that the most difficult ones would be of type (5.1) for Slater-type orbitals. In Chicago, Mulliken gathered a large group of scientists working on this problem, including C.C.J. Roothaan and K. Ruedenberg. At M.I.T., Slater started the famous "Atomic, Molecular and Solid-State Theory Group" (AMSST-group) including such scientists as G. Koster, R.H. Parmenter, G. Pratt, A. Meckler and many others. Mulliken felt the need to try to "standardize" the tables of molecular integrals to make their use as a general as ever possible, and he agreed with Slater and other U.S. leaders to arrange a small conference limited to 25 participants at Ram's Head Inn at Shelter Island at the north tip of Long Island, New York, for this purpose in August, 1951. The Shelter Island Conference became a landmark in the development of quantum chemistry, and one discussed a series of fundamental problems[84]--except perhaps the standardization of molecular integrals. Barnett and Coulson showed their scheme for evaluating molecular integrals using Bessel functions and emphasized the fact that mathematically strict recursion formulas may behave very strangely when the quantities involved are permitted to have "rounding-off errors"-- some stay well-behaved, where others may turn out to be very ill-behaved and practically useless. A great deal of attention was given to the *correlation problem*, i.e., the fact that two electrons i and j ought to avoid each other because of their mutual Coulomb repulsion e^2/r_{ij}. Since electrons with parallel spins avoid each other because of the Pauli exclusion principle, the correlation effect deals mainly with electrons having anti-parallel spins, and, using the hydrogen molecules as an example, Kotani emphasized that its two electrons ought to stay in different molecular orbitals. Instead of doubly filled orbitals, one should have a "correlation splitting" of each orbital and use "different orbitals for different spins"--an approach later coined DODS.

I remember that I reported about various central difference methods for solving systems of ordinary differential equations, which all go back to Gauss-- including a simple derivation of the Numerov method for solving second-order differential equations.[85] The error term could be as small as desired, but, if it was decreased beyond a certain limit, the recursive integration formula turned out to be exceedingly ill-behaved and unstable with respect to rounding-off errors. I remember that I shared a room with a young American scientist, Harrison Shull, and this created a lifelong scientific collaboration and friendship. For a full report about the Shelter Island Conference, the reader is referred to the proceedings.[84]

The Shelter Island Conference was characterized by a great deal of scientific optimism due to the occurrence of the electronic computers: one anticipated that, within the next five years, the electronic structure of the ground state and the excited states of all *diatomic molecules* would be know in great detail. The only problem neglected was the "standardization" of the tables of molecular integrals, and, when the conference was finally over, interested parties had to meet in a hotel room in New York. Mulliken was still very pleased with the outcome of this conference and wrote an excellent introduction to the proceedings giving a survey of the various trends in the field. He pointed out that the earliest example of "correlation splitting" may be found already in Hylleraas' work on the helium atom, where the singlet component of the wave function $(1s'\alpha; 1s''\beta)$ has lower energy than the conventional configuration $(1s)^2$ with a doubly filled 1s-orbital. Somewhat later this problem was investigated in greater detail.[86]

It turned out that Slater had another aspect on this problem. In a study of ${}^2S$ ground state $(1s)^22s$ of the lithium atom[87] in the Hartree-Fock method, he pointed out that, in the conventional configurations $(1s\alpha, 1s\beta, 2s\alpha)$, the 1s-electron with α-spin is influenced by another exchange potential than the 1s-electron with β-spin and that the solution of the Hartree-Fock equations should lead to a configuration $(1s'\alpha, 1s'\beta, 2s\alpha)$, which is subject to a considerable spin-splitting or "spin-polarization" of the 1s-orbital. This configuration has a lower energy than the conventional configuration $(1s)^22s$, but the corresponding Slater determinant D is instead a mixture of the states ${}^2S$ and ${}^4S$ of the total spin.

Slater's work started a discussion about the difference between the "Restricted Hartree-Fock" (RHF) method, in which one had introduced all the proper symmetry constraints which may raise the energy $<H>$ and give rise to non-vanishing non-diagonal Lagrangian multipliers $\lambda_{lk} \neq 0$, and the "Unrestricted Hartree-Fock" (UHF) method, in which one has no symmetry constraints whatsoever and the matrix $\lambda = \{\lambda_{lk}\}$ may be completely diagonalized. There could also be between-stages, in which one had removed one or more of the symmetry conditions in the original RHF-scheme. It was evident that the fully unrestricted UHF-method was identical to the original Hartree-Fock method, but it was somewhat of a surprise to realize that the Slater determinant D corresponding to the minimum of $<H>$ may be a *mixture* of different symmetry types. We will return to this question below.

A big question in connection with the discussion of tables of molecular integrals was what type of atomic orbitals one should use. From the chemical point of view, the natural choice would be the hydrogen-like orbital (STO's) characterized by the symbols 1s, 2s, 2p, 3s, 3p, 3d, ..., etc., but, in such a case, one also had to remember to include the *continuum*, since otherwise the basis would never become complete.[88]

The first tables of molecular integrals and tools for calculating such integrals available in the early 1950's were published by Kotani, et al.[65], and they represented the summary of a gigantic analytic and computational effort. The first European tables were due to Kopineck.[89]

It was evident that, if one wanted to go beyond the Hartree-Fock scheme, one had to approximate the total N-electron wave function not by a single Slater determinant D of type (4.6) but by a series of such determinants:

$$\Psi = \sum_{K} D_K C_K \; , \tag{5.21}$$

or a "superposition of configurations" K. If one starts from a truncated orthonormal one-electron basis of order M, i.e. $\psi = \{\psi_1, \psi_2, \; \dots \; \psi_M\}$, there exists $\binom{M}{N}$ ordered configurations K with the indices $k_1 < k_2 < \dots < k_N$. Using the variation principle (4.5), one obtains the standard set of linear equations for the coefficients C_L:

$$\sum_{L} \; < D_K \mid H - E \cdot 1 \mid D_L > C_L = 0 \tag{5.22}$$

with the secular equation

$$\mid \; < D_K \mid H - E \cdot 1 \mid D_L > \; \mid = 0 \tag{5.23}$$

of order $p = \binom{M}{N}$. For historical reasons, this approach has been called the "configurational interaction" (CI)-method. Since the order p quickly becomes very large for increasing M, the method requires very good computational tools, and these were apparently offered by the new electronic computers.

In the late 1940's, S.F. Boys[90] in Cambridge, England, had taken up the CI-method of calculating the electronic structure of small molecules, and--in order to avoid the problem of the cumbersome evaluation of molecular integrals--he suggested the use of a set of *Gaussian* one-electron functions as a basis. In principle, such a basis may become complete, but it should be remembered that, with a finite basis, it is always difficult to reproduce the correct asymptotic behavior of the wave functions having an exponential decay. Even if Boys, himself, successfully carried out a fairly large number of molecular calculations, it would take a fairly long time before this type of basis was generally accepted.

In the early 1950's, the Hückel method for the conjugated systems was still going strong, and Fukui, et al. had successfully introduced the idea of the "frontier electrons." It was generally considered as some form of semi-empirical Hartree-Fock scheme, and it should then be affected by a "correlation error." In 1953, electronic correlation was introduced into the Hückel-method by Pariser, Parr and Pople,[91] and the PPP-model quickly became one of the most popular methods in organic quantum chemistry. It could be shown that this model could essentially be derived from the standard Hartree-Fock scheme, if one introduces the principle of the "neglect of differential overlap" (NDO), i.e., if one neglects all two-, three- and four-center integrals which contain at least a pair of overlapping orbitals.

In the opinion of the author,[92] the neglect of differential overlap could perhaps be at least partly justified, if one starts from a basis set of orthonormalized orbitals $\varphi = \phi\Delta^{-1/2}$ as defined by (5.3) instead of the ordinary atomic orbitals $\phi = \{\phi_\mu\}$, and this idea has later been further developed.[93]

In computational organic quantum chemistry, the PPP-model using various forms of NDO has become a very forceful semi-empirical tool for treating very large molecules, and one has today the well-known symbols CNDO (= complete neglect of differential overlap), INDO (= intermediate neglect of differential overlap), MINDO (= modified intermediate neglect of differential overlap), NNDO (= neglect of diatomic differential overlap), tc. which characterize different types of standard computer programs available.

The Shelter Island Conference in 1951 started a series of international meetings in quantum chemistry, which became of fundamental importance for the national and international exchange and collaboration in this field. In 1953, a symposium was arranged in Nikko (Japan) by Professor Masao Kotani; in 1955, a symposium was arranged in the spring in Stockholm-Uppsala (Sweden) by I. Fischer-Hjalmars and P.O. Löwdin, and in the fall in Austin (Texas) by F.A. Matsen; in 1958, a symposium was arranged in Paris (France) by A. and B. Pullman and R. Daudel and another in Boulder (Colorado) by the U.S. leaders with R.S. Mulliken as "primus motor." For a complete review of these symposia, the reader is referred to the various proceedings.[94]

Already at the Shelter Island Conference it was agreed that the *correlation problem* connected with the interelectronic Coulomb repulsion e^2/r_{12} was exceedingly important and had to be solved in some way. Except for a classical paper by Wigner[95] in 1933 dealing with the correlation in the electron-gas model, very little was known about this problem. It was evident that the basic Independent- Particle-Model (IPM) as expressed, e.g., in the Hartree-Fock scheme was afflicted with the certain *correlation error*, but for Slater the dominating error was the wrong asypototic behavior of the MO-LCAO-method for separated atoms, for which he was repeatedly warning, whereas for other authors--like Mulliken and Kotani--the error in the HF-scheme for the equilibrium distance ($R = R_0$) would be the most interesting quantity. In this connection, I suggested that the *correlation energy* should be defined as the difference

$$E_{corr} = E_{exact} - E_{HF} \tag{5.24}$$

where E_{exact} is the exact eigenvalue of the Hamiltonian under consideration. This correlation energy is hence essentially a mathematical quantity measuring the error in the approximate method used in solving the problem.

In my study[66] of the cohesive energies of the alkali halides defined through the relation

$$E_{coh} = E_{solid} - E_{free\ ions}\ ,\qquad\qquad\qquad (5.25)$$

I had never encountered any correlation problem, since apparently the correlation energies of the ions in the solid were more or less the same as the correlation energies of the free ions, and the excellent theoretical values for the cohesive energies would hence depend on a systematic cancellation of errors. In a study[96] of the cohesive energy of the sodium metal in 1953 using the same type of approach, I realized that the situation was different and that--even if the preliminary results were in good agreement with experience--a deeper study had to be carried out, and I became seriously interested in the correlation problem.

Taking up the idea of using "different orbitals for different spins," the author[97] suggested at the Nikko Conference in 1953 that, in alterant systems, electrons with different spins should occupy different "alternate molecular orbitals" localized on the two different subsets of atoms of the system, in which case the associated MO-LCAO-wave function may have a correct asymptotic behavior for separated atoms. In an application to the benzene molecule, it was also shown[98] that this method--even in its simplest form--gave an essential part of the correlation energy for $R = R_0$. This approach has become known as the "Alternant- Molecular- Orbital" (AMO) method, and it had later become extended and developed by several authors.[99]

Since I had been working on the alkali halides, I was particularly interested in the correlation energies of the free ions having a rare-gas electron structure. During the years 1953-55,[100] the correlation energies for the He-like ions from H^- to C^{4+} were calculated and found to be remarkably constant around -1.2 eV. At the Texas conference in 1955, it was suggested that, if both the exact energy and the HF-energy were expanded in power series of $1/Z$, then the constant term would be dominating in the difference and explain this phenomenon. In reality, this did not turn out to be true, since the constant term did not have the correct value (-1.2 eV); however, if the power series for $0.1 < 1/Z < 1$ was transformed into an expansion in terms of Chebyshef polynomials, the constant term turned out to be dominating and had the correct value. This is probably a technique which should be applied more often when we try to explain physical and chemical phenomena by means of perturbation theory. During the years 1955-58, the correlation energies for the Ne-like ions from F^- to Si^{4+} were calculated in the Uppsala group;[101] the exact energy of the non-relativistic Hamiltonian H was estimated from the experimental data available with correction for the relativistic effects. Again the correlation energies were found to be remarkably constant around -11 eV ($= -253$ kcal/mole). However, even if this principle seemed to apply to the rare-gas structures in general, it was discovered[102] that it did not apply to other structures because of the energy degeneracy of e.g. the 2s- and 2p-orbitals, etc. In a study of the Be-like ions with the structure $(1s)^2(2s)^2$ by using the $(1/Z)$-expansion, it was found that the correlation energy defined by the difference (5.24) contained a term which was *linear* in the atomic number Z due to the degeneracy in the proper definition of the ground state!

The existence of several HF-schemes (RHF, UHF, GHF, etc.) also made it necessary to specify the nature of the second term E_{HF} in (5.24). Since most calculations, so far, had been carried out in the symmetry restricted Hartree-Fock scheme (RHF), it seemed natural to specify that the correlation energy should be defined as the difference:

$$E_{corr} = E_{exact} - E_{RHF} \tag{5.26}$$

It was also evident that the calculation of the correlation energies would be greatly helped by the occurrence of the new electronic computers.

At the 1955 Texas Conference, there was a general pledge to federal agencies and philanthropic foundations to support quantum chemistry groups all over the world to acquire large-scale electronic computers. In this connection, Peter Debye made the following simple statement as to the electronic computers: "If you have them, use them--if you don't have them, beat them." It is remarkable that, if one transfers this statement to the discussion of the "supercomputers," it is probably valid even today.

During the 1950's, there were some very important theoretical developments in ordinary quantum mechanics originating in the more sophisticated "field theory," upon which I would now like to briefly comment. In the middle of the 1950's, many of the scientists working on the non-relativistic many-particle problem picked up the idea of using so-called *transition-type formulas*. If φ is an arbitrary *reference function* in the L^2 Hilbert space, which is normalized so that $< \varphi \mid \varphi > = 1$, and one multiplies the Schrödinger equation $H\Psi = E\Psi$ to the left by $< \varphi \mid$, one obtains

$$E = \frac{< \varphi \mid H \mid \Psi >}{< \varphi \mid \Psi >} \quad , \tag{5.27}$$

which is a transition-type formula. In contrast to the expectation value (3.15) of the Hamiltonian, which is bounded from below, the transition values (5.27) is completely *unbounded* and goes to $\pm \infty$ as $< \varphi \mid \Psi >$ goes to ± 0, as long as $< \varphi \mid H \mid \Psi > \neq 0$.

One further discovered that, by using a normalized reference function φ, one could reformulate the eigenvalue problem $H\Psi = E\Psi$ in a convenient way by introducing the boundary condition

$$< \varphi \mid \Psi > = 1 \quad , \tag{5.28}$$

which is referred to as the intermediate normalization. It has the advantage that one can now treat closed states and scattering states on the same footing. The associated spectrum $\{E\}$ usually coincides with the ordinary spectrum $\{E\}$; one has only to observe that, if a discrete eigenfunction Ψ happens to be orthogonal to the reference function φ, the corresponding eigenvalue E will not show up in the spectrum $\{E\}_\varphi$, and one speaks of a *lost* eigenvalue. Such an eigenvalue may be recovered by changing the reference function φ.

In connection with scattering theory, one had developed the idea that the solution Ψ should be obtained from the reference function φ by means of a *wave operator* W, so that:

$$\Psi = W\varphi \quad , \tag{5.29}$$

and the relations (5.27) and (5.29) may then be written into the form:

$$E = <\varphi \mid HW\varphi> \quad , <\varphi \mid W \mid \varphi> \; = 1 \quad . \tag{5.30}$$

It should be observed that the wave operator is by no means unique and that simple examples are provided by the multiplicative operators $W = \Psi/\varphi$ and by the ket-bra operator $W = \mid \Psi> <\varphi \mid$. The last one has the additional property of being a one-dimensional projector, satisfying the relations $W^2 = W$, $TrW = 1$. If W'' and W' are two different wave operators, then the difference $\Delta W = W'' - W'$ has the property $\Delta W \cdot \varphi = 0$. If one introduces the projector $O = \mid \varphi> <\varphi \mid$ and the projector $P = 1 - O$ for its complement, one may write ΔW in the form:

$$\Delta W = (O + P)\Delta W(O + P) = (\Delta W)_{11} + (\Delta W)_{12} + (\Delta W)_{21} + (\Delta W)_{22} \quad , \tag{5.31}$$

and the relation $\Delta W \cdot \varphi = 0$ gives then the relation $(\Delta W)_{11} + (\Delta W)_{21} = 0$. Hence ΔW must have the form:

$$\Delta W = (\Delta W)_{12} + (\Delta W)_{22} = AP \quad , \tag{5.32}$$

where A is an arbitrary linear operator. The general wave operator has thus the form $W = W_1$ where W_1 is any special solution, and the problem is now to find some simple form of W, which may depend explicitly on the reference function φ but not on the exact eigenfunction Ψ. In the following, only the wave operators associated with perturbation theory, the resolvent technique or the coupled-cluster method will be of particular interest.

In the study of the electronic structure of atoms, molecules and solid-state, the independent-particle-model and the idea of the existence of "electronic shells" turned out to be exceedingly successful, and one had usually believed that this depended on the fact that the coulomb repulsion between the electrons was a comparatively *weak interaction*. One was hence somewhat surprised when, in the beginning of the 1950's, it was discovered that even inside the atomic nuclei, where the interaction between the nucleons is definitely very strong, there was a clear "shell structure." This phenomenon was theoretically explained by Keigh Brueckner and his collaborators[103] by introducing the idea of the reaction operator into the on-particle self-consistent-field scheme.

If $H = H_0 + V$ and the reference function φ is an eigenfunction to H_0, so that $H_0\varphi$, one has

$$E = <\varphi \mid H \mid \Psi> \; = \; <\varphi \mid H_0 + V \mid \Psi> \; =$$

$$= E_0 + <\varphi \mid VW \mid \varphi> \; = E_0 + <\varphi \mid t \mid \varphi> \; =$$

$$= <\varphi \mid H_0 + t \mid \varphi> \quad , \tag{5.33}$$

where t = VW is the so-called *reaction operator*. In the IPM, the unperturbed Hamiltonian $H_0 = \sum_i(H_i + \mu_i)$ is a one-particle operator, the reference function φ is a single Slater determinant, and one has $E_{HF} = <\varphi \mid H_0 + V \mid \varphi>$. A comparison with (5.33) shows that one can go from the Hartree-Fock energy to the correct energy, if one everywhere in the SCF-scheme replaces the perturbation V by the reaction operator t:

$$V \to t = \sum_{i<j} t_{ij} + \sum_{i<j<k} t_{ijk} + \cdots \quad ; \tag{5.34}$$

Brueckner considered essentially the two-particle part t_{ij} of the reaction operator, but the scheme is easily turned into an exact form[104]. If the original Hamiltonian H has the form:

$$H = \sum_i H_i + \sum_{i<j} H_{ij} \quad , \tag{5.35}$$

it is usually convenient to introduce the notations

$$H_0 - \sum_i (H_i + \mu_i)$$

$$V = -\sum_i \mu_i + \sum_{i<j} H_{ij} \quad , \tag{5.36}$$

where the so-called SCF-potentials μ_i are at the disposal of the investigator. In order to evaluate the wave operator W, Brueckner studied higher-order Schrödinger perturbation theory and derived recursion formulas for calculating perturbation quantities of higher and higher order. Brueckner's ideas deepened not only the entire SCF-scheme but led to a new development in perturbation theory, with many important consequences, as e.g., the linked cluster theorem.[105]

Using (5.33), one get immediately for the correlation energy:

$$E_{cor} = E - E_{HF} = <\varphi \mid t - V \mid \varphi> \quad , \tag{5.37}$$

where the difference (t - ´V) may be interpreted as a non-local "correlation potential." The concept has, so far, been of little interest in the literature, since no one has been able to show that it is transferable from one molecular system to another. It is, however, closely related to a similar application of the transition formula E = $<\varphi \mid H \mid \Psi>$ due to Nesbet.[106]

In the Hartree-Fock scheme, one starts from a single Slater determinant $\varphi = D = (N!)^{-1/2}\mid \psi_k(x_i)\mid$, which is built from N occupied one-electron-functions $\psi_{HF} = \{\psi_1, \psi_2, \dots \psi_N\}$. One can introduce a complete orthonormal basis

$\psi = \{\psi_1, \psi_2, \dots \psi_N, \ \psi_{N+1}, \psi_{N+2} \dots \}$ simply by starting from any complete orthonormal basis and making it orthogonal to Ψ_{HF}, for instance, by the Schmidt procedure. Combining it with Ψ_{HF}, one gets again a complete orthonormal basis, which may be used as a basis for the CI-method. However, in the infinite expansion (5.21), it is now possible to combine terms, so that one gets a finite expansion of the form:

$$\Psi = D + D_{s.e.} + D_{d.e.} + D_{t.e.} + \cdots D_{N.e.} \ . \tag{5.38}$$

when $D_{s.e.}$ (s.e. = singly excited) means a sum of N determinants, which are all obtained from D by means of single excitations, whereas $D_{d.e.}$ means a sum of $\binom{N}{2}$ determinants, which are all obtained from D by means of double excitations ($=$ d.e.), etc. The Hartree-Fock functions are characterized by the fact that they satisfy the so-called Brillouin theorem.[107]

$$< D \mid H \mid D_{s.e.} > \ = 0 \tag{5.39}$$

In substituting the expansion (5.38) into the transition formula $E = < D \mid H \mid \Psi >$, Nesbet observed that all the higher terms would vanish identically because of the orthogonality of the basis so that $< D \mid H \mid D_{t.e.} > \ = \cdots = 0$ Hence he obtained:

$$E = < D \mid H \mid \Psi > \ = \ < D \mid H \mid D > + < D \mid H \mid D_{d.e.} > \ , \tag{5.40}$$

or

$$E_{corr} = < D \mid \sum_{i<j} H_{ij} \mid D_{d.e.} > \tag{5.41}$$

In this connection, it is possible to describe the doubly-excited term $D_{d.e.}$ by means of pairs of particle-hole operators in the terminology of second quantization or by means of $\binom{N}{2}$ pair functions. This starting point has led to a fruitful development due to Nesbet[108] and to Sinanoglu.[109] In terms of the pair functions, it is easy to give an interpretation of formula (5.41) for the correlation energy. It should be remembered, however, that, since the basic formula (5.40) does not define an expectation value but a *transition value*, this interpretation of the correlation energy may have a more mathematical then physical meaning.

Nesbet's formula (5.41) gave the first exact expression for the correlation energy without using the concept of the reaction operator. For a general survey of the treatment of the correlation problem up to 1958, the reader is referred to a review article.[110]

Let us now return to another development in the 1950's, which was related to the theory of the CI-method. For this purpose, we will start from the observation that the number p of determinants in the CI-expansion (5.21) for a particular wave function Ψ is by no means invariant but depends on the choice of basis. As an example, we will consider the HF-scheme in which the wave function Ψ is approximated by a single Slater determinant $D = (N!)^{-1/2} | \psi_1, \psi_2, ... \psi_N |$ and one has $p = 1$. On the other hand, if one introduces a basis of atomic orbitals $\phi = \{\phi_\mu\}$ of order M with M<N, one may expand each one of the Hartree-Fock functions ψ_k according to (5.12), and substituting these expansions into D and expanding, one obtains sud-

denly a determinantal expansion containing $P = \binom{M}{N}$ terms. In this situation, it is then natural to ask the question under what conditions a CI-expansion containing a very large number of terms may be condensed into a single Slater determinant. In 1951, Slater[111] had explicitly asked the more general question what type of one-electron basis $\psi = \{\psi_k(x)\}$ one should use in order to obtain a CI-expansion (5.21) of *fastest possible convergence* when the basis becomes infinite and complete. The problem was solved at M.I.T. in the spring of 1954 by using the concept of *reduced density matrices*.[112]

It was well-known from general quantum statistics that the density matrices $\Gamma(X|X')$ as well as the reduced density matrices[113] are valuable tools in the theory, but it was not believed that they would be of any particular importance in pure quantum mechanics built on the concept of state vectors and wave functions Ψ, where one simply has $\Gamma = | \Psi > < \Psi |$ and

$$\Gamma(X \mid X') = \Psi(X)\Psi^*(X') \ , \tag{5.42}$$

since the wave function Ψ contains all the information. Even in this case, the reduced density matrix of order p is derived by the formula:

$$\Gamma(x_1 x_2 ... x_p \mid x'_1 x'_2 ... x'_p) =$$

$$= \binom{N}{p} \int \Gamma(x_1 x_2 ... x_p x_{p+1} ... x_N \mid x'_1 x'_2 ... x'_p \ x_{p+1} ... x_N) \ dx_{p+1} \cdots \ dx_N \tag{5.43}$$

The first-order density matrices $\gamma (x_1 \mid x'_1)$ for $p = 1$ have a spectral resolution of the form

$$\gamma(x_1 \mid x'_1) = \sum_k n_k \chi_k(x_1)\chi_k^*(x'_1) \ , \tag{5.44}$$

where the one-electron functions $\chi_k(x)$ are called the *natural spin-orbitals*, and the eigenvalues or "occupation numbers" n_k have the general properties

$$0 < n_k < 1, \ \sum_k n_k = N \ . \tag{5.45}$$

It could now be shown[114] that the natural orbital represent the one-electron basis which give the CI-expansion (5.21) of most rapid convergence, which will be referred to as the *"natural expansion"* of the wave function Ψ. In this context, it was also easily shown that a wave function Ψ may be expressed as a single Slater determinant, if and only if all the occupation numbers n_k are either 0 or 1, or if one has

$$\mu = Tr(\gamma - \gamma^2) = 0 \quad . \tag{5.46}$$

Some applications of this approach showed that it led to great simplifications of the wave functions.[115] It would take some time before the natural orbitals became a standard tool in computational quantum chemistry, and Davidson's book[116] was of essential importance in this connection. Since I have given a survey of the development of the theory of reduced density matrices and the so-called representability problem in the parallel symposium in Kingston, Ontario (Canada), the reader is referred to the proceedings[117] of this conference for further details.

During the 1950's, the problem of the *molecular integrals* for Slater-type orbitals (STO's) had been systematically tackled by the scientists in Mulliken's Chicago group, and particularly the contributions by Roothaan and Ruedenberg should be mentioned in this connection.[118] However, there were also many other authors working on the same problem.

In 1955, Matsen[119] published a book containing "Tables of Molecular Integrals," which was very useful in many connections. I still remember how surprised the members of the Soviet delegation in theoretical physics to the 1956 Seattle Conference were, when they found this book in pocket edition at the shelves of general audiences in the bookstore at Idlewild International Airport, and were told by some joker that, "In the United States, a large fraction of the population is doing molecular calculations."

Anyway, it seemed as if Mulliken's dream of breaking "the bottleneck of the molecular integrals" would finally become true, and a great deal of progress was reported at the 1958 Boulder Conference. For details, the reader is referred to the proceedings of this conference.[120]

In 1958, the first five-week Summer Institute in quantum chemistry and solid-state theory was arranged in Vålådalen in the Swedish mountains. It was an effort to save and revive some of the results from the golden years 1928-32 and to educate not only senior students in the quantum theory of matter but also scientists from neighboring fields and to tell them about the strengths and limitations of the theory itself. It was followed by a small international symposium with Linus Pauling and Robert S. Mulliken as leading figures and with participation of many of the internationally well-known younger scientists: B. Bak, G.G. Hall, W. Kern, R. McWeeny, K. Ohno, G. del Re and many others. There was a great deal of discussion also about molecular calculations using electronic computers, and part of the optimism from the Shelter Island Conference in 1951 seemed to be gone--the electronic computers had, perhaps, not delivered what was expected from them, and the proper choice of basis sets and the occurrence of approximate linear dependencies (ALD's) had caused greater problems than anticipated. Still I remember that George Hall made the fairly optimistic forecast that, using large-scale computers, one would in a not too distant future be able to calculate the electronic structure of comparatively large molecules containing 5-8 atoms--not counting the hydrogens.

6. Some Comments on the Developments in the Period 1959-1985

In the description of the developments during the last seventeen years, I will try to be very brief since most of the main lines of thought are already very well know to this audience.

Let me start with some personal remarks. In 1959-60, I started the new Florida Quantum Theory Project as a sister project to the Uppsala Quantum Chemistry Group, and I began spending half an academic year on each side of the Atlantic. The Uppsala Summer Institutes were also supported for a few years by the U.S. National Science Foundation, and--with the exception of 1961 and 1963--they were arranged on a yearly basis somewhere in Scandinavia until 1984. In Florida, we arranged similar five-week Winter Institutes in Quantum Chemistry and Solid-State Theory from 1960-61, and they were for more than a decade supported by the U.S. National Science Foundation. The first Florida international symposium in this field was held on Sanibel Island in the Gulf of Mexico outside the city of Fort Myers, Florida, in January, 1961. From then on the Sanibel Symposia have been organized on a yearly basis, even if, in 1977, they were moved away from the island, which had started to become very crowded. For a brief review of 25 years of Sanibel Symposia, the reader is referred to a recently published survey.[121]

The 1963 Sanibel Symposium was arranged in honor of one of the great pioneers in computational quantum theory--Professor Egil Hylleraas from Oslo (Norway)--and the proceedings were published in *Reviews of Modern Physics*.[123] John A. Coleman gave an excellent review[123] of the mathematical structure of the fermion density matrices. He showed, among other things, that every first-order reduced density matrix $\gamma(x_1 \mid x'_1)$ is ensemble-representable and that the second-order reduced density matrix $\gamma(x_1 x_2 \mid x'_1 x'_2)$ may sometimes have a very large eigenvalue (N/2), which corresponds to some form of "Boss-Einstein condensation" of fermion pairs related to certain superfluidity phenomena in physics.

The relation bewenn the UHF- and RHF-schemes was discussed in greater detail, and the problem of the "symmetry dilemma" in the HF-scheme and similar variational method was formulated.[124] It was shown that the results in the RHF-scheme may correspond to local minima, in which the symmetry constraints become self-consistent, whereas the absolute minimum in the original HF-scheme (or in UHF) may be represented by a Slater determinant D, which is a mixture between different symmetry states. The problem could be resolved by replacing the basic wave function by a proper symmetry projection OD of a single Slater determinant based on general spin-orbitals, and this leads to the so-called "Projected Hartree-Fock" (PHF) method[125] Even if this method has given some interesting results, its possibilities are still far from being explored.

In the studies of the optimal values of the expectation value <H> in the Hartree-Fock scheme, it turned out that <H> had not only maxima and minima of various types but also complicated saddle-points, and the investigation of the *instabilities* in the HF-method soon became a very hot field, in which research is still going on.[126]

In the "General Hartree-Fock" (GHF) scheme, one is using general spin-orbitals which are combinations of α- and β-functions. It is perhaps interesting to observe that the mathematical existence of solutions to the GHF-equations has been proven[127], but that no one has so far calculated any such solutions in practice. In the RHF-scheme, on the other hand, one has calculated a large number of solutions in practice by the iterative self-consistent-field method, but one still does not know whether the exact solutions to the RHF-equations exist in a more strict mathematical sense.

The Florida Quantum Theory Project (QTP) was modeled on the AMSST-group at M.I.T., and, in 1964, Professor John C. Slater joined the project on a half-time basis--and a few years later on a full-time basis. He was particularly interested in the approximation of the exchange operator in the Hartree-Fock effective Hamiltonian (4.10), which has become known as the X_α-method[128] and in which one has

$$V_{X_\alpha}^\uparrow(1) = - 6\alpha(3\rho^\uparrow / 8\pi)^{1/3} \quad , \tag{6.1}$$

where is $\rho^\uparrow$ is the density of the electrons with plus spin. In 1951, Slater had suggested the value $\alpha = 1$, and somewhat later Gaspar[129] found the more appropriate value $\alpha = 2/3$. In the 1960 calculations in Florida and elsewhere, the α-values for the free atoms were introduced in the molecular computations. In the studies of various crystals, the X_α-method was combined with the "Augmented Plane Wave" (APW) method with or without the "muffin-tin approximation," and Slater and his collaborators obtained in this way a series of excellent results in various types of application.[130]

In Slater's X_α-method, the electron density ρ plays a dominant role in explaining the physical properties. This was not a new idea, however, since it goes back to the Thomas-Fermi-Dirac (TDF) method[131] developed in the years 1928-32, and it had later been utilized in many different connections in atomic, molecular and solid-state theory. It was still somewhat surprising when Hohenberg and Kohn[132] in 1965 showed that the ground state energy E of a system with fixed nuclei is a unique functional $E = E(\rho)$ of the electron density ρ, and their theorem forms the basis for the electron density functional (EDF) method. For a review of the development in this area and the current problems involved, the reader is referred to the proceedings of the 5th ICQC in Montreal and the Kingston symposium.

In the 1960's, the *correlation problem* was still of essential interest, and tables of atomic correlation energies were calculated by Clementi.[133] The Nesbet formula (5.41) was still of key importance, but--also in atomic and molecular theory--more and more attention was devoted to Brueckner's wave and reaction operators. Using the partitioning technique, the author wrote a series of papers[134] on the structure of finite--and infinite--order perturbation theory with the aim of deriving "closed expressions" for the wave and reaction operators. In 1968, there was finally a Summer Institute devoted entirely to the correlation problem in Frascati outside Rome, arranged by Carl Moser and Roland Lefebvre, and, for a survey of the development up to this stage, the reader is referred to the proceedings.[135]

In order to study the stability of the genetic code at body temperature (T = 310°K), some fairly large-scale calculations were carried out in the 1960's in Uppsala on the guanine-cytosine and adenine-thymine base pairs of DNA.[136] They were soon superseded by some giant calculations on the guanine-cytosine base pair by Clementi.[137] In both cases, it was found that the Watson-Crick code is remarkably stable even at T = 310°K.

It has been emphasized above that the sums of full CI-expansions of the type (5.21) were essentially invariant under unitary transformations of the truncated basis $\Phi = \{\phi_\mu\}$ of order M. The fact that these expansions were related to the unitary group was probably first discovered by Jordan.[138] The irreducible representations of the unitary group were properly described and classified by Gelfand[139] and co-workers, and the "Gelfand symbols" became useful tools in theoretical physics. In the 1960's, several quantum chemists[140] used the "Unitary Group Approach" (UGA) to simplify the CI-expansion, and Shavitt[141] showed that graphical methods could be highly useful in this connection leading to the "Graphical Unitary Group Approach" (GUGA). In this way, one can today handle CI-expansions which contain more than one million (10^6) configurations.[142] The ultimate goal of this approach is, of course, to be able to reach the *natural expansion* of the total wave function Ψ as directly as ever possible. In his book, Davidson[116] has stressed the importance of the *natural orbitals* in this connection.

Even if one had now essentially solved the problem of the molecular orbitals for the Slater-type orbitals (STO's) *Gaussian* basis sets became more and more popular. For a survey of this change of attitude and of the computational developments, the reader will be referred elsewhere.[143] The exchange of software for molecular calculations was organized by Harrison Shull and Stanley Hagstrom at the University of Indiana in the so-called "Quantum Chemistry Program Exchange": (QCPE), which is still very active. It has been of tremendous importance in the development of our field.

A survey of the standard methods used in quantum chemistry in the late1960's is given in Figure 1.

In connection with the treatment of large biomolecules surrounded by water, Clementi[144] started to use the Monte-Carlo methods once introduced by Metropolis in statistical mechanics to a larger and larger extent. Since this area will be reviewed by several speakers at this conference, I will not go into any details here. It should only be remarked that the Monte Carlo methods as well as the CI-methods require big "number crunchers," and that the development of fast and efficient computers is essential in this connection.

If you do not have access to such big computers, your only chance is to follow Peter Debye's advice from 1955: "Beat them" -- i.e. develop theoretical methods which are so powerful that you do not need so much computational hardware. In this connection, the methods known as *resolvent methods* or propagator techniques seem to be particularly valuable. The latter were developed in quantum field theory and, by way of nuclear physics, they were introduced in quantum chemistry in 1965 by Linderberg and Öhrn[145] They are closely related to the resolvent method, which go even further back in history.

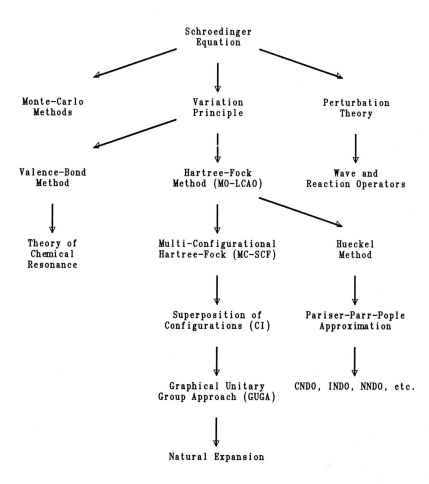

Figure 1. Schematic diagram of the development of some of the methods for solving the many-electron Schrödinger equation. Note that the EDF-method is not included.

Let us for a moment briefly review the basic ideas in the resolvent methods. A Hamiltonian H is said to have the _resolvent_ R(z) defined through the relation

$$R(z) = (z \cdot 1 - H)^{-1} \ , \qquad (6.2)$$

where z is a complex variable. Even if the Hamiltonian H is an unbounded operator, the resolvent is--for $| z - E | > \rho$--a _bounded operator_ satisfying the relation $|| R(Z) \Phi || < (1/\rho) || \Phi ||$. The resolvent R(z) has the same eigenfunctions Ψ as the Hamiltonian H, but with different eigenvalues $r = (z - E)^{-1}$ so that $R \Psi = r \Psi$ and it has further a very simple spectral resolution.

Let us again introduce a normalized reference function φ in the L^2 Hilbert space, and let us consider the eigenvalue problem (3.4) subject to the intermediate normalization (5.28), so that

$$(H - E \cdot 1)\Psi = 0, \; < \varphi \mid \Psi > = 1 \; . \tag{6.3}$$

However, instead of this eigenvalue problem, we will now consider the solution of the inhomogeneous equation:

$$(H - z \cdot 1)\Psi_Z = a \quad , < \varphi \mid \Psi_Z > = 1 \; , \tag{6.4}$$

where the constant a is chosen so that the intermediate normalization is always fulfilled. If the resolvent (6.2) is used to solve the inhomogeneous equation, one obtains

$$\Psi_z = -aR\varphi = \frac{R\varphi}{< \varphi \mid R \mid \varphi >} \; , \tag{6.5}$$

where $a = - < \varphi \mid R \mid \varphi >^{-1}$. We note that the so-called *Weinstein function* $W(z) = < \varphi \mid R \mid \varphi >$ is regular everywhere, except that it has *simple poles* for the eigenvalues $z = E$ of the Hamiltonian H, which correspond to the zero-points $a(z) = 0$. When $z \to E$, the expression (6.5) takes the form ∞/∞, but fortunately it is easy to carry out the limiting procedure. Using the identity $(H - z \cdot 1)R = 1$, one gets the relation:

$$(z \cdot 1 - H)R\varphi = \varphi < \varphi \mid (z \cdot 1 - H)R \mid \varphi > \; . \tag{6.6}$$

Introducing the projector $P = 1 - |\varphi><\varphi|$ and dividing by z, one gets further

$$(1 - \frac{PH}{z})R\varphi = \varphi < \varphi \mid R \mid \varphi > \; , \tag{6.7}$$

or

$$\Psi_z = \frac{R\varphi}{< \varphi \mid R \mid \varphi >} \equiv (1 - \frac{PH}{z})^{-1}\varphi \; . \tag{6.8}$$

This relation, which is valid for all values of the complex variable z and for any choice of the normalized reference function φ, will be referred to as the *resolvent identity* .

If one multiplies the inhomogeneous equation (6.4) to the left by $<\varphi|$, one gets the following alternative form for the quantity a:

$$a = < \varphi \mid H - z \cdot 1 \mid \Psi_Z > =$$

$$= < \varphi \mid H \mid \Psi_z > - z = f(z) - z \quad , \tag{6.9}$$

where

$$f(z) = < \varphi \mid H \mid \Psi_z > =$$

$$= < \varphi \mid H(1 - \frac{PH}{z})^{-1} \mid \varphi > \tag{6.10}$$

is the so-called "bracketing function." The operator

$$W = (1 - \frac{PH}{z})^{-1} \tag{6.11}$$

is referred to as the "wave operator," whereas the operator

$$\Omega = HW = H(1 - \frac{PH}{z})^{-1} \tag{6.12}$$

is referred to as the "bracketing operator."

In order to find the eigenvalues E associated with the equations (6.3), one has now to solve the "algebraic" equation

$$a(z) = f(z) - z = 0 \quad , \tag{6.13}$$

which may be done by means of a first-order iteration procedure or by a second-order Newton-Raphson procedure corresponding to the variation principle. Once an eigenvalue $z = E$ is found, the associated eigenfunction Ψ_E is given by the relation (6.8).

If one puts $H = H_0 + \lambda V$, one can easily derive exact expressions for the wave and reaction operators in Brueckner's theory, and, if one expands these quantities in power series in λ, one obtains the well-known "perturbation expansions" to infinite order. If one instead approximates the wave and reaction operators by means of inner projections, one obtains *rational approximations* to the quantities involved, which are usually much better to handle from the computational point of view. For further details as to this approach, the reader is referred to the review paper[146] I gave in connection with the 5th ICQC in Montreal.

It should further be mentioned, that if one expresses the wave operator W in exponential form, so that $W = e^T$, one gets immediately the connection with the *many-body coupled-cluster theory*.[147]

In the so-called *propagator methods* for evaluating energy differences, ionization energies, electron affinities, etc., one is usually not interested in the Hamiltonian eigenvalue problem but in the *Liouvillian* eigenvalue problem:

$$\hat{L}C = vC \quad , \tag{6.14}$$

with proper boundary conditions. Here the Liouvillian $\hat{L}$ is a *superoperator* defined through the relation

$$\hat{L} = HT - TH \quad , \tag{6.15}$$

for any ordinary operator T; in general any mapping $\hat{M}$ of the operator space $\{T\}$ into or onto itself will be referred to as a superoperator. It is interesting to observe that the eigenvalue problem (6.14) has the formal solution $C = | \Psi_f > < \Psi_i |$ corresponding to a transition from the initial state Ψ_i to the final state Ψ_f and associated with the eigenvalue $v = E_f - E_i$. This approach makes it hence possible to evaluate the energy differences *directly* without solving the Schrödinger equation. However, since the Liouvillian $\hat{L}$ is an unbounded superoperator, it may be more convenient to study the *superresolvent*:

$$\hat{R}(Z) = (z \cdot \hat{1} - \hat{L})^{-1} \quad , \tag{6.16}$$

which is *bounded* whenever $| z - v | > \rho$.

In reality, the special propagator methods are a little bit more complicated than described here, since they usually involve also a "metric" superoperator $\hat{G}$. For more details, the reader is referred elsewhere.[148]

A graphical survey of the developments of the various resultent methods in quantum chemistry is given in Figure 2.

7. Conclusions

During the last few years, we have seen a tremendous development on the computer side from scalar computers, over vector computers, to supercomputers or to large-scale parallel computers á la Clementi. In this process, the computational efficiency of the CI-methods and the Monte-Carlo methods, which both require big "number crunchers," have probably increased by a factor of 100. At the same time, it seems likely that, by going over to more forceful analytical methods--as the resolvent methods--and staying with the scalar computers or moderate vector computers, one can also gain a factor 100. In fact, the resolvent methods seem ideal for large-scale

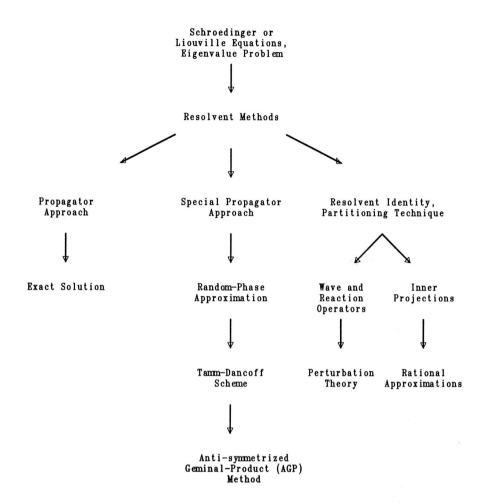

Figure 2. Schematic diagram of the developments of the Resolvent Methods and Propagator Techniques in Modern Quantum Chemistry.

parallel computing, since one may treat various values of the complex variable z simultaneously. The problem is that it may be very difficult to obtain the combined efficiency factor 100 x 100 = 10,000, for the simple reason that there are so many man-years of work invested in the software for the standard methods, that it is exceedingly hard to introduce new, improved methods. This situation presents a very serious dilemma in modern computational quantum chemistry.

For some time, one complained that the computers produced only enormous quantities of numbers, which were hard to understand and which seldom produced new concepts. This situation has been completely changed by the establishment of various types of "*color graphics,*" which have made it possible for ordinary chemists to understand the rather complicated electronic structures of fairly large molecules produced by solving the Schrödinger equation. One has also started to produce "*color movies*" to describe the dynamical phenomena which are obtained by solving the time-dependent Schrödinger equation. Thanks to the computers, it seems hence possible to understand the nature of the solutions to the fundamental equations in pure quantum mechanics.

However, there is another important problem on the horizon connected with the fact that modern theoretical chemistry needs such concepts as temperature, entropy, free energy, etc., in order to be of any real value to the experimentalists. These concepts are not present in pure quantum mechanics based on wave functions and the Schrödinger equation, but are contained in the more general quantum statistics[149] based on system operators Γ and density matrices $\Gamma\,(X \mid X')$. The system operators Γ have the fundamental properties

$$\Gamma = \Gamma^\dagger, \ \ \Gamma > 0, \ \ Tr\Gamma = 1 \ , \tag{7.1}$$

and the set $\{\Gamma\}$ is a *convex set*[150] having limit points satisfying the special relation $\Gamma^2 = \Gamma$. The limit points are hence associated with one-dimensional projectors of the form:

$$\Gamma = \frac{\mid \Psi > < \Psi \mid}{< \Psi \mid \Psi >} \ , \tag{7.2}$$

and they have hence a range in the form of a ray $\{\Psi \cdot \alpha\}$ and are connected with a wave function Ψ. The limit points of the convex set $\{\Gamma\}$ correspond thus to pure quantum mechanics, whereas the interior points are associated with more general physical ensembles. The system operators Γ for a *canonical ensemble* is, e.g., given by the expression

$$\Gamma = \exp \left(\frac{F - H}{kT} \right) \ , \tag{7.3}$$

where the constant F known as the "free energy" is determined by the condition $Tr\Gamma = 1$.

The system operator Γ obeys in general the time-dependent Liouville equation

$$-\frac{h}{2\pi i} \frac{\partial \Gamma}{\partial t} = \hat{L}\Gamma \ , \tag{7.4}$$

where the Liouvillian superoperator $\hat{L}$ is defined by the relation (6.15). We note that the equation (7.4) has exactly the same form as the time-dependent Schrödinger equation and that the methods developed to solve the latter may hence be used also on the former. Since Hilbert-space methods are particularly useful in this connection, it is convenient to introduce a positive binary product

$$\{T_1 \mid T_2\} = Tr \ T_1^\dagger \ T_2 \ , \tag{7.5}$$

in the operator space. This is the famous Hilbert-Schmidt binary product, and the operators which have a finite norm $|| T || = \{T | T\}^{1/2}$ form again a Hilbert space. For the Liouvillian $\hat{L}$, one has the symmetry property:

$$\{T_1 | \hat{L} T_2\} = \{LT_1 | T_2\} \ , \tag{7.6}$$

and the Liouvillian is hence self-adjoint with respect to the binary product (7.5), which is a very important property. Instead of the unbounded superoperator $\hat{L}$, it may then be convenient to study the superresolvent (6.16) which is bounded whenever $| z - v | > \rho$.

In this way, one can develop general quantum statistics in a way which is analogous to pure quantum mechanics, but one can already now foresee that there will be many new computational problems associated with the fact that the binary product (7.5) has a different realization than before. Even in this connection, bigger and better computers will be of essential importance in the future developments.

References

1. M. Planck, Verh. Dt. Phys. Ges. **2**, 202 (1900).
2. A. Einstein, Ann der Physik (Leipzig) **17**, 132 (1905).
3. N. Bohr, Phil. Mag. (6) **26**, 1 (1913).
4. A. Sommerfeld, Sitz. ber. Bayer Akad. Wiss. (Munich), Dec. 1915, 425; see also "Atombau und Spektrallinien," (Vieweg und Sohn, Brauschweig (1919); Z. Phys. **1**, 135 (1920).
5. G.N. Lewis, J. Amer. Chem. Soc. **38**, 762 (1916); I. Langmuir, J. Amer. Chem Soc. **41**, 868 (1919).
6. M.L. Huggins, Doctoral Thesis (1918); see also Am. Scientist **50**, 485 (1962).
7. W. Pauli, Z. Phys. **31**, 765 (1925).
8. N. Bohr, *Drei Aufsätze über Spektren und Atomban* (Wieweg und Sohn, Braunschweig) 1922; see also Collected Works, vol. 4, The Periodic System 1920-1923 (Ed. J. Rud Nielsen, North Holland, Amsterdam, 1977); see also P.O. Löwdin, Int. J. Quantum Chem. **3S**, 331 (1969).
9. R. Courant and D. Hilbert, "Methods of Mathematical Physics," English Edition (Interscience, New York, 1953).
10. E. Schrödinger, Ann. der Physik **79**, 361 (1926).
11. W. Heisenberg, Z.F. Physik **33**, 879 (1925); M. Born, W. Heisenberg and P. Jordan, Z.F. Physik **35,**, 557 (1926).
12. P.A.M. Dirac, Proc. Roy. Soc. London, A113, 621 (1926).
13. E. Schrödinger, Ann. der Physik **79**, 734 (1926); see also J. von Neumann, *Mathematische Grundlagen der Quantemechanik* (Springer, Berlin, 1932).
14. W. Heisenberg, Z. Physik **43**, 172 (1927).
15. W. Heisenberg, Z. Physik **38**, 411 (1926); **39**, 499 (1926).
16. N. Bohr
17. W. Heitler and F. London, Z. Physik **44**, 455 (1927).
18. E.A. Hylleraas, Z. Physik **48**, 81 (1928)
19. E.A. Hylleraas, "Abhandlungen aus den Jahren 1926-37," (Oslo, 1956)
20. Revs. Mod. Phys. **35**, No. 3 (July, 1963).
21. Y. Sugiura, Z. Physik **45** 484 (1927).
22. H.M. James and A.S. Coolidge, J. Chem. Phys. **1**, 825 (1933); **3** 129 (1935).
23. G.E. Uhlenbeck and S. Goudsmit, Naturwiss. **13**, 953 (1925); Nature **117**, 264 (1926).
24. W. Heisenberg, Z. Physik **41**, 239 (1927).

25. P.O. Löwdin, Revs. Mod. Phys. **34**, 80 (1962).
26. J.C. Slater, Phys. Rev. **34**, 1293 (1929).
27. N. Bohr, Proc. Phys. Soc. (London) **35**, 296 (1923).
28. D.R. Hartree, Proc. Cambridge Phil. Soc. **24**, 89 (1928).
29. M. Born and J.R. Oppenheimer, Ann. der Physik (Liepzig) **84**, 457 (1927).
30. J.C. Slater, Phys. Rev. **35**, 210 (1930).
31. V. Fock, Z. Physik **61**, 126 (1930).
32. P.A.M. Dirac, Proc. Cambridge Phil. Soc. **26**, 376 (1930); **27**, 240 (1931).
33. D.R. Hartree, Mem. Proc. Lit. Phil. Soc. Manchester **77**, 91 (1933).
34. V. Fock and M. Petrashen, Phys. Zs. Sow. **6**, 368 (1934); **8**, 547 (1935).
35. T.A. Koopmans, Physica **1**, 104 (1933).
36. D.R. Hartree, "The Calculation of Atomic Structures" (John Wiley and Sons, New York, 1957).
37. P.O. Löwdin, in "NAS-ONR Report Shelter Island Conference 1951," 187 (1951); Quart. Appl. Math. **10**, 97 (1952).
38. R.S. Mulliken, Phys. Rev. **32**, 186 (1928).
39. F. Hund, Z. Physik **51**, 759 (1928); see also J.E. Lennard-Jones, Trans. Faraday Soc. **25**, 668 (1929).
40. F. Bloch, Z. Physik **52**, 355 (1929); **57**, 545 (1929).
41. J.C. Slater, Phys. Rev. **35**, 210 (1930).
42. D.R. Inglis, Phys. Rev. **46**, 135 (1934); J.H. Van Vleck and A. Sherman, Rev. Mod. Phys. **7**, 167 (1935); J.H. Van Vleck, Phys, Rev. **49**, 232 (1936).
43. J.C. Slater, Phys. Rev. **37**, 481 (1931).
44. L. Pauling, Proc. Nat. Acad. SW U.S. **14**, 359 (1928); J. Am. Chem. Soc. **53**, 1367 (1931).
45. P.O. Löwdin, J. Chem. Phys. **21**, 496 (1953).
46. E. Hückel, Z. Physik **70**, 204 (1931); **72**, 310 (1931).
47. E. Hückel, Z. Physik **76**, 628 (1932).
48. G.W. Wheland and L. Pauling, J. Am. Chem. Soc. **57**, 2091 (1939).
49. G.W. Wheland, J. Am. Chem. Soc. **63**, 2025 (1941).
50. K. Fukui, T. Yonezawa and h. Shingu, J. Chem. Phys. **20**, 722 (1952); K. Fukui, T. Yonezawa, C. Nagata and H. Shingu, J. Chem. Phys. **22**, 1433 (1954); K. Fukui, *Theory of Orientation and Stereoselection*, (Springer Verlag, Heidelberg, 1970, 1975).
51. See e.g. A. and B. Pullman, *Quantum Biochemistry*, (John Wiley, New York, 1963), and many later publications.
52. R.B. Woodward and R. Hoffmann, J. Am. Chem. Soc. **87**, 395 (1965).
53. W. Heitler and G. Runier, Nachr. Ges. Wiss. Göttingen **277** (1930).
54. J.C. Slater, Phys. Rev. **37**, 481 (1931), particularly p. 489.
55. L. Pauling J. Chem. Phys. **1**, 280 (1933), and a series of papers in J. Chem. Phys. and J. Am. Chem. Soc.
56. G. Rumer, Nachr. Ges. Wiss. Göttingen **337**, 1932.
57. L. Pauling, J. Chem. Phys. **1**, 280 (1933).
58. F.A. Matsen, A.A. Cantu and R.D. Poshusta, J. Phys. Chem. **70**, 1558 (1966); M. Kotani, K. Ohno and K. Kayama, "Quantum Mechanics of Electronic Structure of Simple Molecules" in *Encyclopedia of Physics*, **37**, 2 (Springer- Verlag, Berlin, 1961) especially pp. 118-142; F.A. Matsen, "Spin-Free Quantum Chemistry" in *Advances in Quantum Chemistry,* **1**, 59-113, (ed. P.O. Löwdin, Academic Press, New York, 1964). Matsen reproduces Kotani's table in an appendix and shows how to derive non-diagonal matrix elements from the diagonal elements given by Kotani; P.O. Löwdin, Colloq. Inter. Centre. Natl. Rech. Sc. (Paris) **82**, 23 (1958).
59. H. Shull, Int. J. Quantum Chem. **3**, 523 (1969).
60. J.H. VanVleck and A. Sherman, Revs. Mod. Phys. **7**, 167 (1935).
61. M. Kotani and co-workers, J. Phys. Soc. Japan **12**, 707, 135 (1957); M. Kotani, *Handbuch der Physik* (ed. S. Flugge, Springer, Berlin, 1961), vol. 37, part II, p. 124; R. McWeeny, Proc. Roy. Soc. (London) A**253**, 242 (1959); Revs. Mod. Phys. **32**, 335 (1960); Phys. Rev. **126** 1028 (1962); etc.; F.A. Matsen, J. Phys. Chem. **68**, 3282 (1964),

F.A. Matsen, A.A. Cantu and R.D. Poshusta, J. Phys. Chem. **70**, 1558 (1966); F.A. Matsen, J. Phys. Chem. **70**, 1568 (1966); F.A. Matsen and A.A. Cantu, J. Phys. Chem. **72**, 21 (1968); M. Simonetta and A. Gavezzotti, Adv. Quantum Chem. (ed. P.O. Löwdin, Academic Press, New York, 1980) **12**, 103; and numerous other references.

62. W.A. Goddard III, Phys. Rev. **157**, 81 (1967), **169**, 120 (1968); J. Chem. Phys. **48**, 1008, 5337 (1968); W.E. Palke and W.A. Goddard III, J. Chem. Phys. **50**, (1969).

63. P.O. Löwdin and O. Goscinski, Int. J. Quantum Chem. **3** S, 533 (1970).

64. L. Pauling, *The Nature of the Chemical Bond* (2nd ed., Cornell University Press, 1941).

65. M. Kotani, Proc. Phys. Math. Soc. Japan **19**, 460 (1937); M. Kotani and M. Siga, Proc. Phys. Math. Soc. Japan **19**, 471 (1937); A. Amemiya, Bull. Phys. Math. Soc. Japan **17**, 67 (1943); M. Kotani, A. Amemiya, E. Ishiguro and T. Kimura, "Table of Molecular Integrals" (Maruzen, Tokyo, 1955).

66. P.O. Löwdin, "A Theoretical Investigation into Some Properties of Ionic Crystals" (Thesis; Almqvist and Wiksels, Uppsala, 1948).

67. A.S. Coolidge, Phys. Rev. **42**, 189 (1932); R. Landshoff, Z. Physik **102**, 201 (1936); Phys. Rev. **52**, 246 (1937).

68. S.O. Lundquist and P.O. Löwdin, Ark. Fys. **3**, 147 (1951) P.O. Löwdin, Adv. Phys. **5**, 1 (1956); R.R. Sharma, Phys. Rev. A13, 517 (1976); H.W. Jones and C.A. Weatherford, Int. J. Quantum Chem. Symp. **12**, 483 (1978); H.W. Jones, Int. J. Quantum Chem. **18**, 709 (1980); H.W. Jones, Int. J. Quantum Chem. **21**, *Conference on ETO Multicenter Integrals* (Reidel, Dordrecht, 1982); E.J. Weniger and E.O. Steinborn, Phys. Rev. A **28**, 2026 (1983); H.W. Jones, Int. J. Quantum Chem. **23**, 953 (1983); H.W. Jones, Phys. Rev. A **30**, 1 (1984).

69. See H.W. Jones, Proc. 5th ICQC, Montreal, Int. J. Quantum Chem. **29**, 177 (1986).

70. P.O. Löwdin, Ark. Mat. Astr. Fys. **30**, 1 (1948); J. Chem Phys. **18**, 365 (1950).

71. J.C. Slater and G.F. Koster, Phys, Rev. **94**, 1498 (1954).

72. R. Pauncz, J. de Heer and P.O. Löwdin, J. Math. Phys. **1**, 461 (1960); J. Chem. Phys. **36**, 2247, 2257 (1962).

73. P.O. Löwdin, Int. J. Quantum Chem. **1S**, 811 (1967).

74. R.H. Parmenter, Phys. Rev. **86**, 552 (1952).

75. P.O. Löwdin, Ann. Rev. Phys. Chem. **11**, 107 (1960); J. Appl. Phys. S33, 251 (1962).

76. P.O. Löwdin, Adv. Phys. **5**, 1 (1956), particularly p. 49.

77. P.O. Löwdin, Adv. Quantum Chem. **5**, 185 (Academic Press, New York, 1970).

78. R. Feynman, "Surely You're Joking, Mr. Feynman" (W. W. Norton, New York, 1985).

79. C.A. Coulson, Proc. Cambridge Phil. Soc. **34**, 204 (1938).

80. R.S. Mulliken, J. Chem. Phys. **46**, 497, 675 (1949).

81. C.A. Coulson and H.C. Longuet-Higgins, proc. Roy. Soc. (London) **A191**, 39; **192**, 16 (1947); **193**, 447, 456 (1948); **195**, 188 (1948).

82. J. Lennard-Jones, Proc. Roy. Soc. London **A202**, 155 (1950); G.G. Hall and C.C.J. Roothaan, Revs. Mod. Phys. **23** 69 (1951).

83. P.O. Löwdin, Phys. Rev. **97**, 1490 (1955); R. McWeeny, Proc. Roy. Soc. London **A223**, 63, 306 (1954).

84. NAS-ONR Report from the Shelter Island Conference in 1951.

85. P.O. Löwdin, ref. 84, p. 187.

86. H. Shull and P.O. Löwdin, J. Chem. Phys. **25**, 1035 (1956)

87. J.C. Slater, Phys. Rev. **81**, 385 (1951); **82** 538 (1951); Revs. Mod. Phys. **25**, 199 (1953).

88. H. Shull and P.O. Löwdin, J. Chem. Phys. **23**, 1362 (1955)

89. J.C. Slater, Phys. Rev. **81**, 385 (1951); **82**, 538 (1951); Revs. Mod. Phys. **25**, 199 (1953).

90. H. Shull and P.O. Löwdin, J. Chem. Phys. **23**, 1362 (1955)

91. S.F. Boys, Proc. Roy. Soc. (London), **A200**, 542 (1950); Svensk Kem. Tidskr. **67**, 367 (1955); Proc. Roy. Soc. (London) **A258**, 402 (1960); Revs. Mod. Phys. **32**, 296 (1960); see also Boys, Cook, Reeves and Shavitt, Nature **178**, 1207 (1956); S.F. Boys and G.B. Cook, Revs. Mod. Phys. **32**, 285 (1960).

92. R.G. Parr, J. Chem. Phys. **20**, 1499 (1952); R.G. Pariser and R.G. Parr, J. Chem. Phys. **21**, 466 (1953); J.A. Pople, Trans. Far. Soc. **49**, 1375 (1953); R.G. Parr and R. Pariser,

J. Chem. Phys. **23**, 711 (1955); J.A. Pople, Proc. Phys. Soc. (London) **A68**, 81 (1955); J. Phys. Chem. **61**, 6 (1957).

93. P.O. Löwdin, Proc. Int. Conf. Theor. Physics Japan in 1953, 13 (1954); Svensk Kem. Tidskr. **67**, 380 (1955).

94. I. Fischer-Hjalmars, Adv. Quantum Chem. **2**, 25 (Academic Press, New York, 1965); K. Ohno, Adv. Quantum Chem. **3**, 240 (Academic Press, New York, 1967).

95. Proc. Symp. Mol. Physics at Nikko, Japan, in 1953; Report Symp. Quantum Theory of Molecules in Stockholm and Uppsala, Svensk Kemisk Tidskr. **67**, 365-398 (1955); Report Molecular Quantum Mechanics Conf. in Austin, Texas, 1955, Texas J. Science 8 (1956); Report Paris Conf. Mol. Wave Mechanics 1957, (Ed. du Centre Nat. Rech. Sci. **82** (1958); Proc. 1958 Boulder Conf. Mol. Physics, Revs. Modern Phys. **32** (1960).

96. E. Wigner, Phys. Rev. **46**, 1002 (1933); Trans Faraday Soc. **34**, 678 (1938).

97. P.O. Löwdin, J. Chem. Phys. **19**, 1570, 1579 (1951).

98. P.O. Löwdin, Proc. Symp. Mol. Physics at Nikko, Japan, in 1953, 13 (1954); Phys. Rev. **97**, 1509 (1955); Proc. 10th Solvay Conf., 1954, p. 71 (Inst. Internat. de Physique Solvay. 10^e Conseil de Physique Tenu a Bruxelles 1954: Les Electrons dans le Metaux, Rapports et Discussions, Bruxelles 1955); Rev. Mod. Phys1 **32**, 328 (1960).

99. T. Itoh and H. Yoshizumi, J. Phys. Soc. Japan **10**, 201 (1955); J. Chem. Phys. **23**, 412 (1955); Busseiron Kenkyu **83**, 13 (1955).

100. R. Lefebvre, H.H. Dearman and H.M. McConnell, J. Chem. Phys. **32**, 176 (1960); P.O. Löwdin, R. Pauncz and J. de Heer, J. Chem. Phys. **36**, 2247, 2257 (1962); J. de Heer, J. Chem. Phys. **37**, 2080 (1962); R. Pauncz, J. Chem. Phys. **37**, 2739 (1962); J. de Heer, Rev. Mod. Phys. **35**, 631 (1963); R. Pauncz, in *Molecular Orbitals in Chemistry, Physics, and Biology*, ed. P.O. Löwdin, Academic Press, New York, 1964, p. 433; Tetrahedron **19**, Suppl. 2, 43 (1963); J. Chem. Phys. **43**, S69 (1965); O. Goscinski and J.L. Calais, Arkiv Fysik **29**, 135 (1965); J. de Heer and R. Pauncz, J. Chem. Phys. **39**, 2314 (1963); R. Pauncz, *Alternant Molecular Orbital Method*, W.B. Saunders, Philadelphia, 1967; J.L. Calais, Arkiv Fysik, **28**, 479, 511, 539 (1965); **29**, 255 (1965); J.L. Calais, Int. J. Quantum Chem. **13**, 661 (1967). For more complete references, see I. Mayer, Adv. Quantum Chem. **12**, 189 (Academic Press, New York, 1980).

101. P.O. Löwdin, Texas J. Science **8**, 163 (1956).

102. A. Fröman, Phys. Rev. **112**, 870 (1958).

103. J. Linderberg and H. Shull, J. Mol. Spec. **4**, 30 (1960).

104. K.A. Brueckner, C.A. Levinson and H.M. Mahmoud, Phys. Rev. **95**, 217 (1954); K.A. Brueckner, Phys. Rev. **96**, 508 (1954); **97**, 1353 (1955); **100**, 36 (1955); K.A. Brueckner and C.A. Levinson, Phys. Rev. **97**, 1344 (1955); L.S. Rodberg, Ann. Phys. (N. Y.) **2**, 199 (1957); to mention only a selection of the rich literature on this subject.

105. P.O. Löwdin, J. Math. Phys. **3** 1171 (1962).

106. H.A. Bethe, Phys. Rev. **103**, 1353 (1956); J. Goldstone, Proc. Roy. Soc. (London), Ser. A, **238**, 511 (1957).

107. R.K. Nesbet, Proc. Roy. Soc. (London), Ser. A, **230**, 312 (1955).

108. L. Brillouin, Actualités Sci. et Ind. **71** (1933); No. **159** (1934); C. Moller and M.S. Plesset, Phys. Rev. **46**, 618 (1934). For a current generalization, see P. O. Löwdin, Proc. Ind. Acad. Sciences (Chem. Sci.) **96** 121 (1986).

109. R.K. Nesbet, Proc. Roy. Soc. (London) **A230**, 312 (1955); R.K. Nesbet, Quarterly Progress Report, Solid State and Molecular Theory Group, MIT, July 15, p. 3, Oct. 15, p. 47, unpublished, (1956); R.K. Nesbet, Phys. Rev. **109**, 1632 (1958); R.K. Nesbet, Phys. Rev. **118**, 681 (1960); R.K. Nesbet, Rev. Mod. Phys. Mod. Phys. **33**, 28 (1961).

110. O. Sinanoglu, Proc. Roy. Soc. (London) **A260**, 379 (1961); O. Sinanoglu, J. Chem. Phys. **36**, 706 and 3198 (1962); O. Sinanoglu, Adv. Chem. Phys. **6**, 315 (1968).

111. P.O. Löwdin, Adv. Chem. Phys. **2**, 207 (ed. I. Prigogine, Interscience, New York, 1959); see also Proc. 1958 Robert A. Welch Foundation Conf. Chem. Research, II. Atomic Structure, 5 (1960).

112. J.C. Slater, Phys. Rev. **91**, 528 (1953).

113. P.O. Löwdin, Phys. Rev. **97**, 1474 (1955).

114. K.Husimi, Proc. Phys. Math. Soc. Japan **22**, 264 (1940).

115. P.O. Löwdin, J. Phys. Chem. **61**, 55 (1957).

116. P.O. Löwdin and H. Shull, Phys. Rev. **101**, 1730 (1956).

117. E. Davidson, *Reduced Density Matrices in Quantum Chemistry* (John Wiley, New York, 1963).

118. See Proc. 1985 Kingston Symposium on Reduced Density Matrices, the Representability Problem, and the Electron Density Functional Method.

119. C.C.J. Roothaan, J. Chem. Phys. 19, 1450 (1951); K. Ruedenberg, J. Chem. Phys. **19**, 1459 (1951); and numerous subsequent papers in J. Chem. Phys.

120. F.A. Matsen, "Tables of Molecular Integrals" (Austin, Texas, 1955); see also H. Preuss, *Integraltafeln zur Quantenchemi*, 4 vols. (Springer, Berline 1956-61).

121. Proc. 1958 Boulder Conference, Revs. Modern Phys. **32**, (1960).

122. P.O. Löwdin, Proc. 1985 Sanibel Symposia (Quantum Chemistry), Int. J. Quantum Chem. S19, (April, (1986).

123. Proc. 1963 Hylleraas Symp., Revs. Mod. Phys. 35 (1963).

124. J.A. Coleman, Revs. Mod. Phys. **35**, 668 (1963).

125. P.O. Löwdin, Revs. Mod. Phys. **35**, 496 (1963), in *Quantum Theory of Atoms, Molecules, and the Solid-State* (Slater Dedicatory Volume, Academic Press, 1966), p. 601.

126. P.O. Löwdin, Rev. Mod. Phys. **97** 1509 (1955); see also paper in Slater volume (see ref. 124), p. 601. For more complete references, see I. Mayer, Adv. Quantum Chem. **12**, 189 (Academic Press, New York, 1980).

127. D.J. Thouless, *The Quantum Mechanics of Many-Body Systems* (Academic Press, New York, 1961); W. Adams, Phys. Rev. **127**, 1650 (1962); J. Cizek and J. Paldus, J. Chem. Phys. **47**, 3976 (1967); J. Paldus and J. Cizek, Prog. Theor. Phys. **42**, 769 (1969); J. Chem. Phys. **52**, 2919 (1970); J. Cizek and J. Paldus, J. Chem. Phys. **53**, 821 (1970); J. Paldus and J. Cizek, J. Polym. Sci., Part C **29**, 199 (1970); J. Paldus, J. Cizek Phys. Rev. A 2 2268 (1970); J. Paldus and J. Cizek, J. Chem. Phys. **54** (1971) J. Paldus, J. Cizek and B. A. Keating, Phys. Rev. A 8, 640 (1973); a. Laforgue, J. Cizek and J. Paldus, J. Chem. Phys. **59**, 2560 (1973); W.G. Laidlaw, Int. J. Quantum Chem. 7, 87 (1973); J. Paldus and A. Veillard, Chem. Phys. Lett. 50 (1977); J. Paldus, J. Cizek, A. Laforgue, Int. J. Quantum Chem. **13**, 41 (1978); J. Paldus and A. Veillard, Mol. Phys. **35**, 445 (1978); M. Benard and J. Paldus, J. Chem. Phys. **72**, 6546 (1980); H. Fukutome, Prog. Theor. Phys. **40**, 1156 (1972); **49**, 22 (1973); **50**, 1433 (1973); **52**, 115 (1974); **52**, 1766 (1974); **53**, 1320 (1975); M. Ozaki and H. Fukutome, Prog. Theor. Phys. **60**, 1322 (1978); M. Ozaki, Progr. Theor. Phys. **62**, 1183 (1979); M. Ozaki, Prog. Theor. Phys. **63**, 84 (1980); H. Fukutome, Int. J. Quantum Chem. 20, 955 (1981); P.O. Löwdin, Proc. Ind. Acad. Sci. (Chem. Sci.) **96**, 121 (1986).

128. E. Lieb and B. Simon, Comm. Math. Phys. **53**, 185 (1977).

129. J.C. Slater, Phys. Rev. **81**, 385 (1951); see also P. O. Löwdin, Phys. Rev. **97**, 1494 (1955), particularly p. 1487.

130. R. Gaspar, Acta Phys. Acad. Sci. Hung. **3**, 263 (1954); W. Kohn and L.J. Sham, Phys. Rev. **140**, A1193 (1965).

131. J.C. Slater, Adv. Quantum Chem. **6**, 1 (Academic Press, New York, 1972), and the proceedings from the Sanibel Symposia, 1965-1975.

132. L.H. Thomas, Proc. Camb. Phil. Soc. **23**, 542 (1927); E. Fermi, Z. Physik 48, 73 (1928); P.A.M. Dirac, Proc. Camb. Phil. Soc. **26**, 376 (1930); P. Gombas, *Die Statistische Theorie des Atoms und Ihre Anwendungen* (Springer, Wein, 1949).

133. P. Hohenberg and W. Kohn, Phys. Rev. **136**, B 864 (1964).

134. E. Clementi, IBM J. Res. Develop. **9**, No. 1 (1965).

135. This series, known as "Studies in perturbation Theory I-XVI," was first communicated as a series of "Technical Notes" from the Uppsala Quantum Chemistry Group and later published as follows: P.O. Löwdin, J. Mol. Spectrosc. **10**, 12 (1963); **13**, 326 (1964); **14**, 112 (1964); 14, 119 (1964), 14, 131 (1964); J. Math. Phys. 3, 969 (1962); 3, 1171 (1962); **6**, 1341 (1965); Phys. Rev. **139**, A357 (1965); J. Chem. Phys. 43, S175 (1965); Int. J. Quantum Chem. **2**, 867 (1968); Int. J. Quantum Chem. S4, 231 (1971); 5, 685 (1971)

(together with O. Goscinski); Phys. Scrip. **21**, 229 (1980); Adv. Quantum Chem. (Academic, New York 1980) **12**; Int. J. Quantum Chem. **21**, 69 (1982).

136. Proc 1968 Frascati Conference on "Correlation Effects in Atoms and Molecules," Adv. Chem. Phys. **14**, (eds. I. Prigogine and S. Rice, Interscience, New York, 1969).

137. P.O. Löwdin, Adv. Quantum Chem. **2**, 213 (1965); Revs. Mod. Phys. **35**, 724 (1963); International Science and Technology (Conover-Mast Publication, New York), May, 1963; Biopolymers Symp. **1**, 161 (1964); *Electronic Aspects of Biochemistry*, p. 167 (ed. B. Pullman, Academic Press, New York, 1964); see also, E. Pollard and M. Lenke, Mutation Research **2** 214 (1965) and P.O. Löwdin, *ibid*, **2**, 218 (1965); R. Rein and J. Ladik, J. Chem. Phys. **40**, 2466 (1964); J. Ladik, Preprint QB 8, Uppsala Quantum Chemistry Group (1963); R. Rein an F. Harris, J. Chem. Phys. **41**, 3393 (1964); R. Rein and F. Harris, J. Chem. Phys, **42**, 2177 (1965); R. Rein and F. Harris, Jr. Chem. Phys. **43**, 4415 (1965); S. Lunell and G. Sperber, Prepring QB 32, Uppsala Quantum Chemistry Group (1966), published in J. Chem. Phys. **46**, 2119 (1967); P.O. Löwdin, Pont Acad. Vatican Scrip. Varia **31**, "Semaine d'Etudy sur les Forces Moleculaires," 637 (1967).

138. E. Clementi, Proc. Natl. Acad. U.S.A. **69**, 2942 (1972).

139. P. Jordan, Z. Phys. **94**, 531 (1955).

140. I. M. Gelfand and M.L. Zetlin, Dokl. Akad. Nauk SSSR **71**, 825, 1017 (1950); I.M. Gelfand and M.I. Graev, Izv. Adak. Nauk SSSR, Ser. Mat. **29**, 1329 (1965) { Amer. math. Soc. Transl. **64**, 116 (1967).}

141. J. Paldus and J. Cizek, Adv. Quantum Chem. **9**, 105 (1975); F.A. Matsen, Adv. Quantum Chem. **11**, 223 (1978); J. Paldus, J. Chem. Phys. **61**, 3321 (1974); Int. J. Quantum Chem. **S9**, 165 (1975).

142. I. Shavitt, Int. J. Quantum Chem. **S11**, 131 (1977); **S12**, 5 (1978).

143. P. Siegbahn, J. Chem. Phys. **72**, 1647 (1980); P. Saxe, D.J. Fox, H.F. Schaeffer and N.C. Handy, J. Chem. Phys. **77** 5584 (1982).

144. H.F. Schaefer III, *Methods of Electronic Structure Theory*, (Plenum Press, New York, 1977); P.O. Löwdin, Adv. Quantum Chem. **12**, 263 (1980).

145. E. Clementi and H.E. Popkie, J. Chem. Phys. **57**, 1077 (1972), and many subsequent papers, e.g. E. Clementi and G. Corongiu, Int. J. Quantum Chem. **S10**, 31 (1983); G. Corongiu and J.H. Detrich, IBM J. Res. Devel. (1983); E. Clementi, G. Corongiu, J.H. Detrich S. Chin and L. Domingo; Int. J. Quantum Chem. **S18**, 601 (1984)

146. J. Linderberg and Y. Öhrn, Proc. Roy. Soc. London Ser. **A285**, 445 (1965); Y. Öhrn and J. Linderberg, Phys. Rev. **139**, A1063 (1965); J. Linderberg and Y. Öhrn, Chem. Phys. Lett. **1**, 295 (1967); J. Linderberg and Y. Öhrn, *Propagators in Quantum Chemistry* (Academic, New York, 1973).

147. P.O. Löwdin, Proc. 5th ICQC in Montreal, Int. J. Quantum Chem. **29** (May, 1986).

148. For a general review of this field, see R.J. Bartlett, Ann. Rev. Phys. Chem. **32**, 359 (1981). See also two forthcoming papers in Adv. Quantum Chem. **18** (Academic Press, 1986) by M.R. Hoffmann and H.F. Schaefer III, respective S.A. Kucharski and R.J. Bartlett.

149. P.O. Löwdin, Adv. Quantum Chem. **17**, 285 (Academic Press, New York, 1985).

150. J. von Neumann, *Mathematische Grundlagen der Quantenmechanik* (Springer, Berlin, 1932); G. Birkhoff and J. von Neumann, Ann. of Mathematics **37**, 823 (1936).

151. P.O. Löwdin, Int. J. Quantum Chem. **12**, Suppl. 1, 197 (1978); **21**, 275 (1982).

The author would like to express his gratitude of Dr. Ceferino Obcemea of the Florida Quantum Theory Project for most valuable help in collecting this bibliography.

JAPANESE SUPERCOMPUTERS AND MOLECULAR ORBITAL CALCULATIONS

Kimio Ohno

Department of Chemistry
Faculty of Science
Hokkaido University
Sapporo, Japan

I. INTRODUCTION

I am a quantum chemist and a user of computers. The following contribution consists of two parts. In the first part, a review of the current status of Japanese supercomputers will be given. In the second part, a new supercomputer system which will be operating from January 1986 at the Institute for Molecular Science (IMS), Okazaki, Japan will be explained and some of its characteristics will be discussed.

II. CURRENT STATUS OF JAPANESE COMPUTERS

It is not easy to report the current status of supercomputers since it it time dependent. There are three Japanese companies which are making supercomputers. They are Hitachi, Fujitsu and NEC (Nippon Electric Co.). In Table 1, brief outlines of their supercomputers are given. The data are collected from commercially available sources during the period of April-July, 1985.

The performance data of the following simple loop

$$A(I) = A(I) + S * V(I)$$

with a loop length 100 are shown in Table 2. In Table 3, the performance data of 29 loops due to Nishimoto are compared. In Table 4, the speed in terms of MFLOPS for execution of the 14 Livermere loops with double precision is shown for the largest machine of the three Japanese makers.

III. NEW SYSTEM AT IMS

A rough outline of the new Hitachi system, which will be installed at IMS, is given in Fig. 1. The utmost importance is attached to reduction of I/O time since most of

Table 1. Features of Japanese Supercomputers

Maker	HITACHI		FUJITSU				NEC	
Name	HITAC S810		FACOM				ACOS	
	10	20	VP50	VP100	VP200	VP400	SX1	SX2
Available since	Oct.1983		Sep.1985	Dec.1983		Dec.1985	Sep.1985	
Peak Speed (MFLOPS)	315	630	140	285	570	1.140	570	1300
Cycle Time (ns)	15		7	7	7	7	7	6
No. Command Scalar	195		195				81	
Vector	80	83	83				70	85
VPU								
Register								
General	16(32b)		16(32b)					
Floating	16(64b)		16(64b)				40	
Control	16(32b)		16(32b)					
Vector (KB)	32	64	32	32	64	128	40	80
Mask (KB)	16	16	0.5	0.5	1	2.0	1	2
Data Format (b)								
Logic	64		1,64				64	
Fixed P.	32		32				32	
Floating P.	32,64		32,64				32,64	
No. Pipeline (log.xphys.)	6x1	6x2	5x1	6x1	6x2	5x4	4x2	4x4
SPU								
Speed (MIPS)	13?		17?				30?	
Buffer Storage (KB)	64	256	64				64	
Mainframe	M280H		M380				ACOS510	
Speed (MIPS)	13?		17?				26?	
Main Memory								
Max. (MB)	128	256	128	128	256	256	256	256
Throughput (G B/s)	2.3	4.6	4				11	
Max. No. Interleave	32	64	128	128	256	256	256,512	512
I/O								
No. Channel	8,16,32		16,32				32	
Max. Throughput (MB/s)	96		48				50	
External Storage								
Max. (GB)	0.5	1	-				2	
Max. Throughput (MB/s)	500	500,1000	-				1300	
Magnetic Disk								
Max. (GB)	5		2.5				2.6	
Throughput (MB/s)	3		3				3	
Av. Access time (ms)	17+8.4		15+8.4				16+8.3	
Semiconductor or Electronic Disk								
Max (GB)	5		2.5				2.6	
Throughput (MB/s)	3		3				3	
Access time (ms)	0.3		0.3				0.3	
Optical disk								
Max (GB)	84		88				48	
Throughput (MB/s)	1.5		0.8				0.8	
Access time (ms)	200		250				8000	

Table 2.

COMPUTER	TYPE	COMPILER	MFLOPS
FACOM VP100	S	FORTRAN77/VP	123.1
FACOM VP100	D	FORTRAN77/VP	123.0
HITAC S810/20	D	FORT77/HAP PP	104.4
HITAC S810/20	S	FORT77/HAP PP	92.3
CRAY X-MP	S	CFT (INLINE)	82.1
CRAY X-MP	S	CFT (BLAS)	74.8
HITAC S810/20	D	FORT77/HAP	62.7
HITAC S810/20	S	FORT77/HAP	59.6
CYBER 205	S	FTN (BLAS)	25.4
HITAC M280H	S	FORT77/IAP	24.2
HITAC M200H	S	FORT77/IAP	23.2
CRAY 1S	S	CFT (INLINE)	19.1
CRAY 1S	S	CFT (BLAS)	18.1
HITAC M200H	D	FORT77/IAP	15.7
HITAC M280H	D	FORT77/IAP	14.7
CRAY 1S	S	CFT	12.3
CYBER 205	S	FTN	8.37

the jobs running at IMS are I/O bound. External storage of 1GB with maximum throughput of 500 MB/s will be helpful although this is a peripheral memory device and will be used as a work file by READ or WRITE statement. Parallel I/O is also adopted to reduce the I/O time. One data set is divided into several parts and each part is stored in a disk volume and parallel WRITE and READ becomes possible. The multiplicity of 8-16 is being contemplated. In addition to this, advance I/O of plural blocks may speed up the I/O speed by factor 3 so that if everything goes well, the I/O speed may increase by factor 24-48. Among total user's disk space of 70 GB, 40 GB will be devoted to parallel I/O and 30 GB to ordinary I/O. The supercomputer S-810/10 will be upgraded in spring 1987 to a new one which has a peak speed of 1.5 GFLOPS and the new scalar processor will have a speed of about 40 MIPS.

Table 3. Nishimoto 29 Loop (Mflops) (Loop Size = 10000)

NO.	S810/20	S810/10	VP100	SX2[1]	EQUATION
	Loop Size = 10000				
1	117.096	53.333	60.864		$SZ(I) = SX(I)$
2	191.939	121.507	86.022		$SZ(I) = SX(I) + SY(I)$
3	184.672	121.507	86.133		$SZ(I) = SX(I)-SY(I)$
4	191.939	119.976	122.775		$SZ(I) = SX(I) = SY(I)$
5	123.077	61.939	32.632		$SZ(I) = SX(I)/SY(I)$
6	228.833	123.077	83.752		$SZ(I) = -SX(I)$
7	240.096	121.507	116.279		$SF = SF + SX(I)$
8	351.288	184.615	119.570		$SF = SF + SX(I) = SY(I)$
9	314.713	243.013	84.638		$SZ(I) = SX(I) + SY(I) = SZ(I)-1$
10	234.192	123.077	115.942		$SF = SF-SX(I)$
11	360.144	182.260	119.570		$SF = SF-SX(I) = SY(I)$
12	261.780	175.644	75.796		$SZ(I) = SZ(I) + SX(I) = SY(I)$
13	257.069	175.644	75.453		$SZ(I) = SZ(I)-SX(I) = SY(I)$
14	123.001	55.804	48.077		$DZ(I) = DX(I)$
15	240.096	123.077	111.173	253.34	$DZ(I) = DX(I) + DY(I)$
16	218.103	123.077	134.590		$DZ(I) = DX(I)-DY(I)$
17	246.002	121.507	134.499	251.78	$DZ(I) = DX(I) = DY(I)$
18	123.077	61.939	31.551	232.86	$DZ(I) = DX(I)/DY(I)$
19	246.002	123.077	147.820		$DZ(I) = -DX(I)$
20	246.002	124.688	195.695	567.24	$DF = DF + DX(I)$
21	360.144	182.260	278.810	972.32	$DF = DF + DX(I) = DY(I)$
22	480.192	225.861	211.193		$DZ(I) = DX(I) + DY(I) = DZ(I)-1$
23	246.002	121.507	196.850		$DF = DF-DX(I)$
24	360.144	184.615	282.220		$DF = DF-DX(I) = DY(I)$
25	334.821	182.260	162.338	415.66	$DZ(I) = DZ(I) + DX(I) = DY(I)$
26	313.152	184.615	161.377		$DZ(I) = DZ(I)-DX(I) = DY(I)$
27	48.008	45.290	53.107		$D(I) = S(I)$
28	92.336	44.863	81.367		$S(I) = D(I)$
29	5.941	5.941	7.404		CONTINUE
Aver.	232.409	128.189	117.845		

Table 4.

	S810/20	VP200	SX2[1]
1	228.0	326.4	745.1
2	239.4	178.1	417.7
3	211.9	331.1	531.4
4	59.2	88.0	127.7
5	5.4	10.0	13.1
6	4.6	9.5	14.2
7	232.7	326.1	794.8
8	48.8	90.4	149.9
9	207.6	257.4	538.0
10	49.0	84.8	87.2
11	9.8	4.8	24.0
12	93.0	114.1	233.2
13	4.2	6.2	8.1
14	8.5	13.9	24.0
Total	1402.1	1840.8	3708.5
Aver.	100.15	131.49	264.89

ACKNOWLEDGEMENT

I am deeply indebted to Professors K. Morokuma and H. Kashiwagi and Dr. U. Nagashima of IMS for giving out information abouth their new system. I am also grateful to Hitachi, Fujitsu and NEC for answering my inquiries.

REFERENCE

1. K. Nishihara, Y. Fukuda, K. Sawai, and C. Yamanaka: Research Report of Institute of Laser Engineering, October 14, 1985.

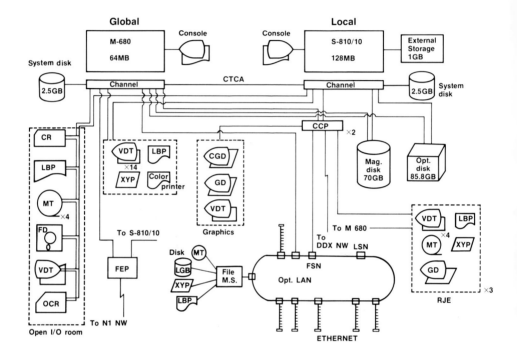

Fig.1 The New System at IMS.

EXPERIENCES WITH THE CYBER 205 FOR QUANTUM CHEMICAL CALCULATIONS

W. Kutzelnigg, M. Schindler, W. Klopper, S. Koch, U. Meier and H. Wallmeier

Lehrstuhl für Theoretische Chemie
Ruhr-Universitat Bochum

D-4630 Bochum, FRG

1. INTRODUCTION

In early 1982 a CYBER 205 has been installed at the Ruhr-Universitat Bochum (RUB). It was then practically a prototype and we had the usual frustrating hardware and software problems that one always has with new computers (probably a bit more than the usual ones), but these have by now been almost eliminated. The CYBER 205 at Bochum belongs to the land (state) Nordrhein-Westfalen and is accessible to all universities of the land. The share of the RUB is decreasing continuously.

We never had the choice between a CYBER 205 and a CRAY 1, our alternative was to get a CYBER 205 or no supercomputer at all. We might have had the possibility to choose a better configuration - as our colleagues in Karlsruhe did, who got a CYBER 205 about a year later. The CYBER 205 in Karlsruhe had 1 Mword central memory and one pipe, while we got 1/2 Mword and 2 pipes. In comparing the two installations with the same programs, we found no case where our two pipes presented an advantage, but our small core caused a number of problems. A warning to all those who might buy a supercomputer: don't save money on core storage! You need a large core for the long vectors that are necessary for a high performance. We have applied for an extension of the core to 2 Mwords and we are still waiting for it. Again our colleagues in Karlsruhe have been luckier. They got this extension already more than half a year ago.

The fact that Bochum and Karlsruhe got the same kind of supercomputer has been fortunate, because this allowed us to exchange ideas and programs with the group of Prof. R. Ahlrichs and to avoid duplication of work.

This lecture will be organized as follows. In Sec. 2 I shall give a short description of the most important features (mainly as far as programming is concerned) of the CYBER 205, in comparison with the somewhat better known CRAY 1 (with which I have no personal experience, so that I can only rely on the literature[1]).

In Sec. 3 I shall summarize our experience with the vectorization of standard quantum chemical programs, while Sec. 4 will deal with some less traditional programs and a discussion of new strategies.

2. THE CYBER 205, IN COMPARISON WITH THE CRAY 1

2.1 Principles of the Architecture; Half-Performance- and Break-Even Lengths

Many scientists tend to prefer the CRAY. It appears in fact that living with the CRAY is a lot easier than with the CYBER 205. On the other hand the CYBER 205 is more of a challenge and demands a much more radical attitude towards coding programs and towards thinking of new strategies.

The comparison of the CRAY 1 (and its successors such as the CRAY 1S and the CRAY XMP) with the CYBER 205 is rather illustrative. More details of this comparison are found in the excellent book by Hockney and Jesshope.[1] Both have pipelined processors (as contrasted to the less common array processors), but they have very different architectures. On the CYBER 205 all vector instructions go from core memory to core memory. This implies a rather large start-up time (at least 50 clock periods of 20 nsec i.e. ~ 1 μsec), but a fast asymptotic speed: $(1/p)$ clock periods are needed for one addition or multiplication, where p is the number of pipes. It also implies that vector instructions can be applied directly only to adjacent locations in core memory (as A(1), A(2), A(3)...). The CRAY 1 uses vector registers and has a much smaller start-up time (3 clock periods of 12.5 nsec i.e. ~ 40 nsec). It can also deal easily with addresses separated by fixed increments. This makes the CRAY very powerful for matrix operations, which are less easily and less efficiently implemented on the CYBER 205.

For the user some characteristic numbers are important.[1]

(a) The ratio R_∞ between the asymptotic vector and scalar speeds. This ratio is roughly 10 both for the CRAY 1 and the CYBER 205. Both computers have rather fast scalar processors of the order of 5-10 Mflops. Note that the CYBER 205 is about as fast as the CRAY 1 in spite of the larger clock period (the clock periods of the CRAY 1S and CRAY XMP are below 10 nsec). The ratio R_∞ is increased for 'linked-together' instructions on the CRAY 1 or for triadic operations like

$$A(I) = B(I) + C * D(I)$$

on the CYBER 205 (in the latter case by a factor of 2). It is also multiplied by the number of pipes for the CYBER 205, i.e. a four-pipe machine is asymptotically twice as fast as a two-pipe machine.

(b) The half-performance vector length $n_{1/2}$, i.e. the vector length for which half the asymptotic speed is reached. Due to the large start-up time, $n_{1/2}$ is ~ 100 for the CYBER 205 (~ 200 for a 4-pipe machine), while for the CRAY 1, $n_{1/2} \sim 10$ ($n_{1/2}$ is much larger for the CRAY XMP than for the CRAY 1).

(c) The break-even length n_b between the scalar and vector processor. For n < n_b the scalar processor is faster, for n > n_b the vector processor. For the CRAY $n_b \sim 2$-3, the vector processor is practically always faster than the scalar processor, while for the CYBER 205 $n_b \sim 10$. This means that if one vectorizes operations for vector lengths < 10 the vectorized program is slower than the scalar program. The break-even length of the CYBER 205 is close to the half performance length of the CRAY 1.

The conclusion is that one needs long vectors if one wants to take advantage of the CYBER 205. Since we are mainly interested in matrices, and matrices can be vectorized only column-wise, we need matrices of at least the dimension 200, which need 40 k words of core storage each; with 10 matrices of this size, the core storage of our installation is full!

Tricks customary from scalar codes, like storing only the upper half of a symmetric matrix, are of only limited help, because they make the vectorization more difficult and imply a substantial decrease of the speed.

It is, by the way, not only important to have sufficiently long vectors on the CYBER 205. It is also essential to check that the maximum vector length of 2^{16}-1 = 65 k (one large page) is not exceeded.

2.2 Data Transfer

It is further rather illusionary to rely on the virtual memory or to transfer information between core and peripheral memory.

In fact the most serious drawback of the CYBER 205 is the relatively slow data transfer to peripheral storage. The transfer of one large page takes as much as 0.2 sec., this is 3 μsec per word. In this time several hundred floating point operations can be performed. To compile first a large amount of data, e.g. two-electron integrals in quantum chemical calculations and to store them for further use on disc storage is very uneconomic even if IO is not charged at the same cost as CPU. It is, namely often unavoidable that the CPU is idle until IO is finished. With an installation such as ours having a relatively small core memory, one is generally in a serious dilemma. Vector lengths necessary for optimum performance cannot be realized. One has either to use small vector lengths, which implies poor performance or one has to transfer part of the data to peripheral storage which yields IO bound programs. Time sharing does not solve these problems, it rather creates new ones. What one must avoid on all account is uncontrolled paging.

2.3 Vectorizing Compiler and Vector FORTRAN

In order to take advantage of the speed of the vector processor one must either program in a language that is directly related to the architecture of the computer (e.g. in assembler code) or in a symbolic language like FORTRAN, which requires a compiler that creates an optimum machine code. Here the difference between the CRAY 1 and the CYBER 205 is that with the CRAY 1 one has only the alternative between standard FORTRAN with a vectorizing option to the compiler, and assembler, whereas the CYBER 205 offers a vector FORTRAN with a whole arsenal of new instructions.

The automatic vectorization by the compiler is moderately effective at the CRAY 1 (usually the innermost DO-loops are vectorized), it is very inefficient at the CYBER 205. On the CRAY 1 it is imperative to formulate the code as much as possible in terms of matrix operations and to use matrix subroutines written in assembler. Then the job of vectorization is nearly finished, while there is a long way to a good vectorized program for the CYBER 205.

The CYBER 205 has many more instructions (206) than the CRAY 1 (~120). Any of these instructions has up to 256 sublevels. However, only a rather small fraction of the instruction set is used by the FORTRAN compiler. In some aspects the CYBER 205 is superior to the CRAY 1, namely for operations like 'gather' and 'scatter', which are genuine vector operations on the CYBER 205, but done in scalar mode by the CRAY 1. (This defect has been repaired on the CRAY XMP).

There is no way to make good use of the CYBER 205 without at least using vector FORTRAN.

The CYBER vector FORTRAN offers (a) vector instructions like

A(1,1;N) = B(1,1;N) * C + D(1,1;N)
E(1;N) = F(1;N) ** G(1;N)   etc.

N is always the length of the vector, A(1,1) or E(1) is the starting address, C is a constant. Use of these instructions makes the code rather compact - though not always very efficient (see later).

There are (b) intrinsic vector functions like

VSQRT(A(1;N); B(1;N))   i.e. $B(I) = SQRT(A(I))$; $I = 1,N$

Q8SDOT(A(1;N),B(1;N)) i.e. $\sum_{I=1}^{N} A(I) * B(I)$

The dot product has a performance of ~100 Mflops, the vector square root is as fast as vector division; it requires about as much time as 7 add or multiply.

Very useful are (c) the so-called bit vectors, the elements of which can only have the value 0 and 1 and on which logical vector operations can be performed. They occupy little space, since 64 components of a bit vector can be stored into one word. With bit vectors the vector generalization of an IF statement is possible

WHERE (BIT(1;N)) A(1;N) = B(1;N) * C(1;N)

Here the instruction A(I) = B(I) * C(I) is only executed for those values of I for which BIT(I) = 1 (rather only for those values of I is the result B(I)*C(I) transferred to A(I)).

A rather strong feature (d) of the CYBER 205 are the instructions that allow one to construct a new vector from the elements of an old vector either by means of a bit control vector as Q8VCMPRS or Q8VXPND or by means of an index vector as Q8VGATHR and Q8VSCATR do.

Q8VCMPRS constructs from a 'long' vector A a 'short' vector B that has only as many elements as the bit vector has elements equal to 1, Q8VXPND reconstitutes from a 'short' vector B a 'long' vector A, filling in zeros where BIT(I) = 0. Q8VGATHR and Q8VSCATR perform B(I) = A(J(I)) and A(J(I)) = B(I) respectively.

These are vector instructions on the CYBER 205 (and the two latter also on the CRAY XMP), but done in scalar code on the CRAY 1.

The first two of these instructions are about as fast as vector floating point multiplications, the latter two need about the time of 2.5 floating point operations.

The CYBER vector FORTRAN furnishes two more classes of instructions: (e) the so-called stack-lib routines, that have to be applied for non-vectorizable DO-loops, e.g. for recursive sequences. They make use of the look-ahead stacks, i.e. of the instruction pipelines rather than data pipelines. (f) the 'special calls' that directly correspond to machine instructions. They are not very warmly recommended in the manuals, the compiler does not make any syntax check, and one cannot use the special calls without referring to the hardware reference manual. However, in order to get efficient codes, one cannot help coding the decisive steps by 'special calls', especially those for operations on sparse vectors, which are not accessible otherwise. An example is the arithmetic compress.

TR = 1.0 E - 10

CALL Q8ACPS(X'0C',,DA,, TR, DBIT, DA)

This causes a vector DA to be compressed on itself by means of a bit control vector DBIT (automatically generated), which is 0 if ABS (DA(I)).LT.TR and 1 otherwise.

There are special calls that allow one to add or multiply two sparse vectors, where either sparse vector (as well as the resulting one) is defined via a "compressed" vector of non-zero components and a bit control vector. These operations are only possible via special calls, like

Q8ADDNS (X'00,BITA,A,BITB,B,BITC,C)

As a beginning vector programmer one gets the impression that vector codes become more compact than scalar codes, mainly since certain DO-loops are replaced by single instructions. However, one realizes soon that good vector programs become more lengthy than the corresponding scalar FORTRAN programs. One reason is that one must not rely on the compiler (which only optimizes the scalar part of the program) but tell the computer exactly what it should do.

So it is e.g. legal to write either

A(1;N) = VSQRT (B(1;N);N)

or

A(1;N) = VSQRT (B(1;N); A(1;N))

In the first case an intermediate dynamic vector of length N is created and the result is stored to this vector before the result is transferred to A (at least so for the existing compiler releases). An expression

A(1;N) = B(1;N)*C(1;N) /(D(1;N)**E(1;N))

is legal, i.e. it will yield correct results, however at least two intermediate vectors will be created. It is therefore recommended to write a program such that one FORTRAN statement consists only of one dyadic or triadic operation, i.e. input: two vectors and optionally a scalar, output: one vector.

If one looks at an efficient vector FORTRAN program it resembles more an assembler program than a traditional FORTRAN program.

There are a few more possibilities to improve the speed of a program for the CYBER 205, which also make the code longer, namely

(a) unrolling of non-vectorizable DO-loops. Instead of

```
DO 1  I = 1, 1024
1 F(I) = ...
```

one writes

```
DO 1 I = 1,1024, 16
  F(I) =
  F(I + 1) =
  .....
1 F(I + 15) =
```

in order to exploit the look-ahead stacks and get an instruction pipeline.

(b) replacement of subroutine calls by inline code (to avoid the long time for subroutine calls ~6 μsec)

Both tricks are rather unelegant, and it would be better if there were a compiler option to generate either of them automatically, say by the instructions DO(UNROLL 16) I = 1,1024 or CALL INLINE ....

Here one remark is in order. When the first computers came, there was no alternative to writing programs in machine code, and this meant that one had to be aware of the architecture of the given computer when writing a program. Later the symbolic languages like FORTRAN took over, and most FORTRAN compilers are by now so efficient that one does usually not gain in using machine language, except sometimes for small parts of the program. Programs in machine independent languages have the big advantage to be transferable from one computer to the other.

Now with the advent of vector processors we are again forced to write programs that will very likely not be transferable to other computers. To vectorize standard program packages requires several man-years of work, and to write entirely new programs requires a similar effort. Who knows what the supercomputer in 10 or 20 years will look like? It is not guaranteed that one does not enter a train which goes in the wrong direction.

3. VECTORIZATION OF CONVENTIONAL QUANTUM CHEMICAL PROGRAMS

3.1 Efficiency of Vectorization

One must first have an idea of what can be gained by vectorization. The ratio R_∞ between the asymptotic vector and scalar speeds is close to 10 (for dyadic operations on a two-pipe machine in single precision - i.e. in 64 bit - arithmetic). So the best that one can achieve is to speed up a given scalar program by a factor ~ 10 (with respect to the scalar code on the same computer; compared to traditional top scalar computers such as the CYBER 175 one has another factor 2-3). Larger speed-up factors are possible - at least in principle - (a) if one has more than two pipes (but this also increases $n_{1/2}$), (b) if one has a large amount of triadic operations involving one scalar, (c) if many operations can be 'linked' together, (d) if one uses half-precision arithmetic - or finally (e) if one has started from a poor scalar code, essentially from any scalar program that does not make efficient use of look-ahead stacks, which applies to most existing programs.

Let us take the asymptotic factor 10 as realistic. Suppose further that 90% of the code is ideally vectorized (which is a lot). Then the overall speed-up-factor is

$$\left(0.1 + \frac{0.9}{10}\right)^{-1} \sim 5$$

A factor 5 must hence be regarded as quite high. If the effective vectorization is 80%, the speed-up-factor is only

$$\left(0.2 + \frac{0.8}{10}\right)^{-1} \sim 3$$

There are mainly four reasons for vectorization rates significantly below 100%:

1. Only a part of the program is vectorizable

2. Vector lengths are too small, either genuinely or due to restrictions on core memory

3. Additional operations (like scatter and gather) are necessary to prepare vectorization, but these operations also cost time

4. It may sometimes be recommended, in order to have efficient vector instructions, to perform dummy operations as in a WHERE block.

3.2 A Simple Example

An example which is somewhat typical for the evaluation of quantum chemical integrals is that of the two-electron one center integrals over Slater functions that arise in atomic theory. The contribution of these integrals to the CI-matrix of two-electron atom in an S-state is

$$V_{ij} = \sum_{k=0}^{k_{\max}(i,j)} c_{ij}(k) \; R_{ij}(k)$$

where the sum over k comes from the angular momentum coupling, where $c_{ij}(k)$ is a Gount (or Condon-Shortley) coefficient and where

$$R_{ij}(k) = \sum_{\alpha=1}^{4} F_{ij}^{(\alpha)} \; (k) \left\{ 1 + \sum_{\mu=1}^{\mu_{\max}(i,j,k)} A_{ij}^{(\alpha)} \; (\mu, k) \right\}$$

The $A_{ij}^{(\alpha)}(\mu, k)$ depend recursively on μ.

In scalar FORTRAN the code would be (for one term in the sum over α)

```
DO 1 I = 1,N
DO 1 J = 1,I

V(I,J) = 0

DO 2 K = 0, KMAX(I,J),2
F(I,J,K) = ...
G(I,J,K) = 1
A(I,J,K,0) = 1

DO 3 MU = 1, MUMAX(I,J,K)

  A(I,J,K,MU) = A(I,J,K,MU-1)*...
3 G(I,J,K) = G(I,J,K) + A(I,J,K,MU)
  F(I,J,K) = F(I,J,K)*G(I,J,K)
2 V(I,J) = V(I,J) + C(I,J,K)*F(I,J,K)
1 CONTINUE
```

The loop over K depends in a complicated way on I and J, and that over MU on I, J, and K. Therefore one cannot simply interchange the DO-loops.

However, direct vectorization is impossible because the inner DO-loop over MU is recursive. Moreover this inner DO-loop is rather short, so that even if it were not recursive, vectorization would not speed up the calculation.

Vectorization is only efficient if one inverts the order of the DO-loops, such that one can vectorize over the indices I and J (combined). One must take care of the fact that the allowed values of K and MU depend on I and J (and that for MU also on K) by defining bit control vectors BK(I,J) and BV(I,J).

A program in vector FORTRAN will then look somehow like

```
V(I,J) = 0.

determine  KAMAX = MAX (KMAX(I,J))
DO 1 K = 0, KAMAX
```

prepare bit vector BK(I,J) to satisfy K = 0, KMAX(I,J),2
compress input matrices with BK(I,J)
construct F(I,J,K)
determine MMAX = MAX (MUMAX(I,J,K))

DO 2 MU = 1, MMAX
    prepare bit vector BV(I,J) to satisfy MU = 1, MUMAX(I,J,K)
    compress with BV(I,J)
    construct A(I,J,K,MU) and G(I,J,K) by means of sparse vector operations

2 CONTINUE

1 V(I,J) = V(I,J) + F(I,J,K) * C(I,J,K) by sparse add.

The program which had ~15 lines in scalar FORTRAN needs ~150 lines in vector FORTRAN.

Most of the vectors have now the ideal lengths, but some of them, especially for large K become rather short, so that one cannot reach 100% vectorization. We got a speed-up factor ~7, corresponding to ~95% vectorization.

It is not too difficult to restructure a relatively short program in this way, but for a standard quantum chemical program package, similar restructuring requires a lot of work, it amounts to reprogramming from scratch.

3.3 Quantum Chemical Program Systems

We have two standard quantum chemical program packages running on the CYBER 205,

(a) The integral program (ZGA) and the SCF part of our old CEPA-PNO program, based on Gaussian lobes and originally written by R. Ahlrichs,[2]

(b) The HONDO program of King and Dupuis[3] in the symmetry adapted version of R. Pitzer[4] together with the rest of the Columbus program system.[5]

It is somehow a luxury to have two vectorized program packages, but they are to some extent complementary. The lobe program ZGA is faster for large molecules e.g. hydrocarbons when only s and p-type basis functions are used (the integrals time for norbornane C_7H_{12} in a double zeta basis is 287 sec. for HONDO, but 96 sec. for ZGA), while the HONDO program is superior when basis functions with high l are included, or for highly symmetric molecules. Some of our less conventional program systems e.g. the IGLO method for magnetic properties[6] are still based on the lobes program, we therefore regarded its vectorization as worthwhile. One advantage of the lobe program is that it is very compact; it consists of ~1000 FORTRAN lines in contrast to some 10,000 FORTRAN lines of the HONDO program. Its simple structure made its vectorization relatively easy and we were able[8] to gain a factor 3 with respect to the scalar code, while for the HONDO program only a factor 2 could be achieved.[7]

The vectorization of the Columbus program package including the HONDO program on the CYBER 205 which has mainly been done in Karlsruhe by R. Ahlrichs and his co-workers, has already been described in detail.[7] The vectorization of the ATMOL program for a CRAY 1S has been outlined by Saunders and Guest,[10] and various other authors have discussed the vectorization of standard quantum chemical program packages.[9,11-14] The main problems seem to be understood by now.

A conventional state-of the art program package consists mainly of

1. An integral program that computes the two-electron integrals (ij|kl) over Gaussian basis functions and stores them on peripheral storage. The number of integrals to be stored is of the order 10^7.

   Optionally a gradient program that computes the derivatives of the integrals with respect to displacements of the nuclei, and stores these derivatives.

2. An SCF-part, that proceeds iteratively. In each iteration cycle all integrals have to be read in from peripheral storage.

3. A 4-index transformation of the two-electron integrals from those over basis functions to those over MO's.

4. A direct CI (with optionally CEPA-like modifications) including single and double substitutions in order to take care of the most important effects of electron correlation.

Although the formalism is to a large part in terms of matrices, vectorization of the code is usually rather difficult. The best vectorization efficiency (both for the CRAY and the CYBER) is obtained for step 3, the 4-index transformation. This has an N^5-dependence on the number of basis functions (while the other steps go as N^4), but it becomes in practice faster than some of the N^4 steps.

An essential ingredient of the vectorization of Ahlrichs et al.[7] for the CYBER 205 is that the transformation of one index is vectorized over a combination of the three remaining indices which implies sufficiently long vectors.

Good vectorization performance is also possible for step 4 (the direct CI). Here it is important to vectorize as much as possible over a combination (ij) of two indices because for vectorization over one index only one gets too short vectors. The efficiency of this step depends on the order in which the integrals are stored.[7]

In both steps 3 and 4 speed-up factors of 5 and more are possible.

The first two program steps resist differently to vectorization on the two supercomputers. Let me illustrate this for simplicity's sake with the lobe program.

The expression for a two-electron integral over Gaussian lobes is simply[2]

$$(ij, kl) = S_{ijkl}\sqrt{B_{ijkl}}\ F_o(B_{ijkl}\ R_{ijkl}^2)$$

where the auxiliary quantities are

$$S_{ijkl} = t_{ij} \cdot t_{kl}$$

$$t_{ij} = \int \chi_i \chi_j \, d\tau = (4\alpha_i \alpha_j)^{3/4} \exp\left\{ -\frac{\alpha_i \alpha_j}{\alpha_i + \alpha_j} \, (\vec{r}_i - \vec{r}_j)^2 \right\}$$

$$B_{ijkl} = (q_{ij} + q_{kl})^{-1}$$

$$q_{ij} = (\alpha_i + \alpha_j)^{-1}$$

$$R_{ijkl}^2 = (\vec{r}_{ij} - \vec{r}_{kl})^2$$

$$\vec{r}_{ij} = \frac{\alpha_i \vec{r}_i + \alpha_j \vec{r}_j}{\alpha_i + \alpha_j}$$

and where the function $F_o(x)$ is defined as

$$F_o(x) = \frac{1}{\sqrt{x}} \, \mathrm{erf}\left(\sqrt{x}\right)$$

$$\mathrm{erf}(x) = \int_0^x \exp(-t^2) \, dt$$

At first glance vectorization seems straightforward. One first computes the two-index quantities t_{ij}, q_{ij} and $\vec{r}_{ij}$ and keeps them in storage. Then one keeps one pair ij fixed and calculates the $B_{ij,kl}$ $S_{ij,kl}$ and $R_{ij,kl}$ for all pairs kl. This can be done by operations on long vectors.

The evaluation of $F_o(x)$ can be rather effectively vectorized. We evaluate $F_o(x)$ as a piecewise rational approximation: $F_o(x) = \dfrac{a_i + b_i x + c_i x^2}{x + d_i}$, the coefficients of this

approximation for the various intervals are stored as vectors with the intervals as vector indices. One determines vectorially to which interval x belongs and gets the coefficients by means of Q8VGATHR, or rather Q8VXTOV (which is the corresponding special call).

For various reasons this apparently straightforward way is not possible.

1. One cannot store all the doubly indexed quantities in core, but one has to transfer part of them to peripheral storage, or - better - to calculate them several times.

2. One does not want the integrals over the lobes, but over groups of 'contracted' lobes, where p-, d- or f-AO's are represented by linear combinations of lobes, i.e. one has an additional inner DO-loop - over a variable, but rather short, length.

3. Usually 8 integrals with different permutations of the same labels are the same, other integrals are identical due to point group symmetry. One cannot afford to compute one distinct integral more than once. It is even more crucial to avoid double storing of the same integral.

4. In order to save time and to make the integral tape short, integrals that vanish or are in absolute value below some given threshold must neither be calculated nor stored.

5. One wants the integrals in a given order, which is usually not that best suited for vectorization.

On a CRAY 1 it is rather obvious how one should vectorize the integral evaluation, namely by vectorizing the summation over the lobes within a group. If one does so on the CYBER 205 one only gains a factor of ~ 1.3, because the vector lengths are too short (in the average ~ 40).

In such a situation one has two possibilities to improve the performance. One of them consists in combining many short vectors to one long vector and apply vector operators to a conglomerate of short vectors of different physical origin. This brute force approach is relatively effective and we get a speed-up factor 3, corresponding to 80% of perfect vectorization.[8] An alternative, which we have not tried so far, since it is much harder, would be to restructure the program, somewhat like in Sec. 3.2.

Step 2 of the program package, the SCF program, contains segments that are readily vectorized, mainly the 'unpacking of the labels.' Other parts are recursive and can be speeded up by loop-unrolling. We got an acceleration factor as high as 8, but unfortunately the SCF part is terribly IO bound. In scalar code one can keep the CPU busy while IO is performed, in the vectorized program the IO/CPU ratio in this step becomes as high as 5/1, although it is done as fast as possible, i.e. asynchronously (by means of Q7BUFIN and Q7BUFOUT) on large-page boundaries.

We had a lot of trouble with the director of our computer center since he does not like jobs which keep the CPU unoccupied while doing IO. He claims that we have not got the right programs for this computer, and he is probably not completely wrong. One way out, though not a fully satisfactory one is to use a 'direct SCF' program (see Sec. 4.2).

The best way to reduce the IO consists in reducing the number of SCF iterations. We have tried various possibilities of convergence acceleration. We got very good results with the method of Pulay[15] and also with a modification of it in which matrix-diagonalizations are replaced by the solution of Bloch equations. The 'Huckel-start' implemented some time ago by H. Kollmar also serves to reduce the number of iterations. We have not yet finished the study of using non-linear acceleration procedures in addition.

It would be nice to compare the performance of various vectorized (and also scalar programs) on various computers (mainly the CRAY 1S and the CYBER 205). Unfortunately the data in the literature leave some confusing impression. Some years ago Guest and Wilson[9] have discussed the vectorization of quantum chemical programs and published timings for calculations of the H_2S molecule with an (8s,7p,2d/4s,2p) basis. However the timings for the CRAY 1 refer to straight implementation of the scalar code without explicit vectorization. We have done calculations with the same basis on the CYBER 205 and collected the results for the integral evaluation in table 1.

The pattern is consistent with the expectation that the properly vectorized HONDO integral program should be somewhat faster on the CRAY 1 than on the CYBER 205.

Table 1.   CPU times (in sec.) for the evaluation of the integrals of H_2S [a]

	IBM 370/104	CYBER 855	CYBER 205 scalar	CYBER 205 vectorized	CRAY 1 [b]
ATMOL	255				41
HONDO	332	(200)[c]	(70)[c)d]	35 [d]	34
ZGA(lobes)		155	64	20	
POLYATOM					113
MOLECULE					66

a) with a (8s,7p,2d/4s,2p) basis as in Ref. 9. All figures except those referring to the CYBER 205 are from Ref. 9.
b) straight implementation without special vectorization
c) the number in parentheses are estimated, no scalar version of HONDO is at present running in Bochum
d) On the CYBER 205 not the original HONDO, but that modified by Pitzer[4] has been used. Roughly 40-50% of the time are needed for the symmetry adaptation.

Recently van Lenthe has published timings for calculations of formic acid HCOOH with the ATMOL program on various scalar and vector computers.[16] We have added CYBER 205 timings from both Karlsruhe and Bochum and also more recent figures of van Lenthe[17] and got so table 2. The claim[16] that the CRAY 1 is faster by a factor 2 or 4 for quantum chemical calculations than the CYBER 205 does not seem to be well-founded. It rather seems that for individually vectorized programs there is not too much difference in the speed of the two computers. On the other hand the CRAY 1 timings of Werner and Reinsch[30] (see also [31] and [7]) are so impressive that it is hard to imagine that the CYBER 205 could be competitive, even if the same internally contracted MCSCF-SCEP method were used. Really conclusive comparisons of the performance of various program systems and of different computers are still to be made.

Table 2. CPU times (in sec.) for calculation of trans-formic acid [a]

Program	ATMOL				HONDO/COLUMBUS		ATMOL
Place	Amsterdam				Karlsruhe[e]	Bochum	Amsterdam[b]
Computer	CYBER 175[b]	CYBER 855[b]	CYBER 205[b)c]	CYBER 205[d]	CYBER 205	CYBER 205	CRAY 1S
Integrals	524	387	191	200	88	93[f]	132
SCF	1118	661	80	66	27	19	55
4-index transf.	877	678	42	34	37	23	33
Direct CI	1744	1825	216	79	73	62	60
Total	4263	3551	529	379	215	196	280

a) as proposed by van Lenthe[16] in a double zeta + polarization basis (58 gaussian groups), keeping 1s frozen in CI.

b) Ref. 16

c) The CPU times are not given in Ref. 16 (only the SBU's), but have been estimated by comparison with other calculations at the same installation.

d) Improved calculations of van Lenthe[17] with a new program. version

e) R. Ahlrichs, private communication.

f) The programs are essentially the same; the calculations in Bochum were done with a newer release.

As the CYBER 205 has a large start-up time for vector operations, it also has a large 'start-up price' namely the high programming effort, and this only pays for 'long vectors' i.e. for programs that are much used.

As far as the costs of calculations on different computers are concerned one can make a few very provisional remarks. Since a speed-up factor of 10 from the CYBER 175 or 855 to the CYBER 205 is realistic and since the price of a CYBER 205 is usually less than twice that of a CYBER 855, computations on the CYBER 205 are cheaper by about a factor 5 (or more) than on the CYBER 855. In view of the higher price of the CRAY 1S, computations on it are more expensive than on the CYBER 205 (possibly by a factor 3-5). Such comparisons are very uncertain, since neither prices of computers are well-defined and reliable quantities, nor does this hold for the prices that users have to pay. More interesting, but still more difficult is the comparison of the price-performance ratios between large-scale computers and minicomputers.

4. LESS CONVENTIONAL PROGRAMS AND NEW STRATEGIES

4.1 Introductory Remark

As I said in the beginning of this lecture it is extremely tedious to 'vectorize' existing program packages, at least for the CYBER 205. The speed-up factors that one gets are often hardly worth the effort. It is certainly better to start from scratch. So we have spent rather little effort into improving existing programs, but we are performing new developments with the architecture of the CYBER 205 in mind.

4.2 The Direct SCF

When the study on direct SCF in this group was started[8] only the scalar direct SCF of Almlof et al.[13] was known. It appears that meanwhile other groups have worked on vectorized versions of direct SCF,[12,14] such that not much new has to be said.

In the direct SCF the two-electron integrals are never stored but are recalculated in every iteration cycle. It is therefore even more important than in conventional SCF to reduce the number of iterations, e.g. by the method of Pulay.[15]

Our direct SCF is substantially slower than the conventional SCF, but it has two advantages (a) it needs almost no IO and the director of our computer center likes such programs (b) it is not limited by the available amount of peripheral storage, which is another serious bottleneck at our installation. We were able to perform an SCF-calculation on a sort of di-norbonane $C_{12}H_{18}$ with a DZ basis at the first try with the direct SCF, while attempts to compute the same molecule with the standard SCF program, have failed for about three months because not enough disc space was available.

One can think of dividing the integrals into those which are computed fast and those that are computed slowly, to recalculate the former in every iteration and to store only the latter. One may store e.g. the integrals over highly contracted groups, or those which involve many AO's with high l. There are certainly still some unexploited possibilities to avoid excessive IO without paying for large amounts of additional CPU times.

We are still wondering whether we should continue to work on these lines, in particular whether we should keep the lobe program or switch to HONDO.

4.3 The IGLO Method (Individual Gauge for Localized Orbitals) for the Calculation of Magnetic Susceptibilities and Chemical Shifts.[6,18-20]

This is essentially a coupled Hartree-Fock (CHF) method in terms of localized MO's with an individual gauge origin for each localized orbital. The equations to be solved are much more complicated than in traditional CHF, but they can be simplified without a significant loss of accuracy.

The CHF like part of the programs consists in the iterative solution of a linear system of equations where each iteration cycle implies a pass through the integrals. Since three operators (for the three spatial directions have to be computed in one cycle, the IO problem is relatively less serious than in the Hartree-Fock part.

For the sake of illustration I give one example, the 2-norbornyl-cation computed with three basis sets, a double zeta (DZ) basis consisting of 92 groups of contracted Gaussians, a DZ+d basis consisting of 125 groups, and a basis of (DZ+d) type, but with uncontracted p-functions, and 148 groups. 4.5 million integrals are stored in a double-zeta calculation and 14.5 million integrals in a DZ+d calculation.

The various program steps need the following CPU times in sec.

	double zeta	double zeta + d	best basis
Integrals (ZGA)	162	1060	1114
SCF (12 iterations)	99	319	515
localization	13	17	18
IGLO -2 el.op	214	640	953
IGLO 1 el.op	109	252	285
IGLO iterations (15)	191	547	818
Miscellaneous	15	37	39
Sum CPU	803	2872	3742
Sum STU	1215	4645	6320

(STU is the accounting unit, that takes care of CPU, IO, SP and LP faults etc.)

The carbon chemical shifts are

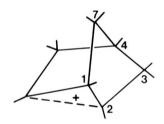

	exp	DZ	DZ + d	best basis
C_1	21.2	37.3	34.0	32.9
C_2	124.5	142.5	131.9	128.0
C_3	36.3	35.7	33.0	32.1
C_4	37.7	33.3	32.3	33.7
C_7	20.4	19.8	19.5	20.4

4.4   Effective Hamiltonians[21-24]

The method of effective Hamiltonians is, in a way, complementary to the more traditional approach in which one starts from an MC-SCF calculation in order to take
care of the non-dynamical correlation effects (near degeneracy and the like) while
in a second step the dynamic correlation is included via CI with doubles. In the
method of effective Hamiltonians one first takes care of the dynamical correlation
by constructing an effective Fock space Hamiltonian in the valence space, and one
includes non-dynamical correlations at the end by diagonalization of the small CI
matrix. One does then not get a single state, but e.g. all states that dissociate to
the same atomic states.

The programs were written in vector code from the very beginning.

The construction of the matrix elements of the effective Hamiltonians involves
essentially contractive multiplication of matrix elements, like e.g.

$$L_{ZU}^{XY} = V_{PQ}^{XY} \ W_{ZU}^{PQ}$$

with an implicit summation over P and Q.

If one vectorizes over one index, say P only, one has vector lengths $\sim$100, which are
too short. For the combined index P,Q one gets vector lengths of 5,000 (which are
reduced by a factor up to 8 by using point group symmetry).

One easily obtains an acceleration factor close to 10. One cannot avoid, though,
that some parts of the program are IO bound.

In the effective Hamiltonian approach we also need the matrix elements of one-,
two-, three- and four-particle operators between the Gelfand states corresponding to
full CI in the model space. We first construct the matrix element of the one particle operator $< \Phi\mu \mid E_Q^P \mid \Phi\nu >$ and then get the matrix elements of operators of
higher particle rank by successive multiplication of extremely sparse matrices
(about 1% non-zero elements). Before the arrival of the CYBER 205 we have
written a scalar code for the operations on these matrices and we have also run it
on the CYBER 205 since we did not regard vectorization as worthwhile. However
the director of our computer center cancelled some of our jobs, because - as he
found out - they did not use vector instructions. So we could not help vectorizing
this program step. The usual sparse matrix instructions are not very effective for
extremely sparse matrices, moreover most operations are on integers and the
CYBER 205 is not well adapted to integer arithmetic. Additionally, the matrices
are too large to be kept in storage. Then we got vector lengths beyond the
maximum allowed ones and had to cut vectors into pieces. With a terrible lot of
effort we were able to get a vector code that was almost as fast as scalar code.

There are probably cheaper ways to learn that for some algorithms the scalar code
is necessarily faster than the vector code.

4.5 Ab initio Pseudopotentials Including Core-Polarization

Currently used pseudopotentials are either of semiempirical or *ab initio* type, in the latter case they are derived from Hartree-Fock calculations (occasionally from relativistic Hartree-Fock calculations) and can hence only describe atoms or molecules in the field of Hartree-Fock cores. If one starts from a rigorous many-body theory also core polarization effects arise, that turn out to be very important in many cases. Müller and Meyer[25] have added these core-polarization corrections in a somewhat semiempirical way, limiting the core-polarization to dipole polarization, using the exact polarizability of the core and adjusting a cut-off-function for the resulting $1/r^4$ potential to empirical data of atomic spectra.

We have developed a formalism, closely related to the theory of effective Hamiltonians, where the core-polarization is obtained in an *ab initio* frame-work. The pseudopotential is constructed by means of the effective Hamiltonian program mentioned in the last section. It arises in non-local form, i.e. as a matrix representation in a given one-electron basis, together with this basis.

As with other pseudopotentials the N^4 dependence is now in terms of the basis for the valence-AO's only, which reduces the necessary computer time considerably.

4.6 Relativistic Quantum Chemistry

Fully relativistic calculations with 4-component spinors for each electron have an enormous requirement of core memory, one needs roughly 32 times as many matrices as in the nonrelativistic case (a two-electron function is a 16-component spinor with complex elements). The problems mentioned earlier in this lecture with the relatively small core memory of our installation become rather disastrous. It is very hard to be substantially faster in the vectorized than in the scalar code and together with the slow IO it is nearly hopeless to get beyond, say the Ne atom or the H_2 molecule although the programs are available (using the back-forth free-particle Fouldy Wouthuysen transformation[26] in order to avoid the variational collapse). For the relativistic work we are very much looking forward to the extension of the core memory to 2 Mwords.

4.7 r_{12}-Dependent Functions

The slow convergence of the CI expansion is, in many respects, frustrating, especially since it is known that much faster convergence is obtained if one uses r_{12}-dependent functions. Applications of r_{12}-dependent functions have so far been limited to very small systems because 'difficult integrals' appeared. We have found a new formulation for wave functions with linear r_{12}-dependent terms in which 'difficult integrals' are avoided.[27]

We have started this approach only very recently, and results beyond He and H_2 are not yet available. It is possible to express all required integrals in terms of two-electron integrals of cartesian gaussians for the basis functions and additional functions with l increased (or decreased) by 1. The most time-consuming step is in the moment the sorting of the integrals from the order furnished by HONDO to that needed for the construction of the modified CI matrix. We shall probably have to write a special integral program. The essential parts of the calculation are done in core and are well vectorized.

An alternative to the use of linear r_{12}-dependent terms consists in computations with Gaussian geminals $e^{-\alpha(r_i - r_j)^2}$. They are less adapted to representing the electronic cusp but all integrals are given in closed form. I should mention that J. G. Zabolitzky and his group have gaussian geminal calculations running on the CYBER 205 in Bochum.[28]

5. CONCLUDING REMARKS

Our attitude towards the CYBER 205 has at first been sceptic. We then found out that with this 'supercomputer' we were able to perform some large scale calculations which never would have been possible on our previous installations, i.e. neither the PE 8/32 nor the CYBER 175. On the other hand we were disappointed for other problems, where the CYBER 205 was of little help, especially for the relativistic calculations for which colleagues who have a minicomputer with a sufficient amount of core memory can do better.

We have not put too much stress on the vectorization of existing programs, but we are performing new developments with the architecture of the 205 in mind. So far we have only occasionally developed new program structures or new algorithms that have been directly inspired by the architecture of the CYBER 205. This aspect of the computational strategy will certainly gain importance. Of course, in developing programs for one special computer, one should know something of the future development of computer architecture, in order not to spend one's effort at the wrong place.

There are a few statements that one can make safely

1. For an efficient vector program all necessary data should be kept in core storage.
2. Transfer of data to peripheral storage should be avoided as much as possible.
3. Programs should be written from scratch for the given vector processor, not by 'vectorizing' existing programs.

The number of two-electron integrals stored on peripheral storage is of the order 10-1000 Megawords. Although computers with this size of core memory, i.e. the CRAY 2S (256 million words) or the ETA[10] (> 300 million words), will become available in the near future, one should find out whether there are other basis sets that yield the same accuracy with half or one third of the basis size such that number of integrals is reduced to < 10 Megawords, even if the evaluation of the integrals is somewhat more expensive. STOs are possible candidates, but they seem to be not competitive so far.[12] Another possibility might be the cusped Gaussians.[29] For explicitly correlated (i.e. r_{12}-dependent) wave functions one also needs a rather small number of basis functions and more complicated integrals. Such functions may become advantageous on a supercomputer. One should also not regard the expansion method as a dogma, and consider completely different approaches, e.g. finite element methods, in view of their implementation on a vector processor. I think it will take a few years until we really know where to go.

ACKNOWLEDGEMENTS

The authors thank Prof. R. Ahlrichs for a fruitful cooperation on the use of the CYBER 205, in which so far we have more received than given. We also thank Prof. V. Staemmler, Dr. R. Jaquet, U. Fleischer and U. Landscheid for their interest. The assistance of the computer center of the Ruhr-University and of the local CDC group, is gratefully acknowledged. Particularly helpful were Dr. G. Schaefer, H. Seeboth, J. Krieger and R. Wojcieszynski.

REFERENCES

1. R.W. Hockney and C.R. Jesshope 'Parallel computers', Hilger, Bristol 1981
2. R. Ahlrichs, Theoret. Chim. Acta **33**, 157 (1974)
3. M. Dupuis, J. Rys and H.F. King, J. Chem. Phys. **65**, 111 (1976); J. Rys, M. Dupuis and H.F. King, J. Comp. Chem. **4**, 154 (1983)
4. R.M. Pitzer, J. Chem. Phys. **58**, 3111 (1973)
5. H. Lischka, R. Shepard, F.B. Brown and I. Shavitt, I.J. Quant. Chem. Symp. **15**, 91 (1981)
6. M. Schindler and W. Kutzelnigg, J. Chem. Phys. **76**, 1919 (1982)
7. R. Ahlrichs, H.-J. Bohm, C. Ehrhardt, P. Scharf, H. Schiffer, H. Lischka and M. Schindler, J. Comp. Chem. **6**, 200 (1985)
8. U. Meier, Diplomarbeit Bochum 1984, directed by V. Staemmler
9. M.F. Guest and S. Wilson in 'Supercomputers in Chemistry,' P. Lykos and I. Shavitt, ed. Wiley, Interscience, New York 1981
10. V.R. Saunders and M.F. Guest, Comp. Phys. Comm. **26**, 389 (1982)
11. V.R. Saunders and J.H. van Lenthe, Mol. Phys. **48**, 923 (1983)
12. D. Hegarty in 'Advanced Theories and Computational Approach to the Electronic Structure of Molecule,' C.E. Dykstra, ed., D. Reidel, Dordrecht 1984, p.39
13. J. Almlof, K. Faegri and K. Korsell, J. Comput. Chem. **3**, 3 (1982)
14. J. Almlof and P.R. Taylor, as Ref. 12, p. 107
15. P. Pulay, Chem. Phys. Letters **73**, 393 (1980); J. Comp. Chem. **3**, 556 (1982).
16. J. van Lenthe in 'supercomputer' **5**, 33 (1985) (P. O. Box 4613, 1009 AD Amsterdam)
17. J. van Lenthe, private communication to R. Ahlrichs
18. W. Kutzelnigg, Israel J. Chem. **19**, 193 (1980)
19. M. Schindler and W. Kutzelnigg, J. Am. Chem. Soc. **105**, 1360 (1983)
20. M. Schindler and W. Kutzelnigg, Mol. Phys. **48**, 781 (1983)
21. W. Kutzelnigg, J. Chem. Phys. **77**, 3081 (1982)
22. W. Kutzelnigg and S. Koch, J. Chem. Phys. **79**, 4315 (1983)
23. W. Kutzelnigg, J. Chem. Phys. **80**, 822 (1984)
24. W. Kutzelnigg, J. Chem. Phys. **82**, 4166 (1985)
25. W. Müller and W. Meyer, J. Chem. Phys. **80**, 3311 (1984)
26. H. Wallmeier and W. Kutzelnigg, Phys. Rev. A 28, 3092 (1983)
27. W. Kutzelnigg, Theoret. Chim. Acta, in press
28. K. Szalewicz, B. Jeziorski, H.J. Monkhorst, J.G. Zabolitzky, J. Chem. Phys. **78**, 1420 (1983); **79**, 5543 (1983); **81**, 368 (1984)
29. W. Klopper and W. Kutzelnigg, J. Mol. Struct. (Theochem) **135**, 339 (1986).
30. H.J. Werner and E.A. Reinsch as Ref. 12, p. 79
31. W. Meyer, R. Ahlrichs and C.E. Dykstra as ref. 12, p. 19

MOLECULAR STRUCTURE CALCULATIONS ON THE CRAY-XMP AND CYBER 205 SUPERCOMPUTERS AT NASA AMES

Charles W. Bauschlicher, Jr. and Stephen R. Langhoff
NASA Ames Research Center
Moffett Field, CA 94035

and

Harry Partridge
Research Institute for Advanced Computer Science
NASA Ames Research Center
Moffett Field, CA 94035

ABSTRACT
Selected molecular structure calculations performed on the NASA Ames CRAY-XMP and CYBER 205 supercomputers are described. We first present an overview of the work underway at Ames. We then present a more detailed discussion of the computation of accurate dissociation energies of ionic diatomic and triatomic molecules, and of studies on small metal clusters that serve as a model for both perfect crystal faces and small gas phase particles. These examples illustrate that supercomputers significantly increase the size of the systems that can be considered, as well as the accuracy to which spectroscopic parameters can be determined.

I. INTRODUCTION

It is widely believed[1] that new generations of computers inspire new algorithms which lead to far greater computational power than one might initially expect. For example, our current capabilities in computational chemistry far exceed the original expectations when our first class VI supercomputer (CRAY-1S) was delivered to NASA Ames. The increased performance is due to a synergism between new algorithm development and programming techniques that explicitly take advantage of the vector capabilities of the supercomputers. We expect similar improvements in computation will arise when multiple instruction multiple data (MIMD) (parallel processing) architectures[2] become available. Hence, these new algorithms, coupled with the capabilities of supercomputers, permit theoretical studies of a wide variety of previously intractable problems. For example, at NASA Ames we are currently studying problems in the general areas of re-entry physics, hypersonic aerodynamics, atmospheric chemistry, materials problems (e.g. fatigue, embrittlement, oxidation), catalysis, combustion, polymer design and non-intrusive spectroscopic diagnostics.

With the availability of supercomputers, it is now possible to generate potential energy surfaces, molecular properties and molecular spectra for diatomic and triatomic systems with sufficient accuracy to resolve problems in electronic and infrared spectroscopy. Calculations are also possible on much larger systems such as $Ni_{13}H_2$ which have shown that the Ni d orbitals are involved in the dissociation of H_2 on Ni surfaces.[3] Also, it is often possible to respond to experimentalists on a very short time scale. For example, in a single day, qualitative potential curves for the lowest states in each of six symmetries for the Ca_2 molecule were run on the CRAY-XMP. Although this calculation could have been run on a VAX it would have taken much longer in both computer and personal time.

In Section II we give a general overview of the work being performed by other members of the Ames computational chemistry group. Sections III and IV give a more detailed discussion of our theoretical studies of ionic molecules and small metal clusters, respectively. Section V contains our conclusions.

II. APPLICATIONS AT AMES

A significant thrust of the computational chemistry effort at Ames is in material science. Work in this area, which ranges from model calculations on solids to *ab initio* calculations on transition metals and large organic molecules, is illustrated below. A more detailed account of our theoretical work on ionic molecules and small metal clusters is given in sections III and IV, respectively. Halicioglu and co-workers.[4-6] are modeling the structures of solids and solid interfaces to answer questions about structural stability, reconstruction and deformation. The unique feature of this work is the use of effective two and three body terms in the expansion of the potential. While this increases the computational work from order of n^2 to n^3, where n is the number of atoms in their model, efficient vector processing has made this quite tractable, and has allowed them to model many processes which cannot be treated using only two body terms. Using these powerful techniques in conjunction with supercomputers, they have been able to model surface reconstruction and bulk structures other than the usual close-packed geometries.

Building on the work for the transition metal atoms[7] and dimers,[8] Walch and co-workers are presently investigating the transition metal trimers. The use of extensive complete-active-space self-consistent field wavefunctions is needed to gain even a qualitative understanding of the bonding. Such calculations are not feasible without a supercomputer. Studies on Sc_3 indicate[9] that 3d bonding is important, leading to a $^2A_2$ ground state, with an equilateral triangle geometry, in agreement with ESR experiments.[10] Preliminary results for Ti_3^+ and V_3^+ indicate that 3d bonding is also important for the transition metal trimers through Cr_3. Calculations[11] for Cu_3 indicate a strongly Jahn-Teller distorted ground state showing 4s bonding derived from three atoms in the $4s^13d^{10}$ state. Calculations on the excited states of Cu_3 have lead to a new assignment of the upper state in the resonant two photon ionization spectrum observed by Morse et al.[12] These studies, in addition to their utility in interpreting experimental studies of these important molecules, provide insight into the bonding in bulk transition metals, since the relative importance of 4s and 3d bonding is controlled by the atomic overlaps, and the metal-metal distances in the trimers are comparable to the nearest neighbor distances in the bulk metals. Such calculations are the building blocks for work on catalysis and other advanced materials problems.

Komornicki and King[13] have implemented a highly efficient vectorized analytic first- and second-derivative code for SCF wavefunctions. This code has been used by Laskowski, Jaffe and Komornicki[14] to compute the rotational barriers and complete vibrational force fields in large (greater than 30 atom) organic molecules, which are then used to model and design improved polymers. In the course of the calibration calculations on their model for polymethylmethacrylate, they found that the experimental data had been incorrectly interpreted.[15] Subsequent experiments support their calculations.[16] With this code, the largest second derivative calculation takes about an hour of CRAY time, and is about 100 times faster than using a finite difference technique.

A long standing research interest at NASA Ames is re-entry physics. The current emphasis is on aeroassisted orbital transfer vehicles (AOTV), which are designed to transport material from low-earth orbit (about 200 km) to higher orbit. Since these vehicles have large heat shields and brake at relatively high altitudes, non-equilibrium radiative heating may be greater than both convective and equilibrium radiative heating.[17] As a result computer simulation of the re-entry process requires knowledge of the detailed chemistry and physics of the very hot air in the shock layer in front of the vehicle. Hence, a major effort is underway to compute accurate chemical and radiative properties of neutral and ionized N_2, O_2 and NO, as well as rates of atomic and molecular excitation and ionization. One example is the study of Partridge and Stallcop[18-19] who have computed the N_2^+ and O_2^+ potential energy curves for the 12 electronic states that dissociate to the ground state asymptotes. Their computed charge-exchange cross sections,[19-20] that are used to compute the binary atom-ion diffusion coefficients, are in excellent agreement with high-energy experimental data.[21-22] Work is currently in progress to compute reliable transport properties (e.g. thermal conductivity and viscosity) for nitrogen and oxygen atom-ion collisions. The transport properties and transition moments are needed to model the flow and radiation in the shock layer in front of the AOTV.

Huo, McKoy and co-workers have implemented the Schwinger multi-channel (SMC) variational method for electron-molecule scattering on the CRAY. Unlike other methods, the SMC code treats the collision process as a multi-centered, $n+1$ electron problem with correct boundary conditions. The requirement of a large basis set to represent both the bound and continuum electrons, as well as the random orientation of the incident plane wave with respect to the molecule, precludes carrying out such calculations on a VAX within a reasonable time scale. The SMC code has been used to calculate elastic and inelastic electron scattering from atmospheric molecules which are ro-vibrationally hot,[23] as well as polyatomic molecules[24] and oriented molecules.[25] Such calculations are important in modeling the chemistry in the shock waves about the AOTV and other hypersonic vehicles.

Another problem of interest at NASA Ames is the line-by-line construction of synthetic spectra[26-27] as the final products of molecular structure calculations. The spectra are generally constructed using a combination of accurate experimental line positions and theoretical line intensities. For example, theoretical studies[28] of the $B^2\Sigma^+ - X^2\Sigma^+$ blue-green band system of AlO showed that spectra derived from shock tube experiments[29] were not fully corrected for problems with self-absorption. Hence, the theoretical line strengths for the blue-green system are considerably more accurate, and have utility in temperature and density measurements in the upper atmosphere. Theoretical studies[30] of the origin of the optical glow observed on the Atmosphere Explorer satellites and the Space Shuttle have also benefited from the construction of molecular spectra. Theoretical spectra for the OH Meinel system[30-31] indicate that OH is a principal contributor to the optical glow on at least the Explorer satellites. Another area of application of molecular spectra is combustion research. We have been able to reproduce almost exactly a 0.1Å resolution OH lamp spectrum containing the 0-0 and 1-1 bands of the $A^2\Sigma^+ - X^2\Pi$ ultraviolet band system.[32] Model studies are underway to determine the attenuation of the lamp spectrum as it passes through a flame containing OH as a means of developing a rapid method of determining the temperature of a flame. Finally, we are extending our study of molecular spectra to triatomic systems using the gaussian wave-packet techniques of Reimers, Wilson and Heller.[33] Applications have been made to the high temperature spectra of the $A(^2\Pi) - X(^2\Sigma^+)$ absorption in the C_2H molecule,[34] which again has applications to combustion and flame diagnostics.

Theoretical studies of the interaction of atoms and molecules with a strong radiation field are needed as input for the development of nonintrusive photodiagnostic techniques in fluid dynamics and combustion research. For this purpose, the "dressed molecule" formulation for the calculation of linear and nonlinear molecular response functions to the laser field has been incorporated into the Ames CI code, SWEDEN. One example of such a study[35] is the calculation of the strong Stark effect observed in the two-photon spectrum of NO, which has been found to be the limiting factor in the temperature diagonostics in cold turbulent flow. These studies are useful for analyzing combustion processes, both cold and hot gaseous flow fields, and stellar spectra.

III. THEORETICAL STUDIES OF IONIC MOLECULES

Using the supercomputers at NASA Ames, the present authors have undertaken a systematic study of the dissociation energies (D_o) of most of the alkali and alkaline-earth fluorides, chlorides, oxides, sulfides, hydroxides and (iso)cyanides. Our recommended D_o values, tabulated in Table I, are believed to be accurate to 0.1 eV and are thus capable of ruling out disparate experimental values, and of permitting a critical evaluation of the various methods used to determine D_o. The systematic study has revealed several previously unrecognized trends among the D_o of these systems (see Refs. 36-43).

In this section we first discuss the methods we employ to study ionic molecules. We next focus on the D_o of the alkaline-earth oxides where theory has been very useful in delineating the correct D_o values. We then discuss the trends observed among the D_o of the various systems revealed by our study. Finally, we discuss the $^2\Pi - ^2\Sigma^+$ separations in the alkali oxides and sulfides. This study provides strong support for a $^2\Sigma^+$ ground state of KO about 200 cm^{-1} below the $^2\Pi$ state.

Our theoretical model[36,38,44] is based upon the observation that the component ions in an ionic system, M^+X^-, closely resemble the free ions, M^+ and X^-. Thus, we separate the molecule into ions and correct to the neutral asymptote using the accurate experimental ionization potential (IP)[45] and electron affinity (EA),[46] and determine D_o from the relation

$$D_0(MX) = E(M^+) + E(X^-) - E(MX, r_e) - IP(M) + EA(X) + \Delta ZPE \qquad (1)$$

Table I. Recommended dissociation energies (D_o) for selected alkali and alkaline-earth fluorides, chlorides, oxides, sulfides, hydroxides and isocyanides.

Molecule	State	D_o (eV)	Molecule	State	D_o (eV)
LiF	$^1\Sigma^+$	6.06	BeF	$^2\Sigma^+$	5.94
NaF	$^1\Sigma^+$	5.00	MgF	$^2\Sigma^+$	4.68
KF	$^1\Sigma^+$	5.10	CaF	$^2\Sigma^+$	5.53
RbF	$^1\Sigma^+$	5.07	SrF	$^2\Sigma^+$	5.62
CsF	$^1\Sigma^+$	5.27			
LiCl	$^1\Sigma^+$	4.89	BeCl	$^2\Sigma^+$	3.90
NaCl	$^1\Sigma^+$	4.22	MgCl	$^2\Sigma^+$	3.26
KCl	$^1\Sigma^+$	4.37	CaCl	$^2\Sigma^+$	4.14
RbCl	$^1\Sigma^+$	4.36	SrCl	$^2\Sigma^+$	4.23
CsCl	$^1\Sigma^+$	4.58			
LiO	$^2\Pi$	3.84	BeO	$^1\Sigma^+$	4.69
NaO	$^2\Pi$	2.83	MgO	$^1\Sigma^+$	2.75
KO	$^2\Sigma^+$	2.86	CaO	$^1\Sigma^+$	4.14
RbO	$^2\Sigma^+$	2.90	SrO	$^1\Sigma^+$	4.36
CsO	$^2\Sigma^+$	3.10	BaO	$^1\Sigma^+$	5.75
LiS	$^2\Pi$	3.30	BeS[a]	$^1\Pi$	2.29
NaS	$^2\Pi$	2.67	MgS[a]	$^1\Pi$	1.70
KS	$^2\Pi$	2.74	CaS[a]	$^1\Pi$	2.41
RbS	$^2\Pi$	2.66			
LiOH	$^1\Sigma^+$	4.64	BeOH	$^2\Sigma^+$	4.70
NaOH	$^1\Sigma^+$	3.51	MgOH	$^2\Sigma^+$	3.31
KOH	$^1\Sigma^+$	3.61	CaOH	$^2\Sigma^+$	4.15
RbOH	$^1\Sigma^+$	3.58	SrOH	$^2\Sigma^+$	4.18
CsOH	$^1\Sigma^+$	3.77	BaOH[b]	$^2\Sigma^+$	4.60
			BeNC	$^2\Sigma^+$	4.15
			MgNC	$^2\Sigma^+$	3.37
			CaNC	$^2\Sigma^+$	4.13
			BaNC[b]	$^2\Sigma^+$	4.50

[a] The dissociation energy (D_e) reported is for the excited $A^1\Pi$ and does not include a zero-point correction.

[b] The results are less accurate (~ 0.2 eV uncertainty) because the core electrons are described by a relativistic effective core potential.

where ΔZPE is the change in the zero point energy between reactants and products. For molecular systems that are well described at the Hartree-Fock (HF) level, this formalism is particularly accurate since most of the differential correlation and relativistic effects are included in the accurate experimental EA and IP. For those systems where the ground state is not well described at the HF level, such as the alkaline-earth oxides, an excited state which is well described by the HF reference is used, and the D_o of the ground state determined by combining the calculations on the excited state with the experimental excitation energy. However, the formalism described by Eq. 1 requires the use of extensive basis sets to fully describe the distortions that occur during bonding. Since the HF limit is approached much more easily for the fragments than the molecule, basis set incompleteness results in D_o values that are systematically low. For the diatomic and triatomic systems studied, we used extensive Slater and gaussian basis sets,[37-38] respectively, that contain at least four sets of polarization functions on each atom. Basis set incompleteness at the SCF level was studied by both basis set saturation studies,[39] and for selected diatomics using an implementation of McCullough's numerical HF code.[47] Our SCF D_o are generally within 0.05eV of the HF limit.

While the formalism in Eq. 1 is potentially very accurate at the SCF level, we have observed[36,38] that correlation effects generally increase D_o slightly. This arises primarily from interfragment correlation effects, i.e. a single excitation on the metal times a single excitation on the negative ion. While such terms are small for the lighter systems, they grow to 0.4eV for the heavier systems such as CsCl. We account for this differential correlation effect through a configuration-interaction (CI) calculation which consists of all single and double excitations relative to the SCF reference, SDCI. These SDCI calculations correlate both the n-1 and n shells on the metal, for example the 3s, 3p and 4s shells of Ca are all correlated. Recent work has shown that even the SDCI calculations may underestimate the D_o, and that a core-valence calculation, which eliminates the nearly constant metal (n-1) shell correlation energy (double excitations out of the n-1 shell), is the optimal procedure for very heavy systems. Basis set superposition errors estimated using the counterpoise method are found to be less than 0.1 eV at the SDCI level, and are at least partially compensated for by basis set incompleteness errors. We have also shown that the size inconsistency of the SDCI method is not a problem if the system is treated as a "supermolecule" at long range. Corrections to D_o for quadruple excitations based on either a Davidson correction[48] or the coupled pair formalism[49] are found to be quite small. Finally, we have applied an empirical "bond-length correction" to the D_o of the heavier systems where we have overestimated r_e. This correction is based on the nearly linear relationship between the energy required to dissociate to ions and r_e that is shown in Fig. 1. This relatively small correction improves the agreement between our theoretical D_o and the accurate thermochemical and flame photometric D_o values for all systems, especially the alkali chlorides.[38,50]

The utility of the theoretical calculations in delineating the correct D_o values is demonstrated here using the alkaline-earth oxides, MgO, CaO and SrO. The recent experimental D_o values[51-52] for MgO of 3.7 ± 2 eV are much larger than our theoretical value[36,40] of 2.75 ± 0.1 eV. However, analogous calculations[38] on MgF give excellent agreement with experiment, and our theoretical value for MgO obeys the nearly linear relationship between D_e (to ions) and r_e illustrated in Fig. 1. In addition, a recent chemiluminescent and laser fluorescence study[53] of the $Mg(^1S) + NO_2$ reaction produces a value consistent with theory assuming a reaction barrier of 0.5 eV.

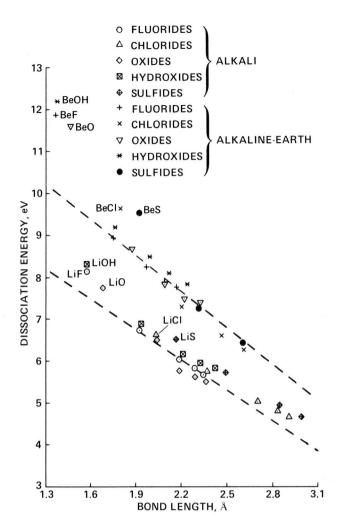

Figure 1. The dissociation energy (without zero-point corrections) with respect to the ionic asymptotes for ionic molecules containing alkali and alkaline-earth metals. For the akline-earth oxides and sulfides, the singly ionic $^1\Pi$ state is used, not the ground state which is a mixture of singly and doubly ionized structures.

For CaO, our theoretical D_o value[36,40] of 4.14 ± 0.1 eV is in excellent agreement with the value of 4.11 ± 0.07 eV obtained by Irvin and Dagdigian from a chemiluminescent study[54] of the $Ca(^1D) + O_2$ reaction. Our value, however, is well below the lower bound of $\geq4.76\pm0.15$ eV deduced by Engelke, Sander and Zare (ESZ) from chemiluminescent studies[55] with ClO_2, which is the value recommended by Huber and Herzberg.[56] Earlier mass spectrometric[57,58] and flame photometric[57,59] studies also support the lower determination. The reliability of the theoretical result gives strong support to the contention[54] that the higher chemiluminescent value is high as a result of interference with CaCl.

For the SrO molecule there is also considerable scatter in the experimental D_o values. The value recommended in the JANAF tables[57] and in the recent review by Pedley and Marshall,[60] $D_o(SrO) = 4.38 \pm 0.17$ eV, is based to a large extent on a reinterpretation[61] of the mass spectrometric determination of Colin et al.[58] and Drowart et al.[51] This is in excellent agreement with our theoretical estimate in Table I of 4.36 ± 0.1 eV. These values are considerably smaller than the value of 4.88 eV recommended by Huber and Herzberg[56] based upon a crossed beam study of Botalli-Cosmovici and Michel,[62] which is in good agreement with the older flame photometric[63-64] and thermochemical values.[65] The higher values were also supported by the lower bound of $\geq 4.67 \pm 0.15$eV obtained in the chemiluminescent study[55] of ESZ. After the calculations were performed, Zare[66] reviewed the original spectra and concluded that the identification of the spectral feature at 17430 cm^{-1} as the (18,0) band of SrO was incorrect. The highest peak that can be distinctly observed is the (11,0) band at 14237 cm^{-1}. If any other peaks are present, they are obscured by SrCl. A revised lower limit of 4.27 ± 0.15 eV was obtained, which is consistent with our theoretical value. Also, the recent work on MgO suggests that the crossed-beam experimental value for SrO could be high if a translation energy barrier were present in the reaction. In fact a barrier of 0.5eV, as found for MgO, would bring this value into agreement with theory as well. Hence, we feel that the theoretical D_o values for MgO, CaO and SrO in Table I are the most accurate and consistent set of values presently available.

The availability of supercomputers has allowed us to extend our method to about 50 ionic molecular systems (see Table I). Calculations on the heavier molecular systems can take up to 2 CPU hours of CYBER 205 time per geometry. Since the calculations would take at least 100 times longer on a VAX, this systematic study would not have been feasible without a supercomputer. In Table II, some of the trends among the D_o of the alkali and alkaline-earth fluorides, chlorides, oxides and hydroxides are illustrated. The trends are always most prominent for the metals Na-Cs and Mg-Sr with Li and Be showing larger differences, probably as a result of either their increased covalent character or the different cores, s^2 versus s^2p^6. These differences are also illustrated by the tendency of molecular systems containing Li and Be to fall above the linear plot of D_e vs r_e in Fig. 1.

One prominent trend is between the D_o of the alkali and alkaline-earth fluorides and hydroxides. The dissociation energies of the fluorides are uniformly larger than the hydroxides by somewhat less than the electron affinity difference of 1.57 eV. We have proposed that the increased stability, over that expected based only on the difference in EA's, of the hydroxides arises from the ability of H to pull charge out of the bond, thereby improving the electrostatic interaction of the metal and oxygen. This effect is somewhat larger for the alkaline-earths, which suggests that the additional valence electron amplifies the difference between the cylindrical OH^- and spherical F^- charge distributions.

Except for the lightest metals, Li and Be, there is also a pronounced correlation between the hydroxides and chlorides. The D_o of the chlorides are larger, however, by far less than the electron affinity difference of 1.57 eV, owing principally to the much larger bond lengths for the chlorides. Also, the relationship is quite different for the alkali and alkaline-earths, probably because the extra valence electron cannot as effectively polarize away from the more spatially extended Cl^- anion. In fact, the D_o of the alkaline-earth hydroxides and chlorides (Mg-Ba) are quite similar as has been noted by Murad.[67]

Table II. Correspondence between the dissociation energies of the alkali and alkaline-earth hydroxides, chlorides, fluorides and oxides.

Metal	D_e(eV) [a]				D_e(MX) $-$ D_e(MOH)		
	MOH	MF	MCl	MO	X = F	X = Cl	X = O
Li	4.74	6.11	4.93	3.90	1.37	0.19	-0.84
Na	3.59	5.02	4.24	2.86	1.43	0.65	-0.73
K	3.68	5.10	4.39	2.89	1.42	0.71	-0.79
Rb	3.64	5.06	4.36	2.90	1.42	0.72	-0.74
Cs	3.83	(5.3)	4.60	(3.1)	1.5	0.77	-0.7
Be	4.80	6.02	3.95	4.78	1.22	-0.85	$-0.02(-1.19)$ [b]
Mg	3.38	4.70	3.29	2.78	1.32	-0.09	$-0.60(-1.04)$
Ca	4.23	5.55	4.16	4.19	1.34	-0.07	$-0.02(-1.07)$
Sr	4.27	5.61	4.25	4.37	1.37	-0.02	$+0.13(-1.04)$
Ba	4.7	(6.0)	(5.8)	1.3	1.3		$+1.1(-1.1)$

[a] Our "best" D_e are given for the ground states. Theoretical D_e for MF and MO discussed in References 36, 38 and 40. Values in parentheses are less certain. For the alkali metals the ground state symmetries are $^1\Sigma^+$ for all MF and MOH, $^2\Pi$ for LiO and NaO and $^2\Sigma^+$ for KO,RbO and CsO. For the alkaline-earth metals the ground-state symmetries are $^2\Sigma^+$ for all MOH and MF and $^1\Sigma^+$ for all MO.

[b] Values in parentheses are computed with respect to the D_e of the excited $^1\Pi$ state of the alkaline-earth oxides. The following T_e values (eV) were used[56]: BeO(1.17), MgO(0.44), CaO(1.05), SrO(1.17) and BaO(2.19).

The D_o of the alkali oxides also correlate well with the hydroxides. The relatively constant difference of about 0.75 eV is again larger than the difference of 0.37 eV in the electron affinities of O and OH. The fact that there is also a correspondence between the D_o of the alkaline-earth hydroxides and oxides is not immediately obvious, because if the $X^1\Sigma^+$ ground states of MO are used for comparison, the difference varies from -0.6 eV for Mg to $+1.1$ eV for Ba. However, the $X^1\Sigma^+$ ground states of the alkaline-earth oxides involve a complicated mixture of both singly and doubly ionized structures. If instead, the $^1\Pi$ or $^3\Pi$ states of MO that are well represented by singly ionized structures are used as a reference, there is again a good correlation with the hydroxides--see Table II. The strong correspondence between the D_o of the different molecular systems lends strong support for the consistency and accuracy of the theoretical results.

The last topic we discuss in this section is the separation (T_e) between the $^2\Pi$ and $^2\Sigma^+$ states of the alkali oxides and sulfides. For the alkali oxides there is a transition from a $^2\Pi$ ground state in LiO and NaO, to a $^2\Sigma^+$ ground state in RbO and CsO. For KO there is conflicting experimental evidence, with the ESR spectrum[68] suggesting a $^2\Pi$ ground state and a magnetic analysis[69] of K + NO$_2$ suggesting a $^2\Sigma^+$ ground state. This transition from a $^2\Pi$ to a $^2\Sigma^+$ ground state has been explained[70] in terms of the competing effects of quadrupole interaction, favoring $^2\Pi$, and Pauli repulsion, favoring $^2\Sigma^+$. Theoretical treatments[40,70,71] have agreed with experiment[68,69,72] for the ground states of LiO, NaO, RbO and CsO, but also give conflicting results for KO. Our theoretical results for T_e are compared with previous theoretical and experimental results in Table III. We observe T_e to be

Table III. $^2\Pi - ^2\Sigma^+$ excitation energies of the alkali oxides and sulfides, in cm^{-1}.

Molecule	Theory		Expt
	This work[a]	Other	
LiO	2359 (2391)	2894(2634)[b],2342(2330)[c]	> 0[g]
NaO	1429 (1701)	2088(2177)[b],1236[d],1785[e]	> 0[h]
KO	$-205(-240)$	233(831)[b], -347[d]	?[i]
RbO	$-516(-650)$	$-138(-114)$[b], -606[d]	< 0[j]
CsO	$-798($ $)$	$-735(-846)$[b], $-497(-726)$[f]	< 0[h,j]
LiS	5107(5035)		
NaS	3814(4033)		
KS	1717(1827)		
RbS	1240(1280)		

a   HF results given first with 15-electron CI results in parentheses.
b   (SCF/8-electron CI) results of Allison et al, Ref 70.
c   (SCF/CI) results of Yoshimine, Ref 101.
d   SCF results of So and Richards, Ref 71.
e   SCF results of O'Hare and Wahl, Ref 102.
f   (SCF/CI) results of Laskowski et al, Ref 103.
g   Freund et al, Ref 72.
h   Herm and Herschbach, Ref. 69.
i   Evidence has been presented separately for both a $^2\Sigma^+$ (Ref 69), and a $^2\Pi$ (Ref 68) ground state.
j   Lindsay et al., Ref 68.

insensitive to correlation, because both states are equally well described by the SCF configuration. This is true of previous theoretical results except for the Allison et al.[70] result for KO, which we believe to be an artifact of their correlating only the O orbitals without first localizing the orbitals to eliminate arbitrary mixings between the K and O orbitals. We have found that basis set quality has a greater influence on T_e than correlation effects. Extensive basis set tests indicate that f functions are more important for the $^2\Sigma^+$ state, whereas d functions are more important for the $^2\Pi$ state. At the numerical Hartree-Fock level the $^2\Sigma^+$ state is predicted to lie 250 cm^{-1} below the $^2\Pi$ state. Since it is unlikely that correlation significantly affects the separation, we feel that our calculations provide the strongest support so far for a $^2\Sigma^+$ ground state for KO. This result is in agreement with the older theoretical study of So and Richards,[71] but in disagreement with Allison et al.[70]

The alkali sulfides have not been characterized experimentally. The results in Table III definitively indicate $^2\Pi$ ground states for Li-Rb. While we have not considered CsS, the trends support a $^2\Pi$ ground state as well. Apparently the longer bond lengths in the alkali sulfides, and the larger size of sulfur anion reduce the importance of Pauli repulsion terms and lead to $^2\Pi$ states. A more detailed account of the study of T_e for the oxides and sulfides is given in Refs. 40 and 41.

IV. THEORETICAL STUDIES OF METAL CLUSTERS

Presently there is considerable effort directed at characterizing the chemistry at metal surfaces. This effort ranges from studies on perfect crystal faces of bulk metal to small metal clusters, both in the gas phase[73-75] and on supports.[76-79] The expanded computational capabilities of supercomputers makes the theoretical study of modest size metal clusters with multiple adsorbates a reality. In this section we describe theoretical work to model dissociation processes and coverage dependent effects of one or several adsorbate atoms or molecules with a metal cluster.

Studies designed to model chemisorption on a bulk metal indicate that it is necessary to use a sufficiently large metal cluster to avoid edge effects,[80-81] yet the environmental atoms[82] around an adsorption site do not participate significantly in the bonding. Hence, we use an all-electron treatment for those atoms directly involved in the bonding, and a one-electron effective core potential (ECP) treatment of the environmental atoms. For transition metal atoms at the all-electron site, we use basis sets which are better than double-zeta quality in the valence 4s, 4p and 3d region. For the environmental atoms we use a valence basis of (4s3p/2s1p) in conjunction with a one-electron ECP, or far from the adsorption site, just (4s/2s). The adsorbate atoms are described by double-zeta quality basis sets in the valence region. Diffuse functions are added to the adsorbate atom if it is ionic. See Ref. 82 for a more complete discussion of basis sets.

Model studies have shown that many of the qualitative features of metal ligand bonding are properly described at the SCF level.[83-85] However, for problems such as the dissociation[3] of H_2 or chemisorption of CO on a Ni(100) surface,[86] correlation must be included for a proper description. Although it may be possible to significantly improve the description through an MCSCF calculation after first localizing the orbitals,[87] the most common approach is to include correlation through a singles plus doubles configuration-interaction, SDCI, procedure. The importance of correlation is illustrated below where we consider in more detail the dissociation of H_2 on a Ni(100) surface.

A. Dissociation of H_2 on a Ni(100) Surface

Both the dissociation site and mechanism of molecular dissociation on a metal surface are difficult problems to resolve experimentally. A classic case is H_2 on a Ni(100) surface. This dissociation process was modeled[3] using a 13 atom cluster with the central Ni atom at the all-electron level, and with the 12 environmental atoms described by a one-electron ECP. Calculations were carried out at the SCF and SDCI levels at a sufficient number of molecular geometries to map out the dissociation pathway. Dissociation was observed to proceed at the on-top site (i.e. directly above a Ni atom- see Fig. 2). The mechanism involves considerable s-d hybridization on Ni, which allows the Ni orbitals to interact with both the H_2 bonding and anti-bonding orbitals, thus breaking the H_2 bond leading to H atoms chemisorbed on the surface.

The above picture of H_2 dissociation on Ni is consistent with the study of Blomberg and Siegbahn[88] on the triatomic system Ni + H_2, where bond formation was observed to occur on the excited singlet surface, and to involve s-d hybridization at the saddle point. It is interesting to note that each calculation on the Ni cluster + H_2 took less time on the CRAY-XMP than a corresponding calculation on the Ni + H_2 triatomic system carried out on a VAX. A significant difference between the Ni cluster and an isolated Ni atom is that the environmental Ni atoms substantially reduce the barrier on the dissociation pathway. A very small barrier of about 4 ± 4 Kcal/mole was found, but the barrier increases to about 50 Kcal/mole if the Ni 3d orbitals are prevented from mixing, and to about 66 Kcal/mole if correlation effects are excluded. Note also that the reaction on the Ni cluster occurs on the electronic ground state surface, whereas the lowest dissociation pathway for the Ni + H_2 system involves the excited ^1D state of Ni atom.

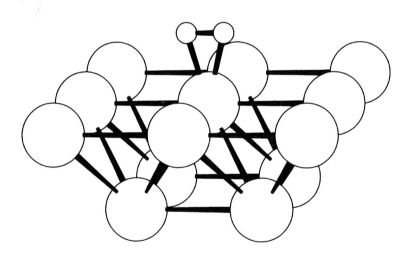

Figure 2. The $Ni_{13}H_2$ cluster used to model dissociation at the on-top site. The H_2 is dissociating towards the bridge site.

Dissociation pathways involving multiple Ni atoms were also investigated. Again, all dissociation pathways that involved only the valence 4s electrons were found to be quite high (< 30 Kcal/mole). Hence, if dissociation pathways involving two Ni atoms exist, they must also involve considerable 3d involvement on Ni. However, we find that once the H_2 is dissociated, there is little d involvement in the bonding.

This is consistent with the small perturbations observed[89] in the metal d bands for chemisorption of H atoms. This gives validity to the observation that the chemistry of the transition metals can appear significantly different from the simple metals, yet at times show little d involvement.

B. Coverage Dependent Effects

Experimentalists commonly study the surface as a function of coverage by monitoring the low-energy electron diffraction (LEED) pattern, the work function or the electron energy loss spectra (EELS). The relatively small changes that occur with coverage in metal adsorbate bond distance are difficult to measure experimentally, and the shifts in vibrational frequency normal to the surface, ω_e, are difficult to interpret based solely on experiment (See Refs 90-95). Hence,

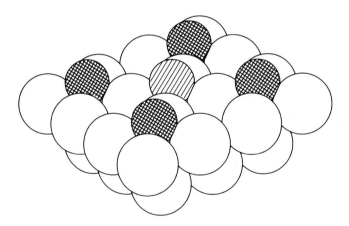

Figure 3. The $Ni_{25}X_5$ cluster used to model the c(2x2)/Ni(100) coverage. The adsorbate atoms are crosshatched. The p(2x2) coverage is modeled by eliminating the four doubly crosshatched adsorbate atoms.

theoretical calculations are valuable if they can establish the correct trends and gain insight into the changes in bonding that occur as coverage is increased. The availability of supercomputers allows one to consider a sufficiently large cluster such that several atoms can be adsorbed. Our chemisorption model uses the 25 atom Ni cluster shown in Fig. 3, with each atom treated at the one-electron ECP level. Four atomic adsorbates were considered, namely O, S, F and Cl. For low coverage, one adsorbate was chemisorbed, which corresponds to 0.25 monolayer (ML)--a primitive (p) 2x2 LEED pattern, while for high coverage, four adsorbates were chemisorbed corresponding to 0.50 ML--a centered (c) 2x2 LEED pattern. The theoretical results are compared with both the rigid Ni lattice dynamics calculation and experiment in Table IV.

Table IV. Summary of Oxygen, Sulfur, Chlorine and Fluorine chemisorbed on Ni(100)[a].

	Sulphur		Oxygen	
	$R_\perp$	ω_e^r	$R_\perp$	ω_e^r
		Cluster Calculation		
$Ni_{25}X$	1.28	345	0.58	280
$Ni_{25}X_5$	1.21	360	0.45	230
Δ	0.07	-15	0.13	50
		Rigid Ni lattice from lattice dynamics		
p(2x2)	294			351
c(2x2)	307			298
Δ	-13			53
		Experiment		
p(2x2)	1.30 ± 0.10[c]	367[b]	0.86 ± 0.07[d]	423[c]
c(2x2)	0.10 ± 0.10[c]	351[b]	0.86 ± 0.07[d]	327[c]
Δ	0.0	16	0.0	96
		Cluster Calculation		
	Chlorine		Fluorine	
	$R_\perp$	ω_e^r	$R_\perp$	ω_e^r
$Ni_{25}X$	1.33	320	0.88	330
$Ni_{25}X_5$	1.36	320	0.94	350
Delta.	-0.03	0	-0.06	-20

[a] $R\perp$ is the height of the absorbate above the metal surface. ω_e^r is the vibrational frequency normal to the surface for a rigid Ni lattice.
[b] Ref. 94.
[c] Ref. 92.
[d] Ref. 104.

Our computed shifts in ω_e are in good agreement with those from lattice dynamics, which should correspond best to our calculation since both exclude the motion of the Ni lattice. Most of the remaining difference with experiment is due to the influence of phonon coupling. The calculations also predict that the distance of the adsorbate from the surface decreases with coverage for O and S atoms. Since a smaller but opposite trend is observed for F and Cl atoms, it is proposed that adsorbed O atoms bond with more than one Ni atom, but as the coverage is increased this multiple bonding is disrupted. S atoms which form weaker double bonds, and Cl and F atoms, which are singly bonded to the surface, are therefore affected less by coverage.

We have also considered[96] the barrier to penetration of an O atom into a Cu lattice as a function of coverage. The barrier is expected to be at the point of closest approach, where the O atom passes through a triangle of metal atoms. Hence, for this study we switched to the (111) face of an 18-atom Cu cluster and considered penetration directly through the three-fold sites (see Fig. 4). Since the closer approach of O atom again makes 3d involvement on the metal possible, the three surface Cu atoms in the triangle are treated at the all-electron level. Since the shift in the barrier with O coverage for Cu and Ni clusters are expected to behave similarly with coverage, we use a Cu cluster to avoid problems with the open-shell d-electrons on Ni.

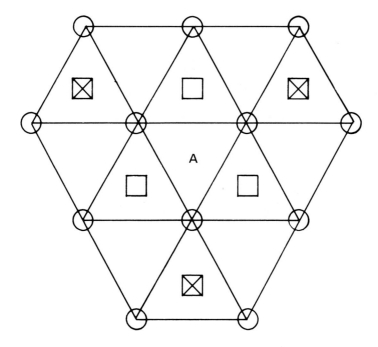

Figure 4. The $Cu_{18}O_n$ cluster used to model the Cu(111) face. The 12 circles represent the top layer copper atoms, and the six squares represent the second layer Cu atoms. The symbol A marks the site of penetration for one O atom at low coverage, while the symbol X marks the sites of the additional O atoms used to model higher coverage.

At low coverage, the penetration barrier is 8.2 eV, but it decreases by almost 3 eV, to 5.3 eV, as the coverage is increased from 0.25 to 0.50 ML. Although this reduction might be somewhat less ın a more elaborate calculation that included relaxation and correlation effects, it is consistent with the observation that oxygen is stable on many metals at low coverage, but penetrates into the metal at high coverages.[97] Apparently, the reduction in the barrier with coverage arises from the electronic effect of more neighbors above the surface, rather than to O atoms inside the lattice making penetration of additional O atoms more favorable.

C. Theoretical Studies of Small Gas Phase Clusters

There has been considerable experimental work on small metal clusters on supports,[76-79] and recently data from gas phase[73-75] experiments on metal clusters and their reactions are becoming available. In these experiments on supports, it is common to measure the change in average metal bond length (of a distribution of clusters) with the coverage of a chemisorbed gas.

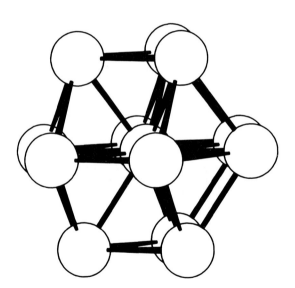

Figure 5. The bare D_{3h} Be_{13} metal cluster. The D_{3d} cluster is formed by rotating the bottom three atoms by 180° about the vertical axis.

Generally only saturation coverage is used and no adsorbate site information is obtained. Hence, theory can give considerable physical insight if it can delineate the trends in cluster bond length with adsorbate coverage and identify the preferred adsorption sites (See Refs 98-100). Unlike our model for chemisorption on bulk metals where we allow one preferred site (e. g. H_2 dissociation on Ni), chemisorption on small metal clusters can occur on all sides of the cluster at once. Hence, all metal atoms must be treated equivalently for a small metal cluster with adsorbates in which all non-equivalent bond lengths are to be optimized. Therefore, we have considered 13 atom Be and Al clusters at the all-electron SCF level in both the D_{3h} (hcp) and D_{3d} (fcc) geometries--see Fig. 5. The results for Be_4 in Table V at the SCF and SDCI level support the contention that geometrical parameters are accurately predicted at the SCF level.

Table V. The Be-Be and Be-adsorbate bond lengths (a_o) for bare Be, $Be_{13}X_2$, and $Be_{13}X_6$ clusters. Also shown is the expansion relative to the bare Be_{13} cluster.

	bare Be Cluster		
Be_4 double zeta SCF	3.97		
Be_4 large basis set CI[a]	3.92		
Be_{13}	4.07		
Bulk[b](hcp)	4.26		

| | | $Be_{13}X_2$ three-fold hollows | |
X	Be-Be	expansion	Be-X
H	4.03	− 0.04	2.97
O	4.24	+ 0.17	2.89
S	4.22	+ 0.15	3.80
F	4.09	+ 0.02	3.24
Cl	4.07	0.00	4.08

		$Be_{13}X_6$ three-fold hollows	
H	4.33	+ 0.26	3.09
O	4.73	+ 0.66	3.10
F	4.19	+ 0.12	3.44

		$Be_{13}X_6$ four-fold hollows	
H	4.14	+ 0.07	3.41
O	4.25	+ 0.18	3.21
F	4.14	+ 0.07	4.03

[a] Ref. 105.
[b] Ref. 106, the average of the two values in the bulk, 4.32 and 4.21.

We first optimized the 13 atom bare clusters with and without the constraint of equal bond lengths. For Be_{13} small distortions were found, but for Al_{13} virtually no distortion occurred.[99] The equilibrium Be-Be bond length of 4.07 Bohr for the Be_{13} bare cluster was found to be significantly less than the average bond length in the hcp bulk metal (4.26 Bohr). Since the distortions away from equal bond lengths in the bare cluster were small, the metal-metal bond lengths were assumed to be equivalent, when adding two and six adsorbates. The clusters with added adsorbates are illustrated in Figs. 6 and 7. We first considered the addition of two

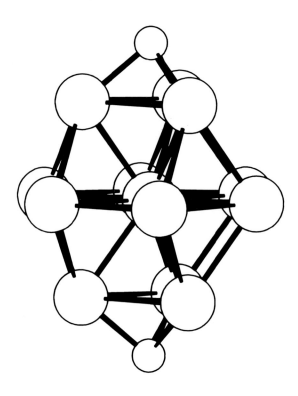

Figure 6. The $Be_{13}X_2$ cluster used to model low coverage chemisorption into the three-fold hollows.

H, O, S, F and Cl atoms into the three-fold hollows. These results are summarized in Table V. Two hydrogen atoms contract the cluster slightly, whereas adding oxygen or sulfur results in a significant expansion. The chemisorption of the more ionic F and Cl atoms has little effect. We then increased the coverage to six adsorbates (high coverage) and examined adsorption at both the three-fold and four-fold hollows. Note that sulfur and chlorine were not considered at high coverage since at low coverage they are similar to oxygen and fluorine, respectively. At high coverage the cluster expands in all cases, but the expansion is far more dramatic for oxygen, and is much larger for the three-fold than the four-fold hollows. Adsorption of hydrogen into the three-fold hollows gives a Be-Be bond length similar to that in the bulk. These results are consistent with the experimental observation for transition metals that at high coverage H atoms expand the cluster to about bulk geometry, whereas oxygen atoms give a much larger expansion. This comparison also suggests that the absorption is into the three-fold sites, whereas one might have expected the electronegative atoms to chemisorb in the larger four-fold hollows. However, unlike the bulk metal, the bond lengths in the cluster can expand making the three-fold hollows more favorable. Preliminary results[100] for the Al_{13} cluster also indicate a large expansion upon adsorption of oxygen. We find that H atoms penetrate the Al_{13} cluster giving a large (0.5 Bohr) expansion. Finally, the good agreement between our results at high coverage and the experimental results involving weak metal support interaction, suggests that the interaction of the metal support with the bare metal cluster is not overly strong, and therefore should give results similar to those in the gas phase.

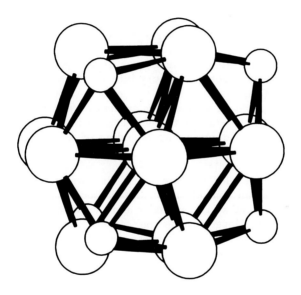

Figure 7. The $Be_{13}X_6$ cluster used to model chemisorption into the four-fold hollows. The $Be_{13}X_6$ cluster used to model chemisorption into the three-fold hollows is generated by a 60° rotation of the six adsorbate atoms about the vertical axis.

V. CONCLUSIONS

The availability of supercomputers has substantially increased the range of problems that we are able to consider in the computational chemistry group at NASA Ames. An overview of our work in the general areas of re-entry physics, atmospheric chemistry, combustion, materials problems, catalysts, polymer design and non-intrusive diagnostics demonstrates the range of interests and capabilities at NASA.

We have presented in more detail our theoretical results for more than 50 ionic molecules. The goal of this work has been to obtain accurate spectroscopic parameters and dissociation energies accurate to 0.1 eV. The large basis sets and the length of the CI expansion of our treatment would not have been feasible without a supercomputer. Our systematic study of the dissociation energies of ionic molecules has revealed several new trends. The theoretical results are sufficiently accurate to rule out disparate experimental values, and to make an overall assessment of the accuracy of the various experimental methods for determining dissociation energies.

Another area of considerable interest to NASA is advanced materials research. Our work in this area was illustrated by the following three examples: (1) the dissociation of H_2 on a Ni(100) surface; (2) chemisorption and coverage dependent effects of atomic adsorbates on a rigid metal lattice; and (3) the effects of coverage on the structure of small metal clusters. Again, these very elaborate calculations would not have been feasible without a supercomputer. The principal results of this study were the following. The dissociation process of H_2 on a Ni(100) surface involves considerable d orbital involvement of the Ni atoms. The average distance of the adsorbates and the frequency normal to the surface of a rigid metal lattice are dependent on coverage, and vary greatly with adsorbate, e.g. O atom is different from S, Cl and F atoms. Finally, we have observed that chemisorption onto small metal clusters probably occurs into the 3-fold hollow sites, thereby resulting in a significant expansion of the cluster. Apparently, the three-fold hollows are more favorable on a metal cluster than on a rigid metal lattice owing to the greater ability of the clusters to deform. Such insights into the bonding at the molecular level are yielding an understanding of such processes as catalysts and oxidation.

Supercomputers have extended our capabilities in three important respects-- the accuracy to which we can compute molecular properties, the size of the clusters and molecular systems that we can consider and the increased speed at which we can respond to problems. Hopefully, the applications presented herein demonstrate the perhaps obvious contention that chemists greatly benefit by the availability of excellent computational capabilities.

ACKNOWLEDGEMENTS

We would like to acknowledge the hard work of Per Siegbahn, Bjorn Roos, Peter Taylor, Jan Almlof and Anders Heiberg who have helped make the MOLECULE-SWEDEN codes a reality. We would like to thank Professors Hans Lischka and Reinhart Ahlrichs for providing us a copy of the Karlsruhe codes for the Cyber 205. We would also like to thank Professor Jan Almlof for supplying us with a copy of his truly novel DISCO code.

REFERENCES

1. This fact is perhaps best illustrated by the computational fluid dynamicists at Ames, who have plotted problem size versus time, and find the growth in problem size exceeds the growth in computer power.

2. "Algorithms vs Architectures for Computational Chemistry", H. Partridge and C. W. Bauschlicher, "Algorithms, Architectures and the Future of Scientific Computation", ed. T. Tajima and F. A. Matsen, University of Texas Press, Austin, Texas; also D. H. Calahan, private communication.

3. P.E.M. Siegbahn, M.R.A. Blomberg, and C.W. Bauschlicher, J. Chem. Phys., **81**, 2103 (1984).

4. T. Takai, T. Halicioglu and W.A. Tiller, Script. Metall., **19**, 709 (1985).

5. T. Halicioglu, H.O. Pamuk and S. Erkoc, Surf. Sci., **143**, 601 (1984).

6. T. Halicioglu and D.M. Cooper, Materials Sci. Engin., **62**, 121 (1984).

7. C.W. Bauschlicher, S.P. Walch, and H. Partridge, J. Chem. Phys., **76**, 1033 (1982).

8. "Theoretical Studies of Transition Metal Dimers", S.P. Walch and C.W. Bauschlicher, "Comparison of *ab initio* Quantum Chemistry with Experiment", ed R. Bartlett, D. Reidel Publishing Company, Boston (1985).

9. S.P. Walch and C.W. Bauschlicher, submitted to J. Chem. Phys.

10. L.B. Knight, R.W. Woodward, R.J. Van Zee, and W. Weltner, J. Chem. Phys., **79**, 5820 (1983).

11. S.P. Walch and B.C. Laskowski, submitted J. Chem. Phys.

12. M.D. Morse, J.B. Hopkins, P.R.R. Langridge-Smith, and R.E. Smalley, J. Chem. Phys., **79**, 5316 (1983).

13. A. Komornicki and H. King, submitted to NATO workshop on gradient methods.

14. B.C. Laskowski, R.L. Jaffe, and A. Komornicki, submitted for publication, proceeding of ACS Hawaii meeting, Dec. 1984.

15. S.D. Hong, S.Y. Chung, R.R. Fedors and J. Moacanin, J. Polymer Sci.: Polymer Physics Ed., **21**, 1647 (1983).

16. A. Gupta, private communication, relayed by R.L. Jaffe.

17. D.M. Cooper, R.L. Jaffe and J.O. Arnold, J. Spacecraft and Rockets, **22**, 60 (1985).

18. H. Partridge, C.W. Bauschlicher, and J.R. Stallcop, J. Quant. Spectrosc. and Radiat. Transfer, **33**, 653 (1985).

19. H. Partridge, and J.R. Stallcop, A paper 85-0917 AIAA 20th Thermophysics Conference, Williamsburg, VA, June 19-21 (1985).

20. J.R. Stallcop and H. Partridge, in press Phys. Rev. Lett.

21. V.A. Belyaev, B.G. Brezhnev and E.M. Erastov, Sov. Phys. JETP, **27**, 924(1968).

22. R.R. Stebbings, A. C.H. Smith and H. Ehrhardt, J. Geophys. Res., **69**, 2349(1964).

23. W.M. Huo, V. McKoy, M.A.P. Lima, and T.L. Gibson, paper 85-1034, AIAA 20th Thermophysics Conference, Williamsburg, VA, June 19-21 (1985).

24. M.A.P. Lima, T.L. Gibson, W.M. Huo and V. McKoy, Phys. Rev. A, to be published.

25. L.M. Brescansin, M.A.P. Lima, W.M. Huo and V. McKoy, Phys. Rev. A, to be published.

26. E.E. Whiting, NASA TN D-7268, Wash. D.C. 1973.

27. E.E. Whiting and R.W. Nicholls, Astrophys. J. Supplement Series. No 235, **27**, 1 (1974).

28. H. Partridge, S.R. Langhoff, B.H. Lengesfield, and B. Liu, J. Quant. Spectrosc. Radiat. Transfer, **30**, 449 (1983).

29. C. Linton and R.W. Nicholls, J. Quant. Spectrosc. Radiat. Transfer, **9**, 1 (1969).

30. S.R. Langhoff, R.L. Jaffe, J.H. Yee and A. Dalgarno, Geophys. Res. Lett., **10**, 896 (1983).

31. S.R. Langhoff, H.J. Werner and P. Rosmus, "Theoretical Transition Probabilities for the OH Meinel System", to be published.

32. S.R. Langhoff, unpublished.

33. J.R. Reimers, K.R. Wilson and E.J. Heller, J. Chem. Phys., **79**, 4749 (1983).

34. J.R. Reimers, K.R. Wilson, E.J. Heller, J. and S. R. Langhoff, J. Chem. Phys., **82**, 5064 (1985).

35. W.M. Huo, K.P. Gross, and R.L. McKenzie, Phys. Rev. Lett., **54**, 1012 (1985).

36. "Theoretical Dissociation Energies for Ionic Molecules", S.R. Langhoff, C.W. Bauschlicher, and H. Partridge, "Comparison of ab initio Quantum Chemistry with Experiment", ed. R. Bartlett, D. Reidel Publishing Company, Boston (1985).

37. C.W. Bauschlicher, S.R. Langhoff, and H. Partridge, submitted J. Chem. Phys.

38. S.R. Langhoff, C.W. Bauschlicher, and H. Partridge, submitted J. Chem. Phys.
39. H. Partridge, S.R. Langhoff, and C.W. Bauschlicher, submitted Chem. Phys. Lett.
40. S.R. Langhoff, C.W. Bauschlicher, and H. Partridge, to be published.
41. H. Partridge, S.R. Langhoff, and C.W. Bauschlicher, to be published.
42. H. Partridge, S.R. Langhoff, and C.W. Bauschlicher, to be published.
43. C.W. Bauschlicher, S.R. Langhoff, and H. Partridge, Chem. Phys. Lett., **115**, 124 (1985).
44. C.W. Bauschlicher, B.H. Lengsfield and B. Liu, J. Chem. Phys., **77**, 4084 (1982).
45. C.E. Moore, Atomic energy levels, Natl. Bur. Stand. (US) circ. **467** (1949).
46. H. Hotop and W.C. Lineberger, J. Phys. Chem Ref. Data, 4, 530 (1975).
47. E.A. McCullough Jr., J. Chem. Phys., **62**, 3991 (1975); L. Adamowicz and E.A. McCullough Jr., J. Chem. Phys. **75**, 2475 (1981); E.A. McCullough Jr., J. Phys. Chem., **86**, 2178 (1982).
48. S.R. Langhoff and E.R. Davidson, Int. J. Quantum Chem., 8, 61 (1974).
49. R. Ahlrichs, P. Scharf and C. Ehrhardt, J. Chem. Phys., **82**, 890 (1985).
50. L. Brewer and E. Brackett, Chem Rev., **61**, 425 (1961).
51. J. Drowart, G. Exsteen and G. Verhaegen, Trans. Farad. Soc., **60**, 1920 (1964).
52. R.D. Srivastava, High Temp. Sci., 8, 225 (1976). Recommended values on review of the experimental literature through 1975.
53. J.W. Cox and P.J. Dagdigian, J. Phys. Chem., **88** (1984) 2455.
54. J.A. Irvin and P.J. Dagdigian, J. Chem. Phys., **73,** 176 (1980).
55. F. Engelke, R.K. Sander and R.N. Zare, J. Chem. Phys., **65**, 1146 (1976).
56. K.P. Huber and G. Herzberg, "Molecular Spectra and Molecular Structure," (Van Nostrand Reinhold, New York ,1979).
57. M.W. Chase Jr., J.L. Curnutt, R.A. McDonald, and A.N. Syverud, J. Phys. and Chem. Ref. Data 7, 793 (1978). JANAF Thermochemical Tables, 1978 supplement.
58. R. Colin, P. Goldfinger and M. Jeunehomme, Trans. Faraday Soc., **60**, 306 (1964).
59. P.J. Kalff and C.Th.J. Alkemade, J. Chem. Phys., **59**, 2572 (1973).
60. J.B. Pedley and E.M. Marshall, J. Phys. Chem. Data, **12**, 967 (1983).
61. L. Brewer, and G.M. Rosenblatt, "Advances in High Temperature Chemistry", edited by L. Eyring (Academic, New York, 1969) p.1.
62. C. Batalli-Cosmovici, and K.W. Michel, Chem. Phys. Lett., **16**, 77 (1972).
63. A. Lagerquist, and L. Huldt, A. Naturforsch., **9a**, 991 (1954).
64. I.V. Veits, and L.V. Gurvich, Zh. Fiz. Khim., **31**, 2306 (1957).
65. G. Drummond, and R.F. Barrow, Trans. Faraday Soc., **47**, 1275 (1951).
66. R.N. Zare, private communication.
67. E.Murad, Chem. Phys. Lett., **72**, 295 (1980).
68. D.M. Lindsay, D.R. Herschbach, and A.L. Kwiram, J. Chem. Phys., **60**, 315 (1974).
69. R.R. Herm and D.R. Herschbach, J. Chem. Phys., 2, 5783 (1970).
70. J.N. Allison and W.A Goddard III, J. Chem. Phys., **77**, 4259 (1982). See also J.N. Allison, R.J. Cave and W.A. Goddard III, J. Phys. Chem., **88**, 1262 (1984).
71. S.P. So and W.G. Richards, Chem. Phys. Lett., **32**, 227 (1975).
72. S.M. Freund, E. Herbst, R.P. Mariella and W. Klemperer, J. Chem. Phys., **56**, 1467 (1972); R.A. Berg, L. Wharton, W. Klemperer, A. Bu and J.L. Stauffer, ibid **43**, 2416 (1965).
73. W.D. Knight, K. Clemenger, W.A. de Heer, W.A. Saunders, M.Y. Chou, and M.L. Cohen, Phys. Rev. Letters, **52** 2141 (1984).
74. M.E. Geusic, M.D. Morsé and R.E. Smalley, J. Chem. Phys., **82**, 590 (1985).
75. E.K. Parks, K. Liu, S.C. Richsmeier, L.G. Pobo, ans S.J. Riley, J. Chem. Phys., **82**, 5470 (1985), and references therein.
76. H. Poppa, Vacuum, **34**, 1081 (1985), and Ultramicroscopy, **11**, 105 (1983).
77. P. Gallezot, Catal.-Rev. Sci. Eng., **20**, 121 (1979).
78. R.D. Moorhead and H. Poppa, Proc. IX Int. Vac. Congress Madrid (1983), Ext. Abstr., P11.
79. G.C. Bond, "Metal Support and Additive Effects in Catalysis", B. Imelik et. al. Edts., p. 1, Elsevier, Amsterdam (1982).
80. P.S. Bagus, H.F. Schaefer, and C.W. Bauschlicher, J. Chem. Phys., **78**, 1390 (1983).

81. T.H. Upton and W.A. Goddard, Phys. Rev. Lett., **46**, 1635 (1981), and T.H. Upton and W.A. Goddard, "Critical Reviews in Solid State and Material Sciences", CRC Press (1981).
82. P.S. Bagus, C.W. Bauschlicher, C.J. Nelin, B.C. Laskowski, and M. Seel, J. Chem. Phys., **81**, 3594 (1984).
83. P.S. Bagus, K. Hermann and C.W. Bauschlicher, J. Chem. Phys. **80**, 4378 (1984).
84. P.S. Bagus, K. Herman, and C.W. Bauschlicher, J. Chem. Phys., **81**, 1966 (1984).
85. C.W. Bauschlicher, P.S. Bagus, C.J. Nelin, and B.O. Roos, submitted for publication.
86. C.W. Bauschlicher, and C.J. Nelin, submitted for publication.
87. P. Madhaven and J.L. Whitten, J. Chem. Phys., **77**, 2673 (1982), and references therein.
88. M. R.A. Blomberg and P.E.M. Siegbahn, J. Chem. Phys., **78** 986, 5682 (1983).
89. F.J. Kimpsel, J.A. Knapp, and W.E. Eastman, Phys. Rev., **B19**, 2872 (1979).
90. C.W. Bauschlicher and P.S. Bagus, Phys. Rev. Letters, **52**, 200 (1984).
91. C.W. Bauschlicher and P.S. Bagus, Phys. Rev. Letters, **54**, 349 (1985).
92. J. Stohr, R. Jeager, and T. Kendelewicz, Phys. Rev. Lett., **49**, 142 (1982) and references therein.
93. J.M. Szeftel, S. Lehwald, H. Ibach, T.S. Rahman, J.E. Black and D.L. Mills, Phys. Rev. Lett., **51**, 268 (1983), and references therein.
94. S. Andersson, P.-A. Karlsson and M. Persson, Phys. Rev. Lett., **51**, 2378 (1983), and references therein.
95. C.W. Bauschlicher, submitted for publication.
96. C.W. Bauschlicher, in press Chem. Phys. Lett.
97. T. Narusawa, W.M. Gibson and E. Tronqvist, Surf. Sci., **114**, 331 (1982).
98. C.W. Bauschlicher, Chem. Phys. Lett., **117**, 33 (1985).
99. C.W. Bauschlicher, and L. Pettersson, to be published.
100. H. Partridge and C.W. Bauschlicher to be published.
101. M. Yoshimine, J. Chem. Phys., **57**, 1108 (1972).
102. P.A.G. O'Hare and A.C. Wahl, J. Chem. Phys., **56**, 4516 (1972).
103. B.C. Laskowski, S.R. Langhoff and P.E.M. Siegbahn, Int. J. Quantum Chem., **23**, 483 (1983).
104. M. Van Hove and S.Y. Tong, J. Vac. Sci. Technol., **12**, 230 (1975).
105. C.W. Bauschlicher, P.S. Bagus and B.N. Cox, J. Chem. Phys., **77**, 4032 (1982).
106. R.G. Wyckoff, "Crystal Structures", Second edition (Interscience, New York, 1964).

THE STUDY OF MOLECULAR ELECTRONIC STRUCTURE ON VECTOR AND ATTACHED PROCESSORS

Correlation Effects in Transition Metal Complexes

Martyn F. Guest

Computational Science Group
SERC Daresbury Laboratory
Daresbury
Warrington WA4 4AD

1. INTRODUCTION

The purpose of this paper is to both review the impact and use in the UK of Vector and Attached Processors in the study of molecular electronic structure and to describe illustrative applications in the study of transition metal complexes.

The review is necessarily selective, and is divided into several sections. In section 2 we provide an outline of the available computational facilities in the UK, describing the role played by the Theory and Computational Science Division at the Daresbury Laboratory, and the SERC Collaborative Computational Projects. Attention is focused on current UK Academic Computing, and the background given to the acquisition in January 1985 of an FPS-164 at Daresbury.

In section 3 we focus on the computationally demanding steps and algorithms featured in the discipline of Quantum Chemistry, where the experience gained in the implementation and vectorization of various packages on both vector and attached processors will be described. The relative merits of the supercomputer and attached processor approach to quantum chemistry calculations is outlined, with a measure provided of the cost-effectiveness of the Cray-1S, CDC Cyber-205 and FPS-164. Performance figures for the FPS-264 will also be presented.

In Sections 4 and 5 we turn to specific applications, considering the qualitatively significant effects of electron correlation in the calculation of molecular ionization spectra and ground state equilibrium geometries for a variety of first-row transition metal complexes.

Correlation effects in the ground states of bis(π-allyl) nickel and $Ni(CN)_4^{2-}$ are described in section 4. The valence ionization energies of $Ni(CN)_4^{2-}$, $Co(CN)_6^{3-}$ and $Fe(CN)_6^{4-}$ measured using X-ray emission spectroscopy, together with the photoelectron spectrum of bis(π-allyl)nickel are compared with the ionization energies calculated using the extended-two-particle-hole Tamm-Damcoff method, which includes some account of both relaxation and correlation effects. For all species studied, excellent agreement between theory and experiment is found which, for bis(π-allyl) nickel is absent for ΔSCF-CI calculations.

Finally in section 5 we present a systematic study of the calculation of the equilibrium ground state geometries of some 15 transition metal complexes at the HF level using an extended basis set and contrast the results with those obtained at the minimal basis set level. CI calculations are reported in an attempt to rationalize the deviations from experiment.

2. COMPUTATIONAL FACILITIES IN THE UK

In 1979 a 1/2 Mword Cray-1 computer was installed at the Science and Engineering Research Council (SERC) Laboratory at Daresbury to provide the opportunity for significant advance in those scientific disciplines which were severely constrained by a lack of computer power. The machine was upgraded to a 1 Mword Cray-1S, and remained at Daresbury for four years, during which time over 400 individual projects selected for their potential for breaking new and worthwhile scientific ground were awarded for Cray use.[1] Following this period of concentrated use by a relatively select community the Cray-1 was transferred to the University of London Computer Centre (ULCC) where it now satisfies a more general computing need for an unrestricted community of users.

A 1 Mword 2-pipe CDC Cyber-205 was installed at the Manchester Regional Computing centre (UMRCC) in May 1983, with a full service commencing some 12 months later, again for an unrestricted community of users. The machine was recently upgraded to 2 Mword.

Although the intention was to replace the Cray by a still more powerful processor dedicated to the most computationally demanding tasks, the limitations on financial resources have so far prevented this. The resulting lack of a computing facility dedicated in this way has prompted one community of users, principally those involved in the Collaborative Computational Projects (CCPs) centred at Daresbury, to fund an FPS-164 facility to fill at least part of the vacuum.

The main responsibility of the Theory Group within the Theory and Computational Science Division (TCS) at Daresbury is to provide theoretical support for the experimental programs on the Synchrotron Radiation Source (SRS) and Nuclear Structure Facility (NSF) at the Laboratory. The Computational Science Group has responsibility for supporting university research in an increasingly broad range of computational subjects, which is largely accomplished by means of the CCPs. Four of the projects currently operational have a Chemistry bias, namely

    CCP1 - Quantum Chemistry
    CCP3 - Computational Studies of Surfaces
    CCP5 - Computer Simulation of Condensed Phases
    CCP6 - Heavy Particle Dynamics

The availability of the Cray-1S in both a production, and perhaps more importantly, a development environment provided the CCPs with a unique opportunity to 'learn the tricks' of the vector-processing 'trade,' and develop software widely used in UK Academic Computing. Unfortunately, severe limitations of the links to ULCC made it difficult for members of TCS (and the academic community) to develop code on the machine after its transfer to London. Furthermore production work could only be carried out with significant delays. Commencement of a full service on the

Cyber at the Manchester Regional Centre revealed that many of the problems in accessing the Cray were evident in trying to use the 205, with good networking facilities far into the future.

It became clear that an attractive option for TCS was provided by the FPS-164 processor, attached to an IBM-compatible AS7000 at Daresbury. This provides 'pseudo' vector processing (VP) capability in a cost-effective way, optimally equivalent to some 3000-4000 IBM 360/195 hours per year, with the potential of being upgraded as required by the addition of MAX boards. Benchmark tests demonstrated that code vectorized for the Cray-1 ran effectively on the FPS-164 with the minimum of effort. With the offer by FPS UK of a machine at approximately half price, a proposal for the acquisition was submitted to, and subsequently accepted by, Science Board. The machine was delivered to Daresbury in October 1984. In its initial configuration it is relatively modest, comprising three 135 Mb disks and one 600 Mb disk and 512 Kwords of main memory, but the intention is to expand the memory and add MAX boards over the course of the next year or so as applications are mounted.

3. COMPUTATIONAL QUANTUM CHEMISTRY

University work in the UK on the computation of the electronic structure of molecules is supported and coordinated through the activities of CCP1. The groups involved in the project include Cambridge, Oxford, Sheffield, Sussex, Queen Elizabeth College (London), York, St. Andrews, Warwick, Bristol, ICI New Science Group and Daresbury.

Computational Quantum Chemistry (QC) is a subject with a history at least as long as that of electronic computers themselves. The subject grew rapidly in the period 1960-1980, resulting in the availability of many reasonably efficient computer programs on scalar machines. These codes represent a considerable investment in man-years (say between 5 and 30) and are relatively large (say between 20,000 and 150,000 lines of FORTRAN), as befits a complicated problem. The advent of the large scale vector processors (eg Cray-1, CDC Cyber-205) and their smaller brothers (eg FPS-164 and the more recent FPS-264 and Convex C1) led to substantial efforts[2-5] to alter the existing codes to conform with the requirements of these new machines. Exploiting the parallelism inherent in electronic structure calculations[5b] will not be considered explicitly in this paper, since this subject will undoubtedly be covered in other contributions at the meeting.

The aim throughout our work with vector machines has been to achieve a code design which was reasonable transportable from one vector (or array or scalar) processing computer to another, with the capability of driving these processors at somewhat near their maximum power. Central to this effort was the the realization that five of the most important steps in a typical quantum chemistry calculation may be structured around the matrix multiply operation (MMO), yielding a relatively machine independent structure, given that the MMO is capable of driving all existing computers at their maximal power. In the following sections we provide a summary of the characteristics of electronic structure calculations and outline the MMO-based algorithms involved.

There are at least three levels of approximation that can be used in the study of the electronic structure of atoms and molecules, and typically the capability to handle all three levels is provided in the current generation of QC packages. These three approximations are SCF (self-consistent field), MCSCF (multi-configuration SCF) and CI (configuration interaction), in order of computational difficulty.

The SCF method is usually derived by assuming a specific form of the solution to the quantum mechanical equation (the Schrodinger equation), leading to a set of coupled equations called the HF (Hartree-Fock) equations. This assumed form is the simplest possible that is in agreement with the physics of the problem. The HF equations are a set of integro-differential equations that could be solved numerically. However it is more common to expand the solutions in a finite set of primitive functions (called the basis set). The set of equations then become a set of coupled homogenous equations and are usually written in matrix form. The eigenvalues and eigenvectors of the matrix (the Fock matrix) are required, and because of the coupling in the matrix (the matrix is defined in terms of its solution) a self-consistent solution is sought. This implies an iterative solution process, iterating until the Fock matrix remains constant from iteration to iteration.

The second level of approximation (MCSCF) assumes a more flexible form of the solution to the Schrodinger equation but follows closely, in so far as is possible, the same development as the simpler SCF method. One major difference is that MCSCF generally requires an orthogonal set of functions as its basis to keep it computationally tractable.

The third level of approximation (CI) attempts to take into account an important physical aspect that is not included in SCF and only to a small extent in MCSCF. This is the instantaneous correlation of the motion of electrons in atoms and molecules. The previous two approximations define a type of average influence of (n-1) electrons on the n'th. Again an orthogonal set of basis functions must be used.

In essence all three approximations simply build a matrix and diagonalize to obtain the solutions. In SCF and MCSCF methods the matrix is of order 10^2 but its elements are rather complicated. In the CI method the matrix is of order 10^5-10^6 but its elements are simpler. The matrix elements are written as a sum of quantities (numbers) each quantity being weighted by some coefficient. The evaluation of these coefficients provides the complexity of the matrix, but there is a deeper complexity in the evaluation of the quantities themselves. All of computational quantum chemistry employs these quantities as constants, although their initial calculation is far from trivial. These constants are in fact integrals of products of either 2 or 4 Cartesian gaussian functions over all space, and are regularly referred to as the 1- and 2-electron integrals.

A. Role of the MMO

As we shall show below, the electronic structure problem can be formulated such that the MMO

$$R = A\,B$$

where perhaps **A** or **B** (but not both) is sparse is of fundamental importance for five steps in a typical calculation

(i) 2-electron integral evaluation over Gaussian Functions[6-7]

(ii) Hartree-Fock construction of a non-correlated wavefunction[8]

(iii) Transformation of the 2-electron integrals from an atomic to a molecular orbital basis[9]

(iv) The construction of a correlated wavefunction using the DIRECT-CI technique[3,10]

(v) The multiconfiguration SCF construction of a correlated wavefunction.[11]

Details of an optimized MMO routine written in assembly language for the Cray-1S, using the 'outer product' formalism and capable of utilizing any sparsity in **A** or **B**, been given previously.[2] Execution rates in excess of 100 Mflop are possible on the Cray with matrix dimensions (vector length, VL) as low as 24, the routine achieving an asymptotic performance of 147 Mflop when a dimension of 64 is used (or any multiple thereof), correlating with a hardware characteristic of the Cray. The performance of a straightforwardly coded vectorized FORTRAN version is in marked contrast, being asymptotically only 37 Mflop with very large matrix dimensions (see figure 1 of ref. 2b). Note that the direction of vectorization in the sparse MMO depends upon which matrix is assumed sparse, so that one should vectorize by rows or columns of the result matrix, depending upon whether **A** or **B**, respectively, is assumed sparse.

It is perhaps worth recalling at this point the Mflop rate for the hierarchy of preferred operations relevant to the machines under discussion:

	Cray-1S	Cyber-205 2-pipe	FPS-164
recursive linked triad (MMO)	147	200	9
linked triadic	49	200	4
linked diadic	38	100	2.5

Effectively the full potential of the FPS-164 in the sparse matrix multiply may be obtained directly through use of the APMATH library routine FMMMV (see reference 4b).

A consequence of the increased vector start-up times on the Cyber-205 is, of course, the requirement for a significantly longer VL if optimal performance is to be achieved. Thus MMO rates of 100 Mflop on a 2-pipe 205 require a matrix dimension of 170, and illustrates the need for an alternative 'long vector' algorithm for small matrices. Such an algorithm,[12] comprising a hybrid scheme of dyadic operations, matrix transposition and bisection techniques outperforms the outer product algorithm by a factor of two for matrices of dimension 20 on a 2-pipe machine, where it rapidly achieves an asymptotic performance of 45 Mflop. Indeed this algorithm can outperform the hardware inner product order on a 4-pipe machine!

B. Gaussian Integral Evaluation

This problem is characterised by stating that one requires to evaluate a large number of six-dimensional integrals, perhaps up to 10^9, by means of essentially analytic (but relatively complicated) formulae. If N is the number of basis functions in the chosen basis set, the number of two-electron integrals is proportional to the fourth power of N. These integrals may be grouped into subsets of essentially identical type, the subsets being denoted < PP/PS > etc, where the labels S, P, D etc refer to a group of functions all having the same exponent and total quantum number $(n + 1 + m)$, but differing in the distribution of n, l, m. Each subset of integrals is evaluated by identically the same algorithm and may be of dimension 10^3 to 10^8. A strategy for vectorization can therefore be devised, whose vector length is the dimension of the subsets. In other words, one attempts to evaluate all (or large groups) of integrals within a given subset 'simultaneously'. The performance of this 'extrinsic' vectorization in terms of Mflop rate will depend little on the details of the algorithm used (on the Cray-1 rates of about 30 Mflop are to be expected), and the 'best' algorithm will be that which minimises the number of floating point operations. Notice that the amount of control required before one decides to perform a floating point operation is not now of primary importance, since the cost of that control can be amortized over long vector lengths. One problem with this 'extrinsic' vector algorithm is memory requirement. Since all auxiliaries for all batches of integrals that are being evaluated 'simultaneously' must be stored, we find the storage requirement rises very rapidly with l, the average l quantum number of the four basis functions of the integral. This problem is exacerbated on the Cyber-205, where much longer vector lengths must be chosen for optimal performance. A comparison of the performance of this algorithm on the Cray-1S, Cyber-205 and the FPS-164 is presented below.

An alternative strategy for vectorization of the integral evaluation process is to make use of the 'intrinsic' vectorization potential of the MMO directly, and this is found possible within the McMurchie-Davidson algorithm.[7] The code we have prepared assumes a real spherical harmonic basis of the form

$$\Xi[A, \alpha, n, l, m] = r_A^{2n}\, r_A^l\, e^{-\alpha r_A^2}\, Y_{lm}$$

and we evaluate all integrals involving all possible m values within the four basis functions of an integral simultaneously. The kernel of the algorithm involves a matrix multiplication of the form

$$R_{pq} = C_p\, G_{pq}\, C_q ,$$

where p and q label the two overlap distribution involved. The matrix elements of G_{pq} are two-electron integrals between the scaled Hermite functions into which the overlap distributions are expanded, while the C matrices comprise the necessary expansion coefficients and are sparse (typically 40-50% of the elements are finite). The time taken to evaluate the C matrix varies as the fifth power of l, the G matrix as the fourth power, and the MMOs as the eighth power, where again l is the average l quantum number of the four basis functions. Use of the intrinsic vectorization potential of the MMO becomes advantageous at sufficiently high l, allowing us to evaluate only one batch of integrals at a time. This 'intrinsic' algorithm in fact becomes more efficient than the 'extrinsic' algorithm for batches of integrals of higher l than $<PP/PP>$.

Neither the extrinsic or intrinsic algorithm outlined above migrates effectively to the 205. The required increase in VL leads to excessive memory requirements in the extrinsic case, while the typical vector lengths in the intrinsic MMO case (eg $<PP/PP>$, VL=10, $<DD/DD>$, VL=21; $<FF/FF>$, VL=35) inhibits optimum performance.

A revised algorithm has recently been developed in which the number of auxiliaries held in memory is significantly reduced, leading to an increase in VL. The number of floating point operations may also be reduced by using the translational invariance properties of **G**. In particular:

(i) The matrix elements of **C** are merely products of x, y and z factors. These factors are now kept separate, leading to a decrease in memory requirements, from l^5 to l^3 Note that the C supermatrix is now structured into blocks of N overlap distributions (N of order 100), effectively removing the generation cost given the subsequent N^2 integral generation.

(ii) The memory requirements for **G** may be significantly reduced if the matrix is stored in compacted form

$$G(pqr,stu) = G(p+s,q+r,t+u,000) \, (-1)^s \, (-1)^t \, (-1)^u$$

with the multiplicative factors subsequently built into the definition of C_q. This leads to an l^3 memory dependence, compared to the previous l^6.

(iii) In the original MMO implementation, stu was collapsed into a single index in both **C** and **G**, and the x,y, and z contractions performed 'simultaneously' in the MMO. In the revised algorithm, with the **C** factors kept separate, and **G** stored in compacted form, the x,y and z contractions are performed separately. The use of the translational invariance properties of **G** leads to a significant reduction in the number of floating point operations.

While the formal structure of **Rpq** remains, the code is presently organized as a mixture of linked diadic and triadic operations. This may be rewritten entirely in linked triadic form, but would require a major reorganization of the data structure, leading to potential out-of-store activities.

The net outcome on the 205 is a significant increase in speed; a factor of 2.5 results from increased VL (from 20 to 50-70 for the $<DD/DD>$ case) permitted by memory reductions, and a further factor of 1.5 from reduced floating point operations. The new code outperforms the original scalar implementation by a factor of 15-20, although the complicated indexing degrades the scalar portion of the code by a factor of 2.

C. HF Evaluation of a non-correlated wavefunction

The most time consuming step in the first-order iteration scheme due to Roothaan[8] is the construction of the Fock matrix as a sum of Coulomb and Exchange operators. The construction of the symmetric exchange matrix, K, from a symmetric density matrix, P, may be represented through the matrix * column vector multiplication

$$K = Q\,P$$

where we define a column vector **K** comprising all distinct elements of K

$$Kij = \Sigma\; <ij/kl>\; Pkl$$

a column vector **P** related to all distinct elements of P by

$$Pkl = Pkl\,(1/2)^{\delta_{kl}}\;(k<l)$$

and a symmetric supermatrix Q, such that

$$Qij,kl = <ik/jl> + <il/jk>\quad(i<j\;,\;k<l\;)$$

The construction of K would thus appear to indicate use of the sparse MMO, where we may take advantage of the sparsity of **P**. Unfortunately **P** is not in general sparse, whereas **Q** may be very sparse, with perhaps less than 5% of its elements finite. To construct the K matrix in this circumstance, taking account of the sparsity of **Q**, will require the GATHER operation, and in some forms of implementation the inner product order and/or the SCATTER operation, the latter being required if one attempts to utilize the symmetry of **Q**. The net outcome is that only 10.3 Mflop is realized in the Cray-1S implementation. However, even at this speed an input/output (I/O) rate of 30 Mbytes/sec is required for reading the Q matrix from backing store, indicating a requirement for a large number of channels plus sophisticated I/O software. Due to its superior SCATTER/GATHER capability, the Cyber-205 produces a rate of 25 Mflops, with a requirement for an I/O rate of 80 Mbytes/sec.

Quantum chemistry is, of course, a subject with a notoriously high demand on I/O devices, both in terms of I/O rate and data capacity. The computational chemist will always be faced with the headache of trying to reduce I/O charges, and the optimum way of achieving this is often a function of both machine and installation. However, the design of any I/O system for quantum chemistry must always feature asynchronous activities with the maximum size of data transfer permitted, and by implication will require a large amount of available memory for coordinating this activity. On the Cyber-205 our present I/O system, utilising the maximum unit of data transfer (12K words using small page transfers), acts to improve the I/O efficiency of certain stages of computation (and subsequent cost) by a factor of ten or more. The development of an efficient I/O system for the FPS-164 has been outlined in reference 4.

D. Integral Transformation

This necessary precursor to the CI process has proved a considerable bottleneck in the past. Our vector version of this program is based on the Yoshimine algorithm,[9] which may be broken down into four steps; we describe here only the first two steps, the second two being closely similar.

Step 1. Sort the atomic integrals $<ij/kl>$ so that for a given ij all kl are available. A disk bin sort method is employed, in which only the SCATTER of integrals into their respective bins proves difficult to vectorize. Here we are only able to sort 1.3×10^5 integrals/sec or 3.0×10^5 integrals/sec on the Cray-1S or Cyber-205 respectively, with associated I/O rates of approximately 6 Mbytes/sec of 15 Mbytes/sec.

Step 2. The transformation of integrals to semi-transformed form consists of nothing more than a sequence of $N(N+1)/2$ matrix multiplications (N is the number of basis functions) of the form

$$Bij = Q^{\cdot}AijQ$$

where Q is the molecular orbital coefficient matrix and Aij is a symmetric matrix of untransformed integrals of common index pair, ij. The matrix Aij is likely to be sparse, given that the integrals are relevant to a large molecule, but the conditions are suitable for the application of our sparse MMO algorithm, with an expected performance of near 147 Mflop on the Cray-1S when basis sets of typical dimension (50 to 200) are employed. This stage of the process is not I/O limited on the Cray when the basis set is larger than 70, and when a single channel is driven at optimal efficiency.

The net outcome of all this is that the integral transformation process is very well adapted to vector processors, producing an overall rate of >100 Mflop on the Cray-1S for large basis sets without being severely I/O bound. A similar performance is observed on the Cyber-205.

E. Configuration Interaction

As the scope of quantum chemistry broadened from the consideration of stable molecules near equilibrium to encompass potential curves and surfaces, transition states, radical ions and excited states, the shortcomings of the HF approximation became increasingly evident. The energy error of the (restricted) HF wavefunction has been termed the correlation energy, and although usually a small fraction of the total energy of a molecular system (0.5% in water), it must be remembered that chemistry is primarily concerned with small energy differences, and these differences may be seriously affected by the correlation energy.

Many schemes have been devised and employed in order to overcome the shortcomings of the HF approach. Of these the configuration interaction (CI) method remains the most generally applicable and most straightforward, though historically the method has suffered from severe computational difficulties. We confine our attention to the construction of a CI wavefunction where the configurations can be generated by single and double excitations from an internal to an external set of molecular orbitals, and where the reference configurations are constructed entirely from the internal orbitals.[3,10]

Within this framework two distinct formulations of the CI problem can be recognised, conventional or 'configuration-driven' CI (involving the calculation of all interactions for each pair of configurations at a time) or 'integral-driven' direct-CI (in which the molecular integrals are examined to determine the matrix elements to which they contribute). The bottleneck of the conventional approach is evident in the number of matrix elements that must be stored, since this is proportional to the square of the number of configurations. With 10^5 configurations, a rather modest CI by present standards, and with only 10% of the matrix elements non-zero, the number of elements is of the order 10^9, an impossible storage and retrieval problem even on large computers. In the direct-CI approach one avoids explicit construction of the intermediate Hamiltonian matrix by constructing the wavefunction directly from the list of molecular integrals.

The main computational step in each iteration of the direct-CI problem is the construction of a vector Z as

$$\mathbf{Z} = \mathbf{H} \, \mathbf{C}$$

where **H** is the Hamiltonian matrix and **C** the wavefunction from the previous iteration. We adopt Siegbahn's classification of integrals and configurations,[10] and also adopt this author's strategy of factorizing the coupling coefficients of the CI method as a product of an internal part (which needs to be computed explicitly, but fortunately there are few in number) and an external part, which takes an extremely simple form. It is then possible to consider every contribution to the **Z** vector above in one of ten interaction types (see Table 2 of ref. 2b), and to formulate for each interaction type a scheme which requires the minimum number of floating point operations and which is implemented through one or more MMOs. The details of this analysis have been presented elsewhere.[3] It suffices here to give the overall conclusions.

(i) An implementation which does minimize the number of floating point operations and proceeds via MMOs is indeed possible, leading to the realization of computation rates of well over 100 Mflops on the Cray-1S. An idea of typical CPU requirements on the Cray-1S is given by timings from a calculation on the iron-formaldehyde complex $Fe(CO)_4H_2CO$. In a 142 basis function treatment, a single reference CISD calculation with 20 internal and 77 external orbitals (leading to 595,091 csfs in C_s symmetry) required 128 secs of CPU time per iteration.

(ii) The calculation can be carried out in a fashion which is not I/O bound, even given there is insufficient memory to hold the whole of the **C** and **Z** vectors in store.

(iii) The extreme importance of vector processing technology in the direct-CI method is evident from timings on the water dimer. A single and double excitation CI calculation, involving 56269 CSFs, revealed a cycle time for the diagonalizer on the Cray-1 approximately 300 times smaller than that of the corresponding calculation on an IBM 360/91,[13] indicating that configuration spaces of the order of 10^6 are now tractable.

The diagonalization phase of the Cyber-205 implementation is observed to run at best 75% Cray-1 speed. Considerable attention has been given to improving the pre-diagonalization activities in the 205 version, so that for 'small' cases at least the overall CPU requirements are roughly comparable to those on the Cray. Thus for a 58 basis function calculation on HCOOH, an 18 electron CISD treatment (40134 csfs) required a total CPU time of 60 seconds and 65 seconds respectively on the Cray and 2-pipe Cyber-205. Moving to a 1-pipe machine increased the CPU time for the corresponding calculation to 94 seconds. The most extensive CI treatment performed to date on the Manchester 205 involved a calculation on the Cr_2 dimer; with 91 basis functions, 12 internal orbitals (frozen 4s) and 61 external orbitals, a total of 2,295,104 csfs are generated from a 365 configuration reference set. The diagonalization cycle time for such a CI is 490 seconds (CPU) and 560 seconds (SBU).

F. Multiconfiguration SCF

The ability to perform accurate multireference CI calculations at rather modest expense, which typically require MCSF wavefunctions as an initial approximation, highlights the need to obtain such wavefunctions reliably and efficiently. The development of the CASSCF (Complete Active Space MCSCF) method[14] proved a significant landmark in the quest for accurate zero-order wavefunctions, particularly in implementations where the need to store the CI coupling coefficients is removed.[15] Furthermore the development of 'direct' second order MCSCF procedures,[16] in which explicit construction of the Hamiltonian and Hessian matrices is avoided, makes possible the full optimization of quite long CI expansions.

A new second order MCSCF method recently described by Werner and Knowles[11] exhibits impressive performance on vector machines, with a formalism based mainly on the MMO. Full details of this method have been presented in reference 11, and it suffices here to give the overall conclusions:

(i) The improved convergence when compared with all previous methods is achieved by minimizing a second order energy approximation which takes into account the orthonormality of the orbitals exactly and is therefore periodic in the orbital rotations. Rotations between internal orbitals and the changes of the CI coefficients are treated effectively to highest possible order without performing a more expensive integral transformation than in any other second-order MCSCF method. The additional computational effort per iteration for the more accurate optimization of the internal orbital rotations is relatively small, and in most cases by far outweighed by the reduction in the number of macroiterations needed.

(ii) The method can be used to minimize an energy average of several states.

(iii) Very large basis sets and configuration expansions can be handled since the Hessian and Hamiltonian matrices are never explicitly calculated and stored. CASSCF calculations with up to 10^5 configurations have been reported using the new method of reevaluating the coupling coefficients very efficiently each time they are needed (see reference 11b, where details of a 178,910 csf CASSCF calculation on the $^5\Delta$ state of FeO are presented).

(iv) In all the calculations reported to date, convergence has been achieved in at most three iterations. The energy has been observed to converge better than quadratically from the first iteration even when the initial Hessian matrix has many negative eigenvalues.

G. Supercomputers and Superprograms

In the above sections we have considered the vector implementation of five of the steps in a typical Quantum Chemistry calculation, leading to the efficient computation of the total energy of a given molecular species at a *fixed* nuclear geometry. Chemistry, however, is not concerned merely with the properties of a molecule at a single point, but with the more general characteristics of multi-dimensional potential energy surfaces, with a quantitative account of the making and breaking of chemical bonds crucial in the study of reaction mechanisms. Ideally we wish to move automatically, and systematically, on a surface from one stable molecular geometry, through one or more transition states, onto a product equilibrium geometry. Such a 'walking process' became viable with the development of efficient methods for calculating gradients of the molecular energy,[17] together with the evolution of robust and efficient algorithms for locating minima and transition states based on first-, and more recently, second-derivative information.

The complexity and sheer size of the programs required in such studies presents a formidable task for the computational chemist. Such a code must include, in addition to the optimized steps above, routines for the evaluation of the energy derivative for a broad class of wavefunctions of increasing complexity, involving computation of the derivatives of the one- and two-electron integrals. All such steps reside under control of optimization routines designed to locate and characterise the stationary points on the potential surface in the minimum number of energy and energy-gradient evaluations.

These programs are potentially vast consumers of both machine cycles and the more general resources of memory, disk space, etc. Much of the 300 hours of Cray-1 time allocated by SERC to users in the QC community in the period 1983-4 was consumed through use of these codes. It is estimated that the equivalent computations on the AS-7000 at Daresbury would have required at least 5000 hours, effectively the entire machine. Some 20-30% of the current VP usage at ULCC and UMRCC is taken up by Quantum Chemistry calculations.

Work in this area at Daresbury has concentrated on the GAMESS program (General Atomic and Molecular Electronic Structure System), a 120,000 line general purpose *ab initio* molecular electronic structure program for performing SCF- and MCSCF-gradient calculations.[18] The program utilizes the Rys Polynomial or Rotation techniques to evaluate repulsion integrals over s,p and d type Cartesian Gaussian orbitals. Open- and closed-shell SCF treatments are available within both the RHF and UHF framework, with convergence controls provided through a hybrid scheme of level shifters and the DIIS method.[19] In addition generalized valence bond (GVB), CASSCF and more general MCSCF[11] calculations may be performed.

The analytic energy gradient is available for each class of wavefunction above. Gradients for s and p Gaussians are evaluated using the algorithm due to Schlegel,[20] while gradients involving d Gaussians utilize the Rys Polynomial Method. Geometry optimization is performed using a quasi-Newton rank-2 update method, while transition state location is available through either a synchronous transit[21] or trust region method.[22] Force constants may be evaluated by numerical differentiation. Large scale multi-reference CI calculations may be performed using the Direct-CI formalism.

A variety of wavefunction analysis methods are available, including population analysis, localized orbitals, graphical analysis and calculation of 1-electron properties.

H. Implementation Strategy on Vector and Attached Processors An Example

Our experience in the development of optimized QC programs on vector processors typified by the Cray-1 and Cyber-205 suggests that optimum performance can only be achieved by resorting to assemble language constructs for many of the vector kernels involved eg MMO, SCATTER, transposition etc. This partly reflects a lack of 'richness' in typical QC software, where for each fetch and store very little floating point arithmetic occurs. This use of assembly code permits, for example, account to be taken of the segmented nature of the scalar and vector functional units, and enhances the MMO on the Cray-1S from some 37 Mflop to an asymptotic performance of 147 Mflop.

The Cray-1S is perhaps not the ideal machine for comparing FORTRAN and assembler performance. Store access conflict problems together with chain-slot loss left much to be desired for the FORTRAN user. Although these effects have been largely remedied on the XMP-1, reliance on FORTRAN still leaves the user unable to take advantage of hiding scalar control activities under vector operations. The situation is less clear on the multiprocessor XMP-n, where store access conflict problems in the case of a general code implemented at the large granularity level would be exacerbated by the generation of redundant store/fetch operations.

While the 205 is potentially an ideal FORTRAN machine, similar problems remain because of deficiencies in the current FORTRAN compiler. Many of the typical loops in QC software involve items from argument lists, and as such inhibit automatic vectorization using the FORTRAN 200 Compiler, given the requirements for a vector length 'known' to be less than 65536. Since use of the 'UNSAFE vectorization' option commonly led to miscompiled code, the strategy of building a META library of FORTRAN callable routines was adopted. Such a philosophy is in line with using the mathematical subroutine library of the FPS-164, which includes FORTRAN callable subroutines written in optimized assembly language to perform, amongst others, vector and matrix operations. The typical improvement figures arising from use of this library have been documented by Dunning and co-workers.[4]

The above discussion raises the obvious question as to what order of operation should this reliance on factored routines be instigated. It would seem clear that factoring of code should occur at the N^2/N^3 level eg MMO, matrix square, diagonalization, etc ie where there is a clear potential for taking advantage of the interplay between loop lengths. Indeed the prospects for large scale granularity may well be at this level. This factoring at order N, however, is more debatable. In cases where one cannot guarantee the magnitude of N (ie where the compiler will use a given algorithm regardless), or where one is dealing with non-vectorized code, then again factoring may appear beneficial, to accommodate alternative optimization techniques eg loop folding. Factoring at this level is not, however, in general recommended, except perhaps where all of the loop may be represented by a single CALL to the library utility. Obviously the larger the library, the more likely this condition would be satisfied. An attempt to widen the range of the BLAS might encourage such factorization.

Implementation of GAMESS on an FPS-164

As an example of the typical problems encountered in migrating code from processor to processor, we consider our implementation of the GAMESS package on the FPS-164. Implementation of a VAX-version of the package had commenced prior to the acquisition of the machine at Daresbury, with the help of staff at FPS, Bracknell. On the arrival of the machine, the author undertook the task of implementing and converting the Cray version of the code. Some of the problems arising in this conversion are outlined below (see also 4):

(i) A potential problem is the use of non-standard data types - INTEGER *2, LOGICAL *1 - in common blocks and equivalence usage. Due account of these effects had been taken during Cray implementation of the code.

(ii) Use of extended DO-loops

(iii) Use of Hollerith data types instead of character type data. Most QC programs are written in FORTRAN-4, but nevertheless compile successfully with FORTRAN-77 compilers, with the aid of various language flag options etc. Yet it was felt timely, given the general requirements of APFTN64, to undertake the task of converting the entire code to a FORTRAN-77 standard, at least as far as character type data was concerned. This conversion took approximately 2 weeks to carry out, involving, for example, major changes to the free-format data input routines.

(iv) Use of dummy arrays that are not initialized on the most recent entry into the subroutine.

(v) A known bug in release E of the APFTN64 compiler meant that many of the 'long' routines with 'complicated' loop structure would not compile correctly at higher optimization levels. One of the most important routines in the 'rotated-axes' 2-electron integral code will not even compile at optimization level 1! It is claimed that this fault will be corrected for in release F of the compiler.

(vi) The most serious problem encountered, and one that took several months to resolve, involved implementation of an efficient direct access asynchronous I/O system. Both the Cray and Cyber versions of the code rely on a multi-buffered I/O system based on the fundamental building block of 512 words (the Cray block). Typically multiple blocks are written under control of a single output statement (using, for example, the Q5 routines on the 205), but may be subsequently processed through multiple read commands. Attempts to conform to this structure using the asynchronous I/O facilities within APFTN64 revealed intolerable elapsed/wait times. The initial solution to this problem involved basing the I/O system on the FILE$ routines (vol. 3 of the Operating System Manual Set), a collection of FORTRAN callable routines providing far greater flexibility than their FORTRAN77 counterparts.

(vii) The 1/2 Mword memory on the Daresbury machine provides a potential constraint on the systems amenable to study. The amount of available memory has been optimized in two ways:

a) In common with most QC codes, GAMESS features a large array which is partitioned and passed to subroutines in segments, the space requirements for each segment depending on the chemical system under investigation. Access to such an array on the FPS is achieved through use of the /SYS$MD/ common block and SYS$ADDMEM routine to define the first usable location in the program workspace.

b) The space requirements of the code itself have been minimised by extensive use of the flexible OVERLAY features of APLINK64.

Cost Effectiveness of the FPS-164

We consider the performance of a Gaussian integrals code implemented on the FPS-164. This code is in standard FORTRAN, and has been designed to vectorize on the Cray-1, where it runs at approximately 30 Mflop. After almost zero effort, and using the APFTN64 (OPT=3) compiler the FPS-164 produced 3 Mflops, whilst the AS7000 yields 1.6 Mflops. It is estimated that after some further polishing of the code using the scientific library, 4.5 Mflops would be realized on the FPS-164.

Timings reported below are for two distinct cases. The 'Vector' case is specifically designed to reflect the vector processing capabilities of the host machine. The 'Scalar' case represents the execution of non-vectorized code.

Performance of a Gaussian Two-Electron Integral Code

| | CPU (times (seconds) | |
Machine	'Vector'	'Scalar'
Cray-1	34.5	15.5
Cyber-205 (scalar)	227.3	16.6
FPS-164	365.0	61.5
AS-7000	616.6	57.3
VAX 11/780	6933.	
Gould 32/9705	1800.	62.0

The timings clearly demonstrate, just as with the Cray-1, that the potential of the machine is only realized when handling 'vectorized code.' Thus in the 'Vector' case, the FPS-164 achieves a performance 1.7 times the speed of the AS-7000 (ie approaching that of an IBM 370/195).

The comparison with the Gould 32/9705 (which is the one of the most cost effective scalar processor known to us) is particularly instructive, where the FPS-164 is superior by a factor of 5 in the 'Vector' case, even though the price of the Gould approaches that of the FPS. In the 'Scalar' case the performance of these two machines is almost identical, again underlining the requirement for vector codes if the FPS-164 is to be used effectively.

Note that in the comparison with the scalar performance of the Cyber-205, the FPS achieves a speed of 0.6 times the Cyber in the 'Vector' case, to be compared with the ratio of 0.27 found in the 'Scalar' benchmark.

These timings are broadly in line with those reported by the Theoretical Chemistry Group at the Argonne National Laboratory, who have implemented a complete set of electronic structure codes on an FPS-164.[4] The generalized valence bond and Hartree Fock programs developed at ANL currently run 28 times faster on the FPS than on the VAX 11-780 host, a figure to be compared with the ratio of 19.0 found above.

Performance of the FPS-264

To provide some idea of the improvement to be expected on the newly-released FPS-264, we include below overall timings recently obtained in optimising the geometrical structure of chromium tetranitrosyl, $Cr(NO)_4$, using a double zeta basis of 110 functions (see section 5). The table below shows the breakdown of this gradient optimization into component parts, and contrasts performance on FPS-164 (release FO.O) and FPS-264.

| | cpu times (seconds) | |
	FPS-164	FPS-264
Input phase	3	1
Vector generation	11	3
1-electron integrals	104	30
2-electron integrals	4221	1192
SCF	2153	622
1-electron gradient integrals	559	167
2-electron gradient integrals	8845	2582
Wavefunction analysis	44	12
Other	12	3
Total cpu seconds	15952	4612

The increased performance of the 264, by a factor of 3.5, suggests that the impact of this machine from Floating Point Systems on computational chemistry will be just as marked as its predecessor. Note that this benchmark involved migrating the executable load module directly from 164 to 264, and does not reflect the possible improvement to be obtained from 264 specific software.

4. CORRELATION EFFECTS IN THE GROUND AND IONIC STATES OF TRANSITION METAL COMPLEXES

There are two, well-proven, approaches to the accurate calculation of the ionization energies of molecules lacking a transition metal atom. In the first, SCF calculations are carried out on both the ground and various ionic states, correlation effects being included by carrying out CI calculations on each state. Ionization energies are then obtained as the energy difference between the ground and ionic states (ΔSCF-CI method).[23] To avoid carrying out individual calculations on both the unionized, and various ionic states, several authors have used the single-particle propagator or Green's function techniques to calculate ionization energies directly taking account of both relaxation and correlation effects. Both approaches have been successful in yielding ionization energies accurate to 0.5 eV for a range of organic molecules,[24,25] and accurate to 0.2 eV for smaller molecules for which extended basis sets can be used.[26]

However, the extension of these methods to realistic transition metal complexes which may be studied experimentally, can present formidable technical difficulties. Thus, a double zeta representation of the valence orbitals, which is generally deemed the minimum size needed for a meaningful discussion of correlation effects, may generate well in excess of 100 basis functions. The 4-index transformation for such a basis size has represented a non-trivial problem, while the size of the CI expansion generated from the large number of valence electrons in the complex, and from a basis of this size, may be upwards of 5×10^5, again representing computational problems. The advent of quantum chemistry codes on supercomputers, combined with the development of direct-CI methods, has significantly increased the ease of conducting calculations on transition metal complexes which include correlation energy. In the present section we describe such calculations for bis(π-allyl)nickel, $Ni(CN)_4^{2-}$, $Co(CN)_6^{3-}$ and $Fe(CN)_6^{4-}$.

Two of the first transition metal complexes studied by *ab initio* methods were bis(π-allyl)nickel, $Ni(C_3H_5.)_2$, and the tetracyanonickelate anion, $Ni(CN)_4^{2-}$.[27,28] Veillard and co-workers recognised the importance of including orbital relaxation effects in the calculations of the valence IEs of these molecules. Since these original calculations, additional experimental studies on the PE spectrum of $Ni(C_3H_5)$ have suggested that ΔSCF calculations lead to an incorrect ordering of the ionic states.

The use of X-ray emission (XRE) spectroscopy, in which the emission spectrum accompanying valence-core transitions is observed, provides valuable data on not only ionization energies, but also information on the nature of the ionized states. In the field of transition metal complexes XRE provides relative energies of those ionized states arising from metal and ligand ionization, information not so readily available from PE spectroscopy. We have recently obtained XRE spectra from the storage ring at the Daresbury Laboratory on the transition metal cyanides, which lead to the relative energies of metal and ligand ionizations (Table 1).

The measured IEs corresponding to the N Kα, C Kα and M L$\alpha_{1,2}$ peaks given in Table 1 suggest that for all three anions, the lowest IEs correspond to metal 3d ionizations, closely followed by those ionic states arising from ligand ionizations.

We now describe the results of calculations of the bonding and IEs of bis(π-allyl)nickel and the three cyanide complexes.

Table 1: Measured Valence IEs (eV) of $Ni(CN)_4^{2-}$, $Co(CN)_6^{3-}$ and $Fe(CN)_6^{4-}$

	$K_2Ni(CN)_4$		$K_3Co(CN)6$	$K_4Fe(CN)_6$
C1s IE	283.7		283.3	284.8
C Kα XRE peak	278.5		278.3	278.6
Valence IE	5.2		5.0	6.2
N1s IE	396.9		396.6	398.1
N Kα XRE peak	392.1		392.1	391.8
Valence IE	4.8		4.5	6.3
$M2p_{3/2}$ IE	855.6		781.3	708.6
M $L\alpha_{1,2}$ XRE peak			851.4	778.2 706.8
Valence IE	4.2		3.1	1.8

Computational Details

The calculations were carried out using the (9s5p) basis of Huzinaga[29] contracted to $(3s2p)$[30] for carbon and nitrogen. The metal bases were constructed from the (12s6p4d) functions of Roos et al,[31] with the two most diffuse s functions replaced by those having exponents of 0.32 and 0.08. An additional p function having an exponent of 0.32, and a diffuse d function were added.[32] These functions were contracted (6s3p2d), giving a close to double zeta representation of the valence region. Calculations were performed at the experimental geometries.

For $Ni(C_3H_5)_2$, RHF calculations were carried out on the ^1A_g ground state of the neutral molecule, and lowest ^2A_g, ^2B_g, ^2A_u and ^2B_u states of the ion. SDCI and MRDCI calculations were performed on all five states with all valence electrons, except the carbon 2s, correlated using the entire virtual manifold. The SDCI computations on the X^1A_g and, for example, the ^2A_g state, involved a total of 206,102 and 472,203 csfs respectively. Subsequent 2-reference CI treatments led to increased configuration spaces of 410,594 and 941,015 csfs respectively.

The second method used to calculate the valence IEs of each species is a Green's Function method. The IEs and their relative intensities (pole strengths) appear as the poles and residues of the one-particle Green's function and can thus be calculated directly, instead of as the difference of the energies resulting from two independent calculations. Two distinct approximations to one-particle Green's function are generally employed (see references 33-36). In the outer valence region where a one-particle picture holds for the physical description of the ionization process (i.e. where ionization can be reasonable well described by the ejection of an electron from a ground state molecular orbital) the outer-valence Green's function (OVGF) may be used.[33-34] If there is a strong relaxation in the sense that the SCF orbitals for the ionized system are a strong mixture of the ground state SCF orbitals, plus the virtual orbitals (as is the case for the complexes considered here), or if in a configurational expansion of the ionic wavefunction the single hole configurations mix strongly with configurations involving ionization plus excitation, as is the case in the inner valence region, then the extended two-particle-hole Tamm Dancoff (2ph-TDA) should be used.[35,36] This latter method is applicable in the entire valence region. Typical for the inner valence region is the appearance of satellite lines which borrow their intensity from the ionization processes which lead to simple hole states, and the possible disappearance of main lines with the intensity being distributed over many lines. This effect has been termed the breakdown of the molecular orbital model of ionization.[37] Both the OVGF and extended 2ph-TDA methods are accurate to third order in the electron interaction in the outer valence region. However, in the inner valence region, when the breakdown of the orbital model of ionization occurs, the structure of the ionization spectrum is only described qualitatively by the extended 2ph-TDA calculations. This is due both to the complicated configurational structure of the states and to the missing higher excitations, and to basis set inadequacies in the neighborhood of a continuum of states.

For $Ni(C_3H_5)_2$, the extended 2ph-TDA calculations included all the filled valence MOs and 20 virtual orbitals. For the cyanide complexes all valence MOs correlating with the 1π, 4σ and 5σ MOs of CN and with the metal d orbitals, together with the lowest 22 virtuals, were included.

Computational Results The Ground States of $Ni(C_3H_5)_2$, $Ni(CN)_4^{2-}$, $Co(CN)_6^{3-}$ and $Fe(CN)_6^{4-}$

RHF calculations of the singlet states of $Ni(C_3H_5)_2$ and $Ni(CN)_4^{2-}$ yielded the same description of the bonding as that obtained previously by Veillard and co-workers,[28,38] although the exact ordering of the closely-spaced MOs is somewhat basis set dependent. For bis(π-allyl)nickel, the highest filled MO, $7a_u$ involves no metal character, and derives from the allyl a_2 π-MOs. The other MO involving these ligand orbitals, $6b_g$, is the second highest filled orbital, and has the greatest contribution to the metal-ligand bonding, via interaction with the nickel $3d_{xz}$ orbital (adopting the conventional axis labelling in C_{2h}. The antisymmetric combination of the ligand b_1 π-MOs, $11b_u$, is non-bonding, whilst the symmetric combination interacts with the metal $3d_{xy}$ orbital, leading to the $9a_g$ and $13a_g$ MOs. The $10a_g$, $5b_g$ and $11a_g$ MOs are mainly non-bonding, consisting of the metal $3d_z2$, $3d_{xy}$ and $3d_{x^2-y^2}$ respectively. The remaining valence orbitals involve mainly the σ-framework of the ligands. The energies of these valence orbitals are used in Table 2, where it can be seen that Koopmans' theorem predicts the mainly metal ionizations to occur at considerably higher energy than the ligand π ionizations.

CI calculations on the ground state of bis(π-allyl)nickel revealed that the HF configuration contributed 83% to the wavefunction, a value considerably less than that usually found for molecules lacking a transition metal atom ($> 95\%$). The major correlating configurations were found to be the single and double excitations $6b_g \rightarrow 7b_g$ and $6b_g^2 \rightarrow 7b_g^2$. The $7b_g$ MO is the metal-ligand anti-bonding counterpart of the $6b_g$ bonding MO. Thus, the important $6b_g^2 \rightarrow 7b_g^2$ configuration yields the left-right correlation of the metal-ligand bonding a correlation effect absent in the isolated metal atom and allyl ligands. Both the $6b_g$ and $7b_g$ MOs are composed of Ni 3d and allyl π-orbitals. However the Ni3d character is greater in the $7b_g$ (43%) than in the $6b_g$ (38%) so that the excitation corresponds to an increase in the population of the $3d(b_g)$ atomic orbitals. The effect of all the correlating configurations additional to the RHF configuration, which make up 18% of the total wavefunction, can be seen from a population analysis of the natural orbitals of the CI wavefunction. The major effects of correlation on the molecular charge distribution are:

Table 2: Ionization energies (eV) of the valence orbitals of bis(π-allyl)nickel.

Orbital	Character	Koopmans' Theorem	ΔSCF	ΔSCF-CI	extended 2ph-TDA	Expt[a]
$7a_u$	$\pi(L)$	7.5	6.7	6.7	6.4	7.7(1)
$6b_g$	$3dxz;\pi(L)$	9.0	5.6	6.6	7.7	8.1(2)
$11b_u$	$\pi(L)$	11.8	11.0	10.8	10.3	10.3(5)
$13a_g$	$3dxy;\pi(L)$	11.7	5.5	6.4	7.6	8.1(2)
$12a_g$	$\sigma(L)$	14.0			13.5	12.7(7)
$5b_g$	$3dyz$	14.0			8.5	
$11a_g$	$3dx^2\text{-}y^2$	14.2			8.2	8.5(3)
$6a_u$	$\sigma(L)$	14.6			13.3	
$10b_u$	$\sigma(L)$	14.6			13.4	12.7(7)
$4b_g$	$\sigma(L)$	15.0			13.7	
$10a_g$	$3dz^2$	15.3			8.8	9.4(4)
$9a_g$	$3dxy;\pi(L)$	16.4			11.5	11.5(6)
$5a_u$	$\sigma(L)$	16.5			14.9	
$3b_g$	$\sigma(L)$	17.3			15.1	14.2(8)
$9b_u$	$\sigma(L)$	18.0			16.2	
$8a_g$	$\sigma(L)$	19.0			16.5	15.6(9).

[a] The band number is given in parenthesis.

(i) an increase in the $3d(b_g)$ populations and an associated decrease in the π-populations of the terminal carbon atoms of the allyl ligands, corresponding to correlation of the metal-ligand bonding electrons.

(ii) a smaller decrease in the population of the $3d(a_g)$ non-bonding electrons and an increase in the π-population of the central carbon atoms of the allyl ligands. This may be associated with the in-out correlation of the non-bonding electrons.

The net result of these effects is an increase in the metal 3d electron density and a decrease in the ligand π density.

The results of the RHF calculations on $Ni(CN)_4^{2-}$, $Co(CN)_6^{3-}$ and $Fe(CN)_6^{4-}$ are summarised below in Tables 3-5.

In the nickel and cobalt species, the highest filled MOs correlate with the 1π and 5σ orbitals of CN^-, and the more strongly bound orbitals with the 4σ orbitals of CN and with the metal d orbitals. By comparison with our reported XRE spectra of these molecules, we see immediate conformation of the well-known inadequacies of Koopmans' theorem for these molecules. In the case of $Fe(CN)_6^{4-}$, the highest filled MO, $2t_{2g}$ is predominantly 3d in character, although the metal-ligand separation here predicted by Koopmans' is much smaller than the experimental value.

Table 3. Calculated Valence Ionization energies (eV) of $Ni(CN)_4^{2-}$.

Orbital	Character	Ionization Energy	
		Koopmans' Theorem	extended 2ph-TDA
$1a_{2g}$	π	3.9	3.1
$8e_u$	π	3.9	2.5
$2e_g$	π	4.2	4.4
$1b_{2u}$	π	4.3	3.6
$2b_{2g}$	π	4.8	2.9
$3a_{2u}$	π	4.9	4.3
$5b_{1g}$	5σ	5.7	4.0
$7e_u$	5σ	5.8	4.2
$9a_{1g}$	5σ	5.9	2.7
$8a_{1g}$	d_z2	7.3	3.4
$1e_g$	d_{xz},d_{yz}	8.4	2.2
$4b_{1g}$	4σ	8.5	6.5
$6e_u$	4σ	8.9	6.7
$1b_{2g}$	d_{xy}	10.2	5.4
$7a_{1g}$	4σ	11.7	8.1

Table 4. Calculated Valence Ionization energies (eV) of $Co(CN)_6^{3-}$.

Orbital	Character	Ionization Energy	
		Koopmans' Theorem	extended 2ph-TDA
$8t_{1u}$	π	0.7	-0.8
$1t_{1g}$	π	1.1	0.4
$1t_{2u}$	π	1.5	0.8
$2t_{2g}$	π	2.0	2.1
$5e_g$	5σ	3.0	0.8
$7t_{1u}$	5σ	3.1	1.5
$8a_{1g}$	5σ	3.6	1.5
$4e_g$	4σ	5.5	3.1
$6t_{1u}$	4σ	5.8	3.5
$1t_{2g}$	$3d$	7.2	-2.1
$7a_{1g}$	4σ	9.1	6.0

Table 5. Calculated Valence Ionization Energies (eV) of $Fe(CN)_6^{4-}$.

Orbital	Character	Ionization Energy	
		Koopmans' Theorem	extended 2ph-TDA
$2t_{2g}$	3d	-4.2	-9.2
$8t_{1u}$	π	-3.8	-5.4
$1t_{1g}$	π	-3.4	-4.1
$1t_{2u}$	π	-3.0	-3.8
$5e_g$	5σ	-2.5	-4.7
$7t_{1u}$	5σ	-1.6	-3.0
$1t_{2g}$	π	-1.1	-2.7
$8a_{1g}$	5σ	-0.8	-2.8
$4e_g$	4σ	0.7	-1.3
$6t_{1u}$	4σ	1.1	-0.9
$7a_{1g}$	4σ	3.7	0.8

The calculated Ionization Energies of $Ni(C_3H_5)_2$

The valence ionization energies calculated by the various methods are summarised in Table 2. As in previous studies,[27,28] Koopmans' theorem predicts the mainly metal 3d MOs ($10a_g$, $11a_g$, $5b_g$) to be considerably more tightly bound than the ligand π MOs ($7a_u$, $11b_u$). However, when ΔSCF calculations are carried out on the ion states, the considerably greater relaxation energy associated with the metal, compared to ligand, ionizations, results in the ground state of the ion being predicted to be ^2A_g, arising from metal 3d ionization, in agreement with the calculations of Veillard and co-workers. The results of our CI calculations on the lowest four ion states of different symmetry show that considerably greater correlation energy is associated with the ^2A_u and ^2B_u arising from ligand ionization than from the ^2A_g and ^2B_g states which correspond to metal ionization. For the ^2A_u and ^2B_u states, the most important correlating configurations are similar to those in the ^1A_g ground state, and involve the $7b_g$ MO, whose characteristic is similar in these ion states and in the ^1A_g ground state. Thus, the important correlation effects which we have identified in the ground state, are also present in the ^2A_u and ^2B_u states, and indeed appear to be slightly more important in the latter two.

Examination of both the metal orbital populations, and the metal ligand overlap populations in the RHF ^2A_u and ^2B_u states reveals similar values to those found in the ^1A_g ground state. Thus, similar correlation energies associated with the metal and metal-ligand bonding electrons in these three states are to be expected, as found by our CI calculations.

However, in the RHF ^2A_g and ^2B_g ion states, there is substantial electron reorganization accompanying metal electron ionization, leading to ligand → metal electron migration. As a result, the total metal 3d population in the ^2A_g and ^2B_g ions are close to 9 electrons. The $3d_{xz}$ (b_g) populations is 1.8 in the ^2A_g state and 1.4 in the ^2B_g state, so that there is no strongly metal-ligand bonding electron pair in either state, as witnessed by the decrease in the bond overlap population in these two states, compared with that in the ^1A_g ground state. There is thus a reduction of correlation energy in the ^2A_g and ^2B_g ion states compared to that found in the ^2B_u and ^2A_u ion states and in the ^1A_g ground state associated mainly in the loss of the correlation energy associated with the metal-ligand bonding electrons in the $6b_g$ MO. The differential correlation energy associated with the (^2A_u, ^2B_u) and (^2A_g, ^2B_g) ion states is reflected in the ΔSCF-CI energies shown in Table 2. Thus, whereas the ^2A_g - ^2A_u separation is 1.3 eV at the ΔSCF level, the introduction of electron correlation reduces this difference to 0.3 eV. However, at both the ΔSCF and ΔSCF-CI level, the ground state is predicted to be ^2A_g.

The ionization energies calculated by the extended 2ph-TDA approximation (Table 2), are, as expected, closed to the ΔSCF-CI than to the ΔSCF results. However, the electron reorganization and correlation given by this method now places the ^2A_u as the ground ionic state, separated by 1.2 eV from the ^2A_g state. This method also allows all the valence ionization energies to be calculated, rather than just the first of each symmetry given by our ΔSCF CI calculations.

We now discuss the valence PE spectrum of bis(π-allyl)nickel in the light of these calculations.

The Photoelectron Spectrum of $Ni(C_3H_5)_2$

The PE spectrum shows nine bands below 17 eV, and was originally assigned by Batich[39] on the basis of methyl substituent results and using both He(I) and He(II) radiation as follows: The first, third and fourth bands are due to nickel 3d ionizations. The second is due to 3d plus a ligand π-orbital, and the fifth and sixth are due to ligand π-orbitals. On the basis of a comparison of the PE spectra of a series of nickel, palladium and platinum bis(π-allyl) derivatives an alternative assignment has been presented.[40] Here the first and fifth bands arise from the $7a_u$ and $11b_u$ ligand MOs respectively, bands two and three from the metal $9a_g$, $10a_g$, $11a_g$ and $5b_g$ MOs, and bands four and six from the $6b_g$ and $13a_g$ orbitals respectively having significant ligand π-character. Bands seven to nine are assigned to orbitals of allyl σ-character. A Green's function approach based upon the semi-empirical INDO method yields[41] an assignment in complete agreement with the above, although it is of interest to note that the INDO method predicts considerably smaller deviations from Koopmans' theorem than are given by *ab initio* methods. Our *ab initio* Green's function calculation yields an assignment of the PE spectrum close to that deduced by Bohm et al.[40] In particular the ground ionic state is predicted to be ^2A_u. The major difference is the interchange of the $9a_g$ and $13a_g$ MOs. However, both MOs have similar atomic orbital characters so that any distinction between them must be somewhat arbitrary. In addition the ordering of the closely spaced states arising from the $6b_g$ and $10a_g$ MOs is inverted in our calculation.

The valence ionization spectrum has been calculated by the extended 2ph-TDA method in the energy range up to about 30 eV. In the energy range up to 11 eV, the lines have a large relative intensity, but frequently the intensity is borrowed from many ground state orbitals, particularly in the case of the a_g and b_g symmetries. This is a reflection of the strong electron reorganization in the electron ejection from the orbitals with a strong nickel d character. Already at about 11 eV satellite lines start to appear and they soon become dense in energy. It is thus clear that the PE spectrum above 10 eV cannot be explained without the consideration of satellite lines. A detailed discussion is however not meaningful as only the qualitative aspects are amenable to calculation for this molecule and because the calculation is only strictly applicable in the limit of high energy exciting radiation. Above about 17 eV the density of lines becomes very large and we observe the effect of breakdown of the molecular orbital model of ionization. There are thus no more simple hole states.

The Valence Ionized States of $Ni(CN)_4^{2-}$, $Co(CN)_6^{3-}$, and $Fe(CN)_6^{4-}$

The valence IEs of these three species, calculated by the extended 2ph-TDA method are given in Table 3 to 5. For $Ni(CN)_4^{2-}$ the first IE at 2.2 eV, and the one at 4.4 eV involve strong mixtures of both Koopmans' theorem configurations arising from the metal $1e_g$ MO and the ligand $2e_g$ MO, reflecting the electron reorganization occurring upon ionization. A similar situation arised for both the b_{2g} and a_{1g} ionizations, where significant mixing between the various Koopmans' theorem configurations occurs. In contrast, the ionic states of other symmetries have a single dominant Koopmans' theorem configuration arising from ligand ionization. When comparing our calculated IEs with the values obtained from the experimental XRE spectrum (Table 1), it should be noted that due to the neglect of the counter-ion, our calculated IEs will be smaller than the experimental values. In our comparison between theory and experiment, we are thus concerned with relative IEs. The single peak in the Ni $L\alpha_{1,2}$ spectrum clearly arises from those ionic states which correlate with $1b_{2g}$, $1e_g$, $8a_{1g}$, $9a_{1g}$ and $2e_g$ ionizations spanning a calculated energy range of 3.2 eV (Table 3). All of these states have significant contributions from configurations arising from ionization from orbitals having substantial Ni3d character. The maximum in the Ni $L\alpha_{1,2}$ emission may be estimated to correspond to an ionization energy of -3.3 eV, since more intensity is likely to be associated with the degenerate ^2E_g states correlating with the $1e_g$ and $2e_g$ ionization. The major peak in both the C and N $K\alpha$ spectra clearly arises from states associated with ligand 1π and 5σ MOs, orbitals having significant 2p character. The corresponding calculated IEs span an energy range 2.5-4.4 eV, placing these peaks very close to the estimated Ni $L\alpha_{1,2}$ maximum at 3.3 eV. Indeed the spectra show substantial overlap of the three major emission bands, in agreement with the results of our calculation. The pronounced shoulder observed in the C $K\alpha$ emission to -3 eV higher IE of the main peak is assigned to transitions from those MOs which correlate with the 4σ orbitals of CN. This observed energy separation is in good agreement with our calculation (Table 3).

The IEs computed for $Co(CN)_6^{3-}$ (Table 4) show a similar trend to those for $Ni(CN)_4^{2-}$. Upon the inclusion of relaxation and correlation effects, the IEs of those MOs of mainly ligand 4σ character decrease by 2-3 eV, whilst there is a much larger decrease in the metal localized IE, here about 9 eV. The 2ph-TDA results predict the first IE of $Co(CN)_6^{3-}$ to arise from the metal localized $1t_{2g}$ MO, with a pronounced gap of greater than 1 eV between it, and the first ligand ionization. In the case of $Fe(CN)_6^{4-}$, the 2ph-TDA calculations yield a splitting of the metal and ligand IEs of more than 3 eV, compared to the value of less than 1 eV at the Koopmans' theorem level of approximation. Thus, along the series

$Ni(CN)_4^{2-}$, $Co(CN)_6^{3-}$, $Fe(CN)_6^{4-}$, the calculated separation of the metal and ligand IEs is <1 eV, 1-2 eV and more than 3 eV. These values are to be compared with our experimental estimates of 0.5 eV, 1.5 eV and 4.5 eV (Table 1).

Summary

The CI calculations for bis(π-allyl)nickel have shown that the important correlation effects, absent in the isolated metal and ligand entities lead to an increase in metal electron density. We have found a similar result for $Ni(CN)_4^{2-}$ where the population of the metal $d_x2_{-y}2$ (b_{1g}) is increased (by 0.06e) upon the inclusion of correlation effects. Thus, in both molecules, electron correlation leads to a less polar environment. A similar decrease in electron density has been shown to occur in ferrocene[42] and leads to a decrease in the metal-ligand bond length. The present example provides further evidence that calculations beyond the Hartree-Fock description are particularly important, even for a semi-quantitative description of the electronic structure of transition metal complexes.

The 2ph-TDA method is particularly attractive method for computing the whole manifold of valence IEs of transition metal complexes. In addition to avoiding the repeated orbital transformations required in ΔSCF-CI methods, the necessity of obtaining a number of roots of very large matrices is also removed. The calculations of IEs of bis(π-allyl)nickel, $Ni(CN)_4^{2-}$, $Co(CN)_6^{3-}$ and $Fe(CN)_6^{4-}$ reported here are the first to employ an ab initio Green's function technique on transition metal complexes. The agreement obtained between such theory and experiment is extremely encouraging. Thus, for $Ni(C_3H_5)_2$ our assignment of the PE spectrum agrees with that obtained experimentally except for some ionizations of similar atomic character which are difficult to distinguish experimentally. For the transition metal cyanides which we have studied, the calculated separation of the metal and ligand ionizations are in excellent agreement with the results from XRE spectra.

To obtain the correct ordering of the closely spaced levels arising from the metal and ligand ionizations requires a balanced calculation of relaxation and correlation effects. The ΔSCF method allows for the calculation of all the relaxation effects which *lower* the metal relative to the ligand IEs. For bis(π-allyl)nickel, correlation effects *increase* the metal relative to the ligand IEs. Due to the difficulty of recovering such differential correlation effects, the ΔSCF-CI results lead to an incorrect ordering of the metal and ligand ionized states. The success of the 2ph-TDA calculations may be ascribed to a more balanced treatment of these relaxation and correlation effects.

5. HARTREE-FOCK GEOMETRIES OF TRANSITION METAL COMPOUNDS

Accurate Hartree Fock geometry optimizations on main group molecules typically yield bond distances too short by some 0.01 - 0.03 A.[43] There is a growing body of evidence that points to far greater errors in the HF geometry of transition metal compounds,[42,44-49] with bond distances involving the metal atom frequently much longer than experimental values or those obtained in extensive

CI treatments. The case of ferrocene and Fe(CO)$_5$ provide examples of this effect. The calculated HF metal-ligand distance in ferrocene is 0.23 A longer than the experimental value,[46,47] while in iron pentacarbonyl the axial carbonyl - iron distance is calculated to be 0.24 A longer than experiment.[49] In contrast the equatorial distance in Fe(CO)$_5$ is found to be in much better agreement, with an error of 0.04 A in the HF bond length.

Calculations by Almlof and co-workers on Ni(CO)$_4$ [45] and a series of sandwich compounds[48] suggest the need for extensive basis sets to achieve stable results which may, nevertheless, remain in rather poor agreement with experiment. Such evidence casts considerable doubt on the value of 'small' basis sets in such studies.[50-51]

The undoubted role of electron correlation in the above systems has been elegantly demonstrated by Siegbahn and co-workers.[49] In a series of large-scale contracted CI calculations, near degeneracy effects were demonstrated as the reason for the failure of the HF model in Fe(CO)$_5$.

In the present section we outline some preliminary results from a study designed to consider further this question of the applicability of the HF model, by examining a broader range of metal-ligand interactions than has been studied to date. In particular we consider the following:

(i) systems that are physically ionic, with few or no d-electrons (ScF$_3$, TiCl$_4$, VF$_5$, VOCl$_3$), where one might not expect problems with the HF model.

(ii) Complexes with both carbonyl and nitrosyl ligands, in particular the isoelectronic series Cr(NO)$_4$, Mn(NO)$_3$CO, Fe(CO)$_2$(NO)$_2$, Co(CO)$_3$NO and Ni(CO)$_4$.

(iii) the substituted manganese carbonyls Mn(CO)$_5$X (X = H,CN)

(iv) bis(π-allyl)nickel, the simplest transition metal sandwich complex, and the nitrosyl (C$_5$H$_5$)NiNO.

All geometrical parameters of the above species have been optimized at the HF level using basis sets of essentially double zeta quality. The s,p basis for the transition metal elements was taken from the (12s6p4d) set of reference 31, with the addition of basis functions to describe the metal 4p orbital[52]. The metal d basis was the reoptimized (5d) set of reference 53, contracted (4/1). For the ligand atoms a (4s/2s) basis was adopted for H and a (9s5p/3s2p) contraction for the first-row elements[30]. The optimized Ni-C bond length of 1.900 A using this (8s6p2d/3s2p/2s) basis is in reasonable agreement with the near Hartree Fock value of 1.921 A[45].

All calculations were performed using the program system GAMESS on the FPS-164 at the Daresbury Laboratory.

Results and Discussion

The optimized geometrical parameters for the species under consideration are presented in Tables 6 and 7, together with the experimental results and parameters derived using a minimal STO-3G basis. The latter results provide the expected agreement with those reported by Pietro and Hehre[53].

Satisfactory agreement between theory and experiment is evident in the transition metal fluorides (Table 6), with the calculations correctly depicting the unequal axial and equatorial bond distances in VF_5. In the formally multiply bonded $VOCl_3$, the DZ description predicts too short a V-O distance and too long a V-Cl distance compared to the experimental values, by 0.05 A and 0.03 A respectively.

Far greater discrepancies are found, however, in the isoelectronic series of carbonyl and nitrosyl complexes. With the exception of $Fe(CO)_2(NO)_2$, we find in each case that the metal-nitrogen bond distance is significantly underestimated, and the metal-carbon distance significantly overestimated, compared to the experimental values. Thus the metal-nitrosyl distance is calculated to be *shorter* than the experimental values by 0.101 A, 0.102 A and 0.167 A in $Cr(NO)_4$, $Mn(NO)_3CO$ and $Co(CO)_3NO$ respectively. As in the case of nickel tetracarbonyl, the metal-carbon distance in $Co(CO)_3NO$ and $Mn(NO)_3CO$ is calculated to be *larger* than the experimental values, by 0.108 and 0.091 A respectively.

The qualitative origin of this error in the HF N-O bond distance may be readily understood from a consideration of the associated CI wavefunction[54,55]. In the case of $Co(CO)_3NO$, the HF configuration may be written thus:

[core] $11e^4$ $17a_1^2$ $12e^4$

where the 11e (75% cobalt 3d) and $17a_1$ (66% cobalt 3d) are essentially metal 3d orbitals, and the 12e orbital is the cobalt-nitrosyl π-bond. The dominant terms in the CI wavefunction involve excitations from the 12e to 13e orbital, where 13e is of cobalt-nitrosyl π-antibonding character ie correlation of electrons in the metal-nitrosyl π-bond. The significant occupation number of the 13e orbitals (0.52) reveals the importance of these π bond-antibond excitations, and highlights the poor description of the metal-nitrosyl interaction at the HF level. Clearly the admixture of these terms will act to increase the HF bond length. This leads rather naturally to the corresponding iron complex, $Fe(CO)_2(NO)_2$. A previous CI study aimed at illucidating the photoelectron spectrum of the complex[55] pointed to the exceedingly poor description of the iron-nitrosyl π-bond. The HF configuration

$9a_1^2$ $6b_2^2$ $3a_2^2$ $6b_1^2$ $10a_1^2$

comprises the Fe 3d orbitals ($9a_1$, $6b_2$ and $3a_2$) and the Fe-NO π-bonding orbitals ($10a_1$, $6b_1$). Excitations to the formally vacant metal-nitrosyl π-antibonding orbitals ($4a_2$ and $7b_2$) are again found to exhibit considerable weight in a CASSCF wavefunction (15,416 csfs with the above 10 valence electrons in 11 orbitals), with occupation numbers of 0.17 and 0.23 electrons respectively. In fact the HF geometry for the iron complex is seen to be alarmingly in error; the

Fe-C bond distance is overestimated by 0.35 A, an effect even larger than that in $Fe(CO)_5$. The outcome of a CASSCF geometry optimization currently in progress will hopefully shed further light on this matter.

Returning to the results of Table 7, we find a metal → ring distance of 1.85 A in $Ni(C_5H_5)NO$, to be compared with the experimental value of 1.72 A, and a previous HF-DZ value of 1.82 A (reference 48, in which other geometrical parameters were held fixed at the experimental values). The Ni-N distance is again underestimated, by 0.06 A.

Sizable errors are evident in both $HMn(CO)_5$ and bis(π-allyl)nickel. In the manganese compound both axial and equatorial Mn-C bond distances are overestimated, by 0.137 and 0.159 A respectively. In $Ni(C_3H_5)_2$ the magnitude of the error in the predicted Ni-C distance is a function of the carbon position in the allyl moiety; the distance to the central carbon of the allyl group is overestimated by 0.10 A, while the error in the Ni - terminal carbon bond length is 0.24 A, a value comparable to that found in ferrocene.

Table 6. Calculated and experimental equilibrium geometries for compounds of Sc, Ti, V and Cr.

Molecule	Point Group	Geometrical Parameter[a]	STO-3G	DZ	Expt.
ScF_3	D_{3h}	r(Sc-F)	1.845	1.879	1.91
$TiCl_4$	T_d	r(Ti-Cl)	2.167	2.214	2.17
VF_5	D_{3h}	r(V-Fax)	1.641	1.744	1.734
		r(V-Feq)	1.608	1.702	1.708
$VOCl_3$	C_{3v}	r(V-O)	1.468	1.518	1.570
		r(V-Cl)	2.107	2.177	2.142
		< (ClVCl)	109.9	110.5	111.3
$Cr(NO)_4$	T_d	r(Cr-N)	1.576	1.689	1.79
		r(N-O)	1.218	1.160	1.16
$Cr(CO)_6$	O_h	r(Cr-C)	1.786	1.975	1.92
		r(C-O)	1.167	1.142	1.16

a) All bond lengths in angstroms, angles in degrees

Table 7. Calculated and experimental equilibrium geometries for compounds of Fe, Co, Mn and Ni.

Molecule	Point Group	Geometrical Parameter	STO-3G	DZ	Expt.
$Ni(CO)_4$	T_d	r(Ni-C)	1.579	1.900	1.836
		r(C-O)	1.160	1.138	1.142
$Fe(CO)_2(NO)_2$	C_{2v}	r(Fe-C)	1.708	2.198	1.8
		r(Fe-N)	1.518	1.822	1.77
		r(C-O)	1.154	1.127	1.15
		r(N-O)	1.221	1.202	1.12
		< (C-Fe-C)	107.7	92.2	
		< (N-Fe-N)	111.8	129.6	
$Co(CO)_3NO$	C_{3v}	r(Co-N)	1.478	1.593	1.76
		r(Co-C)	1.664	1.938	1.83
		r(N-O)	1.231	1.197	1.10
		r(C-O)	1.155	1.132	1.14
		< (C-Co-N)	111.9	114.3	
$Mn(NO)_3CO$	C_{3v}	r(Mn-C)	1.751	1.921	1.83
		r(Mn-N)	1.513	1.658	1.76
		r(C-O)	1.154	1.133	1.14
		r(N-O)	1.222	1.164	1.10
		< (N-Mn-C)	106.1	104.8	
$(C_5H_5)NiNO$	C_{5v}	r(Ni-C)	2.084	2.211	2.11
		r(Ni-N)	1.420	1.571	1.626
		r(C-C)	1.420	1.424	1.43
		r(N-O)	1.271	1.165	1.165
		r(C-H)	1.078	1.067	1.09
$HMn(CO)_5$	C_{4v}	r(Mn-Cax)	1.725	1.960	1.823
		r(Mn-Ceq)	1.717	1.982	1.823
		r(Mn-H)	1.628	1.684	1.50
		r(C-Oax)	1.162	1.137	1.139
		r(C-Oeq)	1.163	1.137	1.139
		< (H-Mn-Ceq)	72.3	82.3	83.6
		< (Mn-C-Oeq)	171.2	172.8	
$Mn(CO)_5CN$	C_{4v}	r(Mn-CCN)	2.045		1.98
		r(Mn-COax)	1.825		1.822
		r(Mn-COeq)	1.804		1.853
		r(C-Oeq)	1.162		1.134
		r(C-Oax)	1.152		1.134
		r(C-N)	1.156		1.16
$Ni(C_3H_5)_2$	C_{2h}	r(Ni-C)	1.744	2.088	
		r(Ni-Ct)	2.183	2.253	2.10
		r(Ct-Cc)	1.405	1.399	1.41
		r(Ct-H)	1.074	1.076	1.08
		r(C-H)	1.094	1.073	1.08
		< (Ct-C-Ct)	128.5	124.0	

Summary

The systematic study described here provides further evidence of the limitations of the HF method in structure optimizations of transition metal complexes. The metal-carbonyl bond length is consistently overestimated in all environments, and it would seem that improvements to our basis (eg the addition of d-polarization functions on the ligands) will act to increase the metal-carbon distance still further (see reference 45). The shortening of the metal-nitrosyl bond predicted at the HF level may be attributed to near degeneracy effects, and hence should be well described within an MCSCF treatment. However, the more general need to recover dynamical correlation contributions through MR-CI treatments will, at least for systems of the size under discussion, require the full capabilities of the next generation of supercomputers.

6. CONCLUSIONS

We have outlined the strategy employed in implementing and optimising quantum chemistry codes on the present generation of vector and attached processors. While this exercise has been of value in its own right, perhaps the most important outcome is that it enables us to view the prospect of realizing the full potential of the next generation of machine with some degree of confidence. If computational chemistry is to grow in stature as a subject, it must be able to respond and adjust rapidly to new developments in computing. This it can undoubtedly do given access to computer systems on which the full power of the system is realisable, and not constrained by either operating system, vendor software, or, for that matter, local management.

ACKNOWLEDGEMENTS

The work described in Sections 1-3 was carried out in collaboration with V.R. Saunders. The calculations of Section 4 and 5 were conducted with I.H. Hillier, M. Rossi and A. Sgammelotti. The untiring efforts of M. Rossi are particularly acknowledged.

The author also wishes to thank FPS (UK) for all their assistance provided to date.

REFERENCES

1. 'Some Research Applications on the Cray-1 Computer at the Daresbury Laboratory 1979-1981', Daresbury Laboratory, 1982.
2. M.F. Guest and S. Wilson in Proc. American Chem. Soc. Meeting, Las Vegas, August 1980 (Wiley-Interscience, New York, 1981); 2b] V.R. Saunders and M.F. Guest, Comput. Phys. Commun. **26**, 389 (1982).
3. V.R. Saunders and J.H. Lenthe, Molec. Phys. **48**, 923 (1983).
4. R. Shepard, R.A. Bair, R.A. Eades, A.F. Wagner, M.J. Davis, C.B. Harding and T.H. Dunning, Jr., Int. J. Quant. Chem. **17**, (1983); R.A. Bair and T.H. Dunning, Jr., J. Comp. Chem. **5** 44 (1984); 4b] 'FPS-164 Matrix Multiplication Subroutine Guide', R. Bair, Argonne National Lab. (1984).

5. R. Ahlrichs, H.J. Bohm, C. Ehrnardt, P. Scharf, H. Schiffer, H. Lischka and M. Schindler, 6th Seminar on Computational Methods in Quantum Chemistry **31**, (1984). 5b] for recent applications in parallelism see the work of E. Clementi and co-workers, Department 48B, Kingston, NY; 'Algorithms vs Architectures for Computational Chemistry', H. Partridge and C. Bauschlicher, in 'Austin Conference on Algorithms, Architectures and the Future of Scientific Computation', Austin, 1985.

6. M. Dupuis, J. Rys and H.F. King, J. Chem. Phys. **65** 111 (1976); H.F. King and M. Dupuis, J. Comput. Phys. **21**, 144 (1976).

7. L.E. McMurchie and E.R. Davidson, J. Comput. Phys. **26**, 218 (1978).

8. C.C.J. Roothan, Rev. Mod. Phys. **32**, 179 (1960).

9. M. Yoshimine, IBM Technical Report, RJ-555, San Jose, USA (1969).

10. P.E.M. Siegbahn, J. Chem. Phys. **72**, 1647 (1980).

11. H.-J. Werner and P.J. Knowles, J. Chem. Phys. **82** 5053 (1985); 11b] P.J. Knowles and H.-J. Werner, Chem. Phys. Letts. **115**, 259 (1985)

12. V.R. Saunders, VAMP on the Cyber-205 (1985).

13. G.H.F. Diercksen, W.P. Kraemer and B.O. Roos, Theor. Chim. Acta (Berlin) **36**, 249 (1975).

14. B.O. Roos, P.R. Taylor and P.E.M. Siegbahn, Chem. Phys. **48**, 157 (1980).

15. P.E.M. Siegbahn, Chem. Phys. Letts. **109**, 417 (1984); P.J. Knowles and N.C. Handy, Chem. Phys. Letts. **111**, 315 (1984).

16. B.H. Lengsfield III and B. Liu, J. Chem. Phys. **75**, 478 (1981); B.H. Lengsfield III, J. Chem. Phys. **77**, 4073 (1982); J. Olsen, D.L. Yaeger and P. Jorgensen, Adv. Chem. Phys. **54**, 1 (1983).

17. P. Pulay, Molec. Phys. **17**, 197 (1969); **18**, 473 (1970); **21**, 329 (1971).

18. M. Dupuis, D. Spangler and J. Wendolowski, NRCC Software Catalog, Vol. 1, Program No. QG01 (GAMESS), 1980; M.F. Guest and J. Kendrick, GAMESS User Manual, Daresbury Technical Memorandum (1985).

19. P. Pulay. J. Comp. Chem. **3**, 556 (1982).

20. H.B. Schlegel, J. Chem. Phys. **77**, 3676 (1982).

21. S. Bell and J.S. Crighton, J. Chem. Phys. **80**, 2464 (1984).

22. C.J. Cerjan and W.H. Miller, J. Chem. Phys. **75**, 2800 (1981).

23. I.H. Hillier, Pure and Appl. Chem. **51**, 2183 (1979).

24. W. von Niessen, L.S. Cederbaum, W.P. Kraemer, and G.H.F. Diercksen, J. Am. Chem. Soc. **98**, 2066 (1978).

25. W. von Niessen, W.P. Kraemer and L.S. Cederbaum, J. Elec. Spec. **8**, 179 (1976); J. Schirmer, W. Domcke, L.S. Cederbaum and W. von Niessen, J. Phys. **B11**, 1901 (1978).

26. W. von Niessen, L.S. Cederbaum, G.H.F. Diercksen, J. Chem. Phys. **67**, 4124 (1977); and references cited therein.

27. M.-M. Rohmer and A. Veillard, J. Chem. Soc. (Chem. Comm.) 250 1973.

28. M.-M. Rohmer, J. Demuynck and A. Veillard, Theor. Chim. Acta (Berl.) **36**, 93 (1974).

29. S. Huzinaga, J. Chem. Phys. **42**, 1293 (1965).

30. T.H. Dunning, J. Chem. Phys. **53**, 2823 (1970); 30b] T.H. Dunning and P.J. Hay, in 'Modern Theoretical Chemistry', ed. H.F. Schaefer, Plenum (NY) Vol. 4, p1 (1977).

31. R. Roos, A. Veillard and G. Vinot, Theor. Chim. Acta (Berl.) **20**, 1 (1971).

32. P.J. Hay, J. Chem. Phys. **66**, 4377 (1977).

33. L.S. Cederbaum and W. Domcke, Adv. Chem. Phys. **36**, 205 (1977).

34. L.S. Cederbaum, Theor. Chim. Acta (Berl.) **31**, 239 (1973); J. Phys. **B8**, 290 (1975).

35. J. Schirmer and L.S. Cederbaum, J. Phys. **B11**, 1889 (1978).

36. J. Schirmer, L.S. Cederbaum and O. Walter, Phys. Rev. **28**, 1237 (1983).

37. W. von Niessen, J. Schirmer and L.S. Cederbaum, Comp. Phys. Rep. **1**, 57 (1984); and references cited therein.

38. A. Veillard and J. Demuynck, in 'Modern Theoretical Chemistry', ed. H.F. Schaefer, Plenum (NY) Vol. 4, p187 (1977).

39. C.D. Batich, J. Am. Chem. Soc. **98**, 7585 (1976).

40. M.C. Bohm, R. Gleiter and C.D. Batich, Helv. Chim. Acta. **63**, 990 (1980).

41. M.C. Bohm and R. Gleiter, Theor. Chim. Acta (Berl.) **57**, 315 (1980).

42. T.E. Taylor and M.B. Hall, Chem. Phys. Letts. **114**, 338 (1985).

43. W.A. Lathan, L.A. Curtiss, W.J. Hehre, J.B. Lisle and J.A. Pople, Prog. Phys. Org. Chem. **11**, 175 (1974).

44. J. Demuynck, A. Strich and A. Veillard, Nouv. J. Chim. **1**, 217 (1977).

45. K. Faegri and J. Almlof, Chem. Phys. Letts **107**, 121 (1984).

46. H.P. Luthi, J.H. Ammeter, J. Almlof and K. Korsell, Chem. Phys. Letts. **69**, 540 (1980).

47. H.P. Luthi, J.H. Ammeter, J. Almlof and K. Faegri, J. Chem. Phys. **77**, 2002 (1982).

48. J. Almlof, K. Faegri, B.E.R. Schilling and H.P. Luthi, Chem. Phys. Letts. **106**, 266 (1984).

49. H.P. Luthi, P.E. Siegbahn and J. Almlof, J. Phys. Chem. **89**, 2156 (1985).

50. W.J. Pietro and W.J. Hehre, J. Comp. Chem. 4, 241 (1983).

51. L. Seijo, Z. Barandiaran, M. Klobukowski and S. Huzinaga, Chem. Phys. Letts. **117**, 151 (1985).

52. D.M. Hood, R.M. Pitzer and H.F. Schaefer, J. Chem. Phys. **71**, 705 (1979).

53. A.K. Rappe, T.A. Smedley and W.A. Goddard, J. Phys. Chem. **85**, 2607 (1981).

54. R.F. Fenske and J.R. Jensen, J. Chem. Phys. **71**, 3374 (1979).

55. M.F. Guest, I. H. Hillier, A.A. MacDowell and M. Berry, Molec. Phys. **41**, 519 (1980).

LARGE SCALE COMPUTATIONS ON THE LOOSELY COUPLED ARRAY OF PROCESSORS

E. Clementi, S. Chin, and D. Logan

IBM Corporation
Data Systems Division
Dept. 48B / MS 428
Kingston, New York 12401

ABSTRACT

An experimental parallel computer system which is expected to achieve supercomputing performance across the entire spectrum of scientific and engineering applications is described. This system allows the execution of single large scale scientific and engineering applications on multiple processors. The system hardwares and softwares will be briefly described, as well as the programming strategies used to migrate codes from sequential to parallel. The validity of this approach to solving large scale problems is verified by analyzing the performance results of a variety of application programs. The type of scientific/engineering applications which may be investigated using this type of system is demonstrated by discussing one of our applications in biochemistry; namely the statistical and quantum mechanical study of DNA. Finally, ongoing and future extensions to this system are presented.

I. INTRODUCTION

The concept of super-computing, while relative at best, is complicated by consideration of vast areas of problems that require solution. One area that has traditionally been associated with the need of super-computing, and in which our primary interest lies, is that of scientific and engineering problems. Success in these applications has invariably given impetus to the development of more elaborate calculations. Thus there has always been constant pressure to expand the limits of currently available computer resources and beyond. However the extent to which traditional sequential processors can be pushed into higher realms of computational power is limited by pragmatic or fundamental constraints. Of general importance is the constant need of developing more efficient algorithms and languages. Of a more fundamental nature are limitations that deal with density of packing integrated circuits, heat dissipation and finally the speed of light. Such considerations have lead to alternative strategies in attacking the problem.

One of the first and most successful approaches has been the development of vector oriented processors such as IBM 3838, the CRAY series computers or the CDC CYBER 205. However the useful application of this technique has to a large extent

Lecture Notes in Chemistry, Vol. 44
Supercomputer Simulations in Chemistry. Edited by M. Dupuis
© Springer-Verlag Berlin Heidelberg 1986

been limited to applications that are inherently vectorizable. While such applications are significant in number they do not include many areas of scientific computing. An alternative approach has been the development of parallel structures i.e. systems of many independent processors that may be concurrently applied to the solution of a single problem. It is now generally conceded that both approaches are complementary. Tomorrow's super computer likely will have the properties of massive parallelism incorporating processors that retain highest possible performance in vector and scalar computation; certainly a high degree of parallelism and vector features will be typical in future architectures.

With this comment on the necessary attributes of super-computing it is logical that one of the first concerns should be that of the parallel structure. In this work we shall expand on some of our previous review papers, particularly the one in reference 1e. The idea of parallel computers is certainly not new. It has already been the subject of numerous research projects and a very vast literature. For a representative subset see Ref. 1 and references therein. Even a system intended specifically for computational chemistry has been talked about.[2] Our interest in such systems has been spurred by the need of extending calculations in theoretical chemistry and biophysics, with which our laboratory has been traditionally involved, well beyond that obtainable on current systems.

As a a first step in this direction we have defined our objectives as follows : 1) to develop a kernel architecture that at minimum is the equivalent of a CRAY 1S or CYBER 205 2) is easily extended well beyond this limit 3) has much more flexibility and versatility 4) permits the quick migration of large scientific applications to parallel execution 5) does not cost too much.

Many of the characteristics of our parallel strategy follow from these priorities. These characteristics are: 1) not thousands, nor hundreds of processing units (PU's) are considered, but very few, less than 20; 2) each PU is a well-developed engine which executes asynchronously or even independently as a stand alone from the others; 3) the system softwares needed to execute application programs are as much as possible those commonly available for sequential programming; 4) to start with we constrain ourselves to FORTRAN, since this is the most widely used scientific application language and 5) we restrict ourselves to the minimum level of programming variations relative to the old sequential codes; 6) because of the applications we are interested in, 64-bit hardware precision is required; and 7) if a larger number of PU's is advisable then a hierarchical system will be considered from the very beginning.

With respect to point 1) we wish to implement a pragmatic approach which is in no way critical of more ambitious attempts that are likely to become standard and available for general application programming, but only in the next decade. Concerning point 2) we use IBM host or hosts (for example, IBM-4341, IBM-4381 or IBM-308X) with several FPS-164 or FPS-264 attached array processors (AP). Since the latter could in principle be either standard scalar CPU's (Central Processing Units) or AP's, in the following we shall refer either to "slave CPU's" or to "attached AP's" as equivalent approaches to obtain parallelism. Concerning point 3) we note that we have implemented our architecture in two physical systems, referred to as "loosely coupled array of processors", 1CAP-1 and lCAP-2. They differ in two important respects in the identity of the IBM host and the operating systems employed. The systems softwares that are needed for parallel programming, as opposed to sequential programming, are those concerned with communication of commands and files. These issues, which are operating system imperatives, will thus be explored from two approaches. Concerning point 4) one would eventually like to have a compiler and/or an optimizer which would include the above communication facilities. As a first step we have developed precompilers that permit the insertion of directives that may be interpreted as extensions to the FORTRAN language. These directives, described in a later section, incorporate basic functions such as "fork" and "join" needed in expressing parallelism. Concerning point 7) we note that we are currently exploring ways to connect lCAP-1 with lCAP-2 in a hierarchical way.

The idea for the lCAP-type architecture was first conceived by Clementi and his group in May of 1983. At that time there was one FPS-164 attached to an IBM 4341. By July of that same year the original one had grown to three FPS-164s. Even at this time experimentation with parallelism had begun. In December of 1983 the three processors were expanded to six, and parallelism was here to stay. In May 1984 the full complement of ten FPS-164s were installed in IBM Kingston and became operational as the parallel computer now known as lCAP-1. The initial success of lCAP-1 encouraged additional ventures into the lCAP architecture. Indeed, in the middle part of 1985 the IBM European Center for Scientific and Engineering Computing, ECSEC, began operation with a copy of the lCAP-1 system. In May of 1985 an interim lCAP system for the Cornell Production Supercomputer Facility, PSF, was installed and operational. In August of 1985 the first three FPS-264 machines were delivered to the lCAP-2 system at IBM Kingston. In October 1985 the Cornell PSF became fully operational with an IBM 3084 QX and 4 FPS-264s. Almost at the same time, three additional FPS-264 machines were added to the lCAP-2 system at Kingston, bringing the total to six.

Additional description of our hardware configuration is provided in Section II below. In Section IV we present the strategies we have developed to modify our applications programs for effective parallel execution on our system. In order to more precisely understand how these strategies are implemented, we describe the operating system communication considerations in Section III. Section V outlines basic considerations of performance in a parallel processing environment, while Section VI illustrates performance results across a wide spectrum of scientific codes. In Section VII we discuss one of the specific applications in biophysics that is amenable to parallel processing. section VIII discusses future extensions to the architecture that will be an aid in attacking a larger variety of such applications. Finally, in Section IX we wrap up our current experience with our system.

II. PRESENT CONFIGURATION FOR lCAPs

As mentioned previously there are at present two parallel processing systems working in our laboratory. Both share the same fundamental architecture of a distributed system with a front end CPU and attached PU's (or AP's). In the original implementation of this system there was a limited ability for slave to slave communication; for this reason we term our architecture a loosely coupled array of processors (lCAP). The first of these systems, called lCAP-1, is hosted either by an IBM 4341 or 4381 and attaches to 10 FPS-164 processors. The second and more powerful system, called lCAP-2, employs as host an IBM 3084 and presently as slaves 6 FPS-264s processors; it is currently being expanded to include 10 of these processors. In spite of these differences the two systems are very similar (aside from operating system considerations which will be discussed in the following section).

The lCAP-1 system is structured such that seven FPS-164s are connected to an IBM 4381 host and the remaining three are attached to an IBM 3814 switching unit so they can be switched between an IBM 4341 host or the IBM 4381 host. The FPS-164 processors are attached to the IBM hosts through IBM 3 Mbyte/Sec channels available on these hosts. A third IBM 4341, connected to a graphics station, completes the host processor pool. The graphics station include an Evans and Sutherland PS300 and an IBM 5080 graphics terminal and a large set of graphics packages for such diverse uses as cad/cam applications or molecular modelling. The three IBM systems are interconnected, channel to channel, via an IBM 3088 connector. A schematic diagram of the configuration appears in Fig. 1.

loosely Coupled Array of Processors (lCAP)

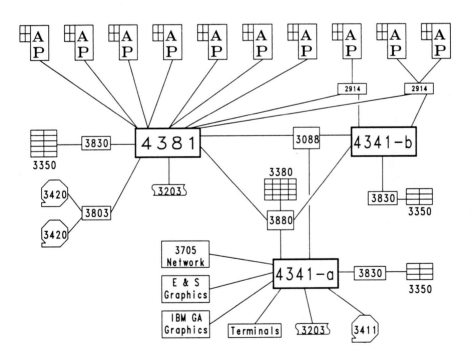

Figure 1. Initial configuration for lCAP-1. Included are the 10 FPS-164's, the three IBM hosts (two 4341's and one 4381), printers, tape drives, and graphics substations.

Each FPS-164 contains an independent PU and its own memory and disk drives for either temporary or permanent data sets (the latter is a rather seldom occurrence). The CPU on the FPS-164 runs at 5.5 million instructions per second (MIPS), and several concurrent operations (up to ten) can take place on each instruction cycle. In particular, one 64-bit floating-point addition and one 64-bit floating-point multiplication can be initiated each cycle, so that peak performance is about 11 million floating point operations per second (11 Mflops). Of course, one must make the distinction between peak performance (a characteristic of the machine hardware) and sustained performance (depending on the application and the code which implements it as well as the hardware).

Each of the FPS-164's has at least 4 Mbytes of real random access memory; two have more: 8 Mbytes and 10 Mbytes, respectively. The memory on the IBM 4341 model M2 and model P2 are 8 Mbytes and 16 Mbytes, respectively. The IBM 4381/3 has 32 Mbytes. Thus, taken as a whole, there is 90 Mbytes real storage available in lCAP-1.

Each FPS-164 also has four 135 Mbyte disks, for a total of 5.4 Gigabytes. In addition there are banks of IBM 3350 and IBM 3380 disks accessible to the host computers, totalling about 25 Gigabytes of disk storage.

Floating Point Systems also supplies the FPS-164/MAX; this is a special-purpose board that can be added to the FPS-164 to augment performance, particularly on vector operations. Each MAX board contains two additional adders and two additional multipliers, and so adds 22 Mflops to peak attainable performance. Up to 15 boards can be placed in a single FPS-164, converting it to a machine with a peak performance of 341 Mflops. At present each of our AP's has been equipped with two MAX boards. This has upgraded our peak performance from 110 to 550 Mflops. Ultimately our system could grow to 3410 Mflops peak capability, but (recalling the distinction between peak performance and realized performance) it is clearly desirable to first explore the gains that one can realistically obtain with only a few 164MAX boards per AP, so we have settled at 550 Mflops. Tape drives, printers, and a communication network interface complete the lCAP-1 configuration.

One especially attractive feature of the above system is that a variable number of AP's may be attached to either the 4381 or the 4341; the former configuration is used for primarily production jobs while the latter serves for debugging and experimentation.

The lCAP-2 system is hosted by an IBM 3084 QX and has at present 6 FPS-264's as slave processors. The FPS-264 is compatible with the FPS-164, and codes developed for either machine run on the other machine without modification. The CPU of the FPS-264 runs at a peak performance of 38 Mflops, or 3.5 times faster than the FPS-164. However, because of improved memory interleaving and larger program cache, we have observed the performance of the FPS-264 to be between 3.5-4.0 times that of the FPS-164. With the intended expansion to include 10 of the FPS-264's the peak performance of lCAP-2 will be 380 Mflops. However, again we stress the distinction between peak performance and observed performance. We expect the average performance of the lCAP-2 system to surpass that of the lCAP-1/MAX, as the MAX boards are special purpose boards with limited use. This has been verified by some preliminary tests on lCAP-2 that shall be discussed later. Each of the FPS-264's have 8 Mbytes of real memory and two of the FD64 disk drives; totalling 1.2 Gbytes of disk storage on each machine. The IBM 3084 has 128 Mbytes of real memory and IBM 3350 and 3380 disk packs totalling 50 Gbytes of disk storage. Again, tape drives, printers and a communication network complete the lCAP-2 configuration.

Data conversion and communication between the host computer and the attached FPS-X64s are handled by hardware and software that is provided and supported by Floating Point Systems as a standard feature. An optimizing FORTRAN compiler and supporting utilities (including disk I/O) are also standard products for the

FPS-X64. The compiler produces reasonably compact pipelined machine code that takes advantage of the unique multiple independent functional units of the FPS architecture. An extensive library of subroutines is also provided. Much of the latter have been optimized to the maximum of performance in turning on error checking code during such procedures as pipeline initiation and termination within our laboratory.[3]

A library of mathematical routines will be available for use on the MAX boards. When properly employed they may achieve a gain in performance that is impressive. For large matrix multiplications the processing speed increases by approximately 22 Mflops i.e. the rated peak performance of the supplemental vector board. The applicability of the MAX boards in general application programs is now under investigation.

It should be noted that such upgrades have no effect on the parallel programming strategy to be discussed later. The strategy is equally effective for AP's of any architecture or computational speed. In principle, we could substitute 10 vector-oriented supercomputers for our 10 FPS-X64's. However given the notable differences in cost between these options the latter one is unrealistically high.

Presently, when we run a given job on two or more AP's (parallel mode) we attempt to ensure that the amount of data flowing from one AP to another AP, via the host processor, is kept to a minimum. Indeed, jobs requiring transfers of long files from AP to AP are not optimal on our configuration because of channel transfer rate limitations. To offset this condition both systems are currently being expanded through the incorporation of both a fast common bus as well as a number of shared memories that will permit direct AP to AP communication. These extensions shall initially be installed and tested on lCAP-1, but we plan to expand both systems to include both shared memories and the fast bus.

The shared memory systems and associated softwares were designed and developed by Scientific Computing Associates, Inc. (SCA). Each shared memory is at minimum 32 Mbytes in size, and may be multiply connected to up to 4 of the FPS-X64 processors. Each unit is capable of sustained data transfer rates of 64 Mbytes/Sec, but when attached to the FPS-164 will operate at a rate of 44 Mbyte/Sec (this is the maximum achievable rate given the cycle time of the FPS-164). When attached to the FPS-264 the data transfer rate may possibly run at full bandwidth (64 Mbyte/Sec). The addition of these shared memory systems provides the ability to perform quickly large asynchronous transfers of data between processors. This constitutes a departure from lCAP and a step towards a tightly Coupled Array of Processors (tCAP). However, we note that the system is flexible and reconfigurable between lCAP and tCAP.

At the present time two of these shared memory systems are installed and operational on lCAP-1 and are connected to 6 of the FPS-164's. Two are connected to the IBM 4341 and the remaining 4 are connected to the IBM 4381 production machine. We are thus able to test the use of the SCA memories without disrupting the production on the 4381. In addition, we have the flexibility to use all 6 FPS-164's processors in a ring configuration. Some preliminary results with this configuration shall be given later. By year end a total of 5 of these memories shall be installed on lCAP-1 configured as shown in Fig. 2. This configuration gives the flexibility of allowing one ring with 10 processors, or alternatively a number of smaller rings. The multiple connections between processors allow added fault tolerance, and the 5 shared memories give an aggregate data transfer rate between APs of 110 Mbyte/Sec (5 x 22 Mbyte/Sec). Additional extensions to the shared memories, as well as the connectivity on lCAP-2 will be discussed later.

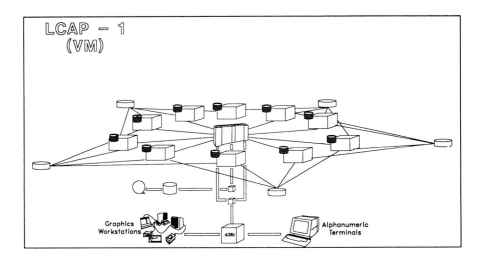

Figure 2. lCAP-1 configuration with five 32 Mbyte shared memories. Notice that each FPS-164 has two connections to both neighbors. This connectivity has the flexibility of allowing one to debug with the shared memory and 2 FPS-164's on the 4341, but also to run tests/production with five shared memories and 10 FPS-164's on the 4381.

III. SYSTEM CONSIDERATIONS

Our system is essentially a Multi Instruction Stream Multi Data Stream (MIMD) system[4], in the form of a distributed network of nodes. The distributed nature of the system allows appropriate utilization of the parallelism of the code; i.e. the "degree of parallelism"[1b] of the software is matched by the hardware. An important advantage of high-level parallelism, as in high-level programming languages, is its portability. A high-level parallel code may be executed on any system with a rudimentary communication protocol. In addition, improvements at the instruction level can be independently pursued without disturbing the setup of the parallel algorithm.

We have implemented our basic architecture in two physically different systems. They differ in one important aspect in the operating systems that the host employs. This directly dictates the mode by which each system achieves host to slave communication. In the following we describe this issue for both systems. It will be seen that, aside from some details of implementation, the overall communication structures are logically equivalent. Thus parallel programs written for either system may with ease migrate to the other.

The lCAP-1 system, hosted by either an IBM 4341 or 4381, runs under the IBM Virtual Machines/ System Product (VM/SP) operating system.[5] For the AP's, we use the software provided for hosts running under this system as by Floating Point Systems. We have not found it necessary to modify either set of software in order to run our applications in parallel.

VM/SP is a time-sharing system in which jobs run on virtual machines (VM) created by the system; these VM's simulate real computing systems. The standard software provided by Floating Point Systems to use the FPS-164's embodies the restriction that only one AP can be attached to a VM. Of course, for a task running in parallel, more than one AP is required. Our solution to this is to introduce extra "slave" VM's to handle the extra AP's we need. To make this work, one must have a way to communicate between different VM's; this is provided by the Virtual Machine Communication Facility (VMCF), which is a standard feature of VM/SP.[5]

A parallel task will consist of several FORTRAN programs, each running on a separate VM in the host system, and each controlling a particular AP on which additional FORTRAN code runs. On one of the VM's, the "master", is the part of the original FORTRAN code intended to be run on the host, combined with utility subroutines that handle communication with the "slave" VM's and with the AP attached to the "master" VM (if any). The logical structure of this system is illustrated in Fig. 3. The programs running on the "slave" VM's are nothing more than transfer points for communication between the "master" program and the AP's attached to the "slaves". Since each VM is attached only to a single AP, the standard utilities provided by FPS.[6] for communication between host and AP can be used without modification.

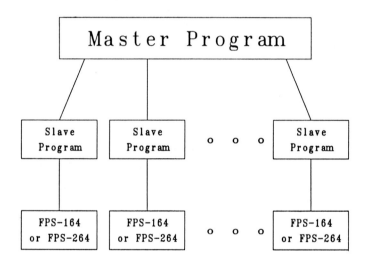

Figure 3.   Basic structure for parallel execution of a program with the lCAP master/slave topology. This structure is true for both VM and MVS. In VM the slaves are actually secondary VM's, and in MVS the slaves are sub-tasks.

The lCAP-2 system is hosted by a 4 processor IBM 3084 running Multiple Virtual Storage (MVS) operating system.[7] In contrast to VM/SP, this operating system has typically been employed as a batch oriented system. The operational counterpart of the slave VM of the former system is represented by a subtask in this system. These subtasks may share the same address space as the master, and therefore communication of data between master and subtask consists of a simple address translation followed by single memory references. In our implementation we associate with each FPS-X64 a different subtask. The software for the data conversion and subsequent transfer to the AP are used as provided by FPS without modification.

It remains to describe the utilities that handle communication between "master" and "slaves". As mentioned, under VM, the vehicle for communication between VM's is provided as part of the VM/SP system, namely VMCF, thus reducing the dependency on system programming. Since use of VMCF requires calls to the system from assembler code, it is desirable to package this code in utility subroutines that can be invoked from normal FORTRAN code. The development of the first such set of utilities, called VMFACS,[1e] (Virtual Machine FORTRAN-Accessible Communications Subroutines), was one of the first steps in implementing our parallel system. Since that time we have experimented with other facilities such as the Inter-User Communication Vehicle (IUCV),[5] and have also written other sets of communication softwares intended to make the implementation of the communication protocol more automatic and user friendly.[8a] Under MVS, the address translation and single memory references have been packaged into a number of different communication programs, some from this laboratory, and others from within IBM. Among these are DNL, Paradigm, and the IBM Multi-Tasking Facility (MTF).[8b] Similar to these but still somewhat different are the routines we have implemented. Both our VM and MVS software are identical in function, though of course they differ in actual implementation. However, the similarities in function have enabled us to incorporate the same set of pre-compiler directives into our pre-processing program for both operating systems. This is extremely important in maintaining portability of code between the two systems. As implemented, parallel codes which run on the VM system may run equally well on the MVS system simply by pre-compiling the same source. Further details of our pre-compiler shall be described in the next section.

Another difference in executing parallel code in the two environments is concerned with subtask or VM initiation and synchronization. The latter, while operationally different on the two systems, are roughly equivalent in terms of efficiency and execution overhead. Moreover existing software in either case required little or no modification to achieve the desired result.

IV. PARALLEL PROGRAMMING STRATEGY AND CONSIDERATIONS

We begin with the observation that large-scale, typically CPU-bound, calculations almost invariably involve loops, either explicitly or in an equivalent sense, that are traversed many times. Most of the CPU time is consumed in such loops, so that if we adapt the tasks contained in these loops to parallel execution, we shall find that we actually have most of the code (as measured by execution time) running in parallel.

This is easy enough to accomplish. Let us suppose that our sequential FORTRAN code has an explicit or a logically inferred DO loop of the form

    DO 500 I = 1,N
    . . . .
    . . . . ,

with some computational kernel inside the loop (up to statement 500). Then, if we suppose that NCPU is the number of AP's available for parallel execution, we can keep the same computational kernel and modify the loop to read

    DO 500 I = ICPU,N,NCPU
    . . . .
    . . . .

This portion of the program, with the computational kernel and modified loop, is dispatched to each of the NCPU AP's. Each AP must of course have a different value for the index ICPU, with $1 \leq ICPU \leq NCPU$.

This fundamental scheme has been applied to most of the application programs we have migrated to parallel execution, and was effective in every case. Thus a typical program flow would consist of an initial sequential part handling initial input, setup, etc., followed by a parallel part running simultaneously on several AP's. At the end of this portion, the results from the parallel execution must be gathered up and processed by another sequential portion. This may be a prelude to another period of parallel execution, or, ultimately, to development of final results and the end of the run.

There is an obvious limit on this scheme: the computational kernel for a particular value of I in the loop example above must not depend on results computed in earlier passes through the loop with a different value of I. Our experience so far indicates that this is not a severe restriction; indeed, we find many codes tend to fall naturally into such a form. There are some "exceptions", of course, and, to start with, we have simply left the ones we have encountered in the sequential part of the code.

The procedure above constitutes a classical fork and join process whereby a master forks off a number of subtasks to an equal number of slaves and upon their completion joins the results. Accordingly our extensions to the FORTRAN language for parallel programming have centered around this concept. These language extensions have taken the form of precompiler directives which the programmer inserts in the original body of code.[9.] The fork directive takes the general form of: EXECUTE ON ALL/SLAVE: SUBROUTINE (arguments). It effectively initiates the execution of the subroutine named on either all the attached processors or a particular slave. Accordingly the partitioning of the DO loop described above necessitates the loop being restructured as a callable subroutine. Data to be passed to the routine (such as the loop index of the previous example) may be passed either as arguments of the routine or through common blocks. The join directive takes the general form of: WAIT FOR ALL/SLAVE. It effects a synchronization point within the master's sequential code for task completion either on all the attached processors or optionally on a particular slave. This directive is usually followed by one or several directives which specify the data to be returned as well as their subsequent treatment or merging conditions.

The precompiler, through which the modified code is processed, then generates the transfer or communication programs that will run on the slaves. This is effected through the translation of the above directives into the desired sequence of primitive communication routines alluded to in Section III. Its output is a pair of FORTRAN programs each of which is compiled and executed on the host and slaves accordingly. The module that runs on the attached processors, which contain the parallel subtask, are compiled and downloaded in a separate phase. As previously mentioned, all the directives are supported both under VM and MVS with almost total compatibility and portability.

V. GENERAL PERFORMANCE CONSIDERATIONS

Here we consider the performance of application programs on the lCAP systems modelled by using a rather simple but useful representation of a parallel program flow. In this model we assume that execution consists of three parts; namely a sequential part, a parallel part, and a part due to the overhead which is incurred during the parallelization process. The sequential portion of the program usually involves reading the input data, setting initial conditions etc. In our cases this time is usually minimal while the parallel routines are the most CPU consuming part of the program. The parallel routines usually involve loops or loop structures that are executed many times. The overhead refers to the time necessary to process the communication between processors. This communication is necessary to execute programs in parallel, but is not present in the sequential code. Therefore we may represent the time for parallel execution as the sum of three terms:

$$P = T_s + T_p + T_o$$

where T_s is the sequential time, T_p is the maximum of the times for the N processors to compute the parallel portion of the program, and T_o is the time needed for the overhead. In general, T_s is a constant regardless of the number of processors used, while T_p should decrease linearly with the number of processors, and T_o increases as some function of the number of processors.

The speed-up obtained by executing in parallel can then be defined as

$$S_p = S / P$$

where S is the time for sequential execution of the same program. The efficiency of parallelization can then be written as

$$E = S_p / N$$

where N is the number of processors employed.

In the ideal case, the speed-up will equal the number of processors, and exhibit an efficiency of unity. In practice, however, this is never realized. The speed-up that is actually observed is dependent upon the algorithm used, the parallel implementation of the algorithm, and the architecture of the parallel system. This is evident from the definition of parallel execution time. Optimal speed-up can only obtained when both T_s and T_o are zero. In addition, T_p must be exactly equal on all processors, and be 1/N of the total sequential time. In general, T_s is dependent upon the algorithm chosen, T_p is dependent upon the parallel implementation of the algorithm, and T_o is architecture dependent.

The reasons then that our observed speed-up curves will deviate from the ideal case may be summarized as a combination of three effects. First, any program has associated with it an inherent sequential time which cannot be parallelized which prevents obtaining 100% efficiency. This effect may be quite significant. If we recall Amdahl's Law for vector processing,[10] a code which is 50% vectorizable only runs 2X faster in vector mode than in scalar mode with a vector unit which is infinitely faster than the scalar unit. The same type of relationship holds for parallel processing also. The maximum speed-up that can be obtained with N processors is

$$\text{Max } S_p = N / (N - PAR(N-1))$$

where N is the number of processors and PAR is the fraction of the code which is parallelizable. If a program is 90% parallelizable, with 10 processors the maximum speed-up that can be obtained is 5.26, or 52.6% efficient.

Second, it is common that in executing a code in parallel certain processors will finish their assigned subtasks and be required to wait for the other processors to finish. This is commonly referred to as load balancing. The magnitude of this effect is a function of both the algorithm and the methods used in implementing it in parallel. Certain algorithms may be inherently load balanced while others may not (an independent loop which executes exactly the same number of instructions each iteration is perfectly load balanced, while a loop structure which contains a lot of branching dependent upon the input data may not be load balanced). If many of the processors spend a significant amount of time waiting for other processors, then this algorithm is not well load balanced. A related factor concerns the operation count. By this we mean that if we sum the number of operations on all of the slave processors, in general, this sum is at a minimum greater than or equal to the corresponding number of operations in the sequential program. Within these definitions a given parallel algorithm may be well load balanced, but not highly efficient due to the operation count.

Third, are the factors due to the parallel architecture employed and or the overhead associated with the parallelization process. Under the lCAP architecture this overhead includes the time necessary to fork and join the subtasks, and to transfer data among the processors. Also included in this time is the collection and merging of the data from the different processors. The overhead depends upon both the system softwares, hardwares, and the methods used to perform the fork and join. In the lCAP system this is accomplished with the Pre-Compiler as previously described. These operations are mainly of two types; host operating system overhead, and the data transfer from the host to the AP.

Thus far we have considered principles of a parallel implementation of general programs or algorithms. This type of approach may be useful in understanding the given characteristics of a particular job, but is limited in the information one can draw concerning the system in general. Ideally, we would like to have a mathematical model of our system which could be used to predict its performance. This model would be parameterized by such factors as the speed of the host CPU, the data transfer rate from the master to the slaves, the data transfer rate from the slaves to the attached processors, the speed of the host I/O, the speed of the AP I/O, the speed of the AP processor, the speed of the shared memories, etc. The information one might expect to draw from these models would be the amount of time it takes a job to go from one resource to another, the amount of time a particular resource is busy, the lengths of the queues at a given resource etc. This type of analysis is pertinent not only lCAP, but rather to the large body of systems now being built or under consideration. It can be useful in designing new systems (such as lCAP-3, which will be mentioned later), in upgrading older systems, or in understanding current system performance.

The detailed analysis of a parallel processing system is a very complex undertaking. Usually either analytic or simulation techniques are employed to investigate the manner in which resources are scheduled within the general operating environment.[11] In the first method equations which relate the system parameters to the performance measures are solved. There are several important restrictions in this scheme, namely the inter-arrival time distributions must be assumed exponential, simultaneous resource possession is not possible, priority queuing disciplines are not allowed, each waiting line must have an infinite capacity, the service times in a multi server queue must have the same service time distribution, the routing decisions must be specified by a set of branching probabilities and others. Further details on this approach are discussed in reference 11 and the references therein.

The second approach of simulation is the one we shall adopt to model lCAP. This method of solution is a statistical experiment which observes the behavior of the system as it evolves over time. The simulation may select arrival and service times and routing decisions through either tabulated data or random selection from relevant probability distributions. Complex queueing disciplines are permitted, simultaneous resource possession is allowed, multiple copies of a job can be created, job, chain and global variables are allowed. These features may permit an accurate representation of the system to be modelled when an analytic solution is not viable. More information on this method of solution can also be found in reference 11.

In order to begin to set up a model for our system we first must identify the "key" components of the system. From a basic model, more elaborate models may be built to help in understanding more complex and intricate details of the system. We begin by identifying the key components of our lCAP systems. We note the following important resources in lCAP:

1. The speed of the host CPU in performing the sequential portion of the code.
2. The operating system overhead of the host for dispatching and initiating the slave programs.
3. The rate of data communication among the master and slave tasks on the host CPU.
4. The operating system overhead involved in initiating and executing a channel program.
5. The data communication rate from the slave subtasks along the IBM channel to the AP.
6. The speed of the AP CPU.
7. The communication of data from the AP back to the slaves, and from the slaves back to the master for processing.

Notice here that we have not considered the shared memory systems, and that the description of the host CPU and the AP CPU are quite general. Here our objective is to begin with a simple model, and to test this model as it is being built by comparing the simulation results with those actually obtained. When these comparisons prove accurate, we then plan to attempt to model the system in a more detailed fashion. These studies shall help not only in understanding our current system, but we expect that they shall become an invaluable tool in helping us decide how to design future extensions and modifications to our system.

VI. APPLICATIONS

In the context of the above considerations, one can begin to better understand a parallel system and the bottlenecks associated between the system and a particular application program. As examples we consider several application programs that have been converted to parallel on the lCAP system. We shall briefly describe the programs and their performance results on the lCAP system.

We begin with several codes developed within this laboratory. Many are concerned with the statistical and quantum mechanical study of biological or chemical systems. More recently, we have expanded our interests to include engineering applications, and one such application will be briefly discussed. Finally, we will examine several programs that were acquired through our visitor's program.

In general most of the programs to be discussed are well documented in the literature. We shall make no attempt to describe these programs in detail, but simply give enough of a description to allow the reader to understand the parallelization process and the results obtained. For more details concerning both the theory and programs, the reader is referred to the references.

The first group of programs are concerned with determining fundamental properties of matter from first principles. This is approached from a molecular point of view at two levels of detail; microscopic and macroscopic. The first is represented by our quantum mechanics code which attempts to describe the properties of a single molecule by performing self-consistent field (SCF) calculations under the Roothaan-Hartree-Fock method.[12] For the purpose of this discussion the program can be partitioned into two time consuming parts. The first calculates the values of the electron-electron interactions and the second part the SCF wavefunction.

In the integrals program the calculation of any one integral is independent upon the remaining integrals. The computed integrals are stored on disk files which can be larger than a Gbyte in size. While the evaluation of any one integral is fast, the computation of a medium sized molecule involves the evaluation of millions of integrals. Depending upon the structure of the program usually the loops run over either the atoms in the molecule or over the contracted functions. Therefore, by partitioning the loops as previously described, different groups of integrals are computed on different processors. The only difficulty associated with this process is that each atom or contracted function usually involves the computation of different numbers of integrals. Therefore, the problem is not load balanced. The measured speed-up of the parallel integrals program on the lCAP system is shown is Fig. 4a. This figure indicates also the degradation effects due to the sequential portion of the code, the communication overhead, and the effect of load balancing the parallelization process. It is the latter effect that is most dominant in this application. There does not exist any overhead associated with the parallelization process. This is due to the fact that each of the subtasks never need to communicate with the other subtasks. The program is more than 99% parallelizable, thus little degradation is observed due to the sequential portion of the code.

In the SCF program the majority of the time is spent in constructing the Fock matrix from the integral file(s). Therefore, the parallelization is obtained by processing each integral file in parallel, with each processor computing his own Fock matrix; all of which are then added together to get the total Fock matrix. Note here that since each AP has its own local disk with its own local integral file that the I/O is also performed in parallel. The diagonalization of the Fock matrix is left in the sequential portion of the program. The results are given in Fig. 4b, and are a bit different from those with the integral program. The deviation of the observed curve is again due in most part, to the load balancing of the problem. This follows since the integral files differ in size, thus the compute load on each processor varies. However, there are also significant contributions due to the sequential portion of the program (3%) and to the overhead. Here the majority of the sequential part of the program is the diagonalization of the Fock matrix, and the overhead encompasses the transfer of the Fock matrix back and forth to the subtasks for each iteration. Note that even though data transfer must be incurred, each iteration that this overhead is not the most significant contribution to the degradation. The diagonalization of the Fock matrix is relatively short compared to its construction; thus there is little loss by leaving the diagonalization in the sequential portion of the code.

The integrals and SCF programs described above are generally used to investigate the ground state electronic configuration of a molecule without including electron correlation. Various procedures are commonly used to include to effects of electron correlation into calculations, one of them being configuration interaction (CI).[13] In the next few paragraphs we wish to discuss the parallelization of the Multi-Reference single and Double Configuration interaction (MRDCI) program package as developed by R.J. Buenker. The theory of the CI technique has been discussed elsewhere; here we just wish to present a brief description of the program and the scheme used for parallelization.

The MRDCI program is divided in several steps; 1). the integrals calculation, 2). symmetry detection and transformation to the symmetry basis, 3). SCF calculation, 4). molecular integral transformation, 5). configuration generation, reference CI and configuration selection, 6). CI with the selected configurations, and 7). diagonalization of CI matrices. In spite of the fact that there are many components to the CI calculation, there are only three basic algorithms that have been parallelized. Note that these algorithms are in addition to the integrals and SCF program previously described.

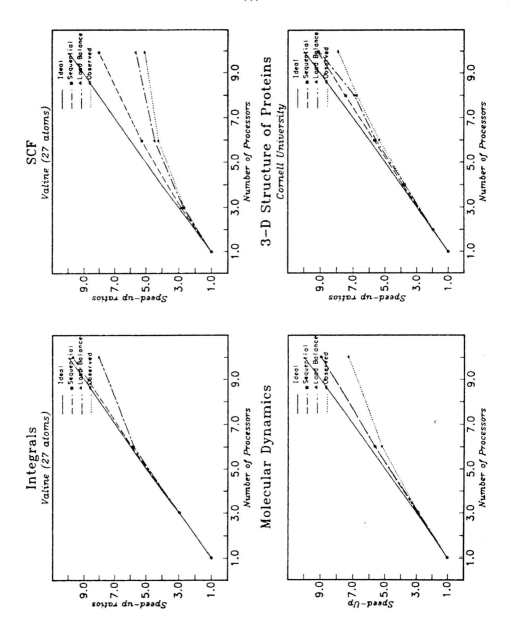

Figure 4.   a. Speed-up curves for the integral portion of <u>ab initio</u> molecular program. The results are shown for valine (27 atoms) with a minimal basis set.   b. Speed-up curves for the SCF portion of the <u>ab initio</u>

molecular program. The results are shown for valine (27 atoms) with a minimal basis set. c. Speed-up curves for a 100 steps of a Molecular Dynamics simulation of 512 water molecules. d. Speed-up curves for 1000 minimizations of the 17 dimensional surface characterizing the glutamine-leucine conformational space.

The first of these algorithms is a two electron integral transformation. This algorithm is used both in the symmetry and molecular transformations; the only difference being the nature of the coefficients used for the transformation. In the symmetry transformation a few number of integrals multiplied by unity factors (plus or minus one: only dihedral symmetry is allowed) contribute to the value of a symmetry adapted integral. On the other hand in the molecular transformation all the integrals in a given symmetry contribute to each molecular integral with real number weights, namely, the MO vectors from the SCF.

The transformation of the two-electron integrals is a two step process with each step transforming two of the four indices that label each integral. At the beginning of each of these two steps the integrals can be considered as randomly distributed over several files each on one of the processor's local disks. Then, the steps for the transformation are as follows; - each processor performs a "bin sort" with the integrals it has on its disk. The purpose of this sort is to put together all the integrals with a common pair of indices, and prepare the "core loads". A "core load" will contain all the integrals of the form (ij/...) for several pairs of indices ij; for each of these pairs the integrals are arranged in canonical order, i.e. (ij/11),(ij/21),(ij/22),...,(ij/nn). As each processor has only a part of the complete set of integrals, the core loads obtained by each processor are only partially full. Each processor fills its core load with the integrals it has, leaving empty holes (zeros) where other processors will put their integral values. In the following step the master gets a given "core load" from every slave, and obtains the complete "core load" by adding the contributions from all the processors. Once a core load is completed it is assigned to one processor for the transformation step. The master repeats this process for all the "core loads", and every time all the processors have a "core load" to transform, the transformation step is performed on all processors in the usual fashion, but in parallel.

The transformed integrals are not sent back to the master, but are kept on the disk on the processor ready for the next integral transformation, except for the last step in the molecular transformation when each slave sends his transformed integrals back to the master and a file containing all the integrals is generated.

The reason for this file is that the CI matrix computation requires the integrals in canonical order. In this way a four index label identifies the position of the integral in the file. When the computation of the CI matrices requires that the file be split among all the processors, the master then knows where in the file to find the integral, and which processor has control of that integral.

The second of these algorithms is the CI matrix calculation. In the MRDCI configurations are classified according to their excitation degree into "supercategories". The CI elements corresponding to the interaction between two supercategories are computed together, so that interactions of the same degree of complexity are grouped together. Therefore the parallelization is achieved by distributing the computation of interaction elements between two supercategories among all the processors. When computing complete CI matrices, i.e. the reference space and the selected space CI matrices, each processor receives all the configuration but it only computes the interactions of one out of every "N" configurations where N is the number of processors. On the other hand, when computing the interaction of all the generated configurations with the reference space, all processors receive the reference configurations and one contiguous sub-range of configurations.

The actual element computation proceeds in three stages that are parallelized independently. First the configurations are examined and two files are generated: the formula tape with information of the type of interaction and the label tape with the labels of the integrals required to compute that interaction. This part is completely parallel by itself and does not require any communication between processors. In the next step, the label tapes have to be read and the integral file searched for the corresponding integrals. As every processors needs integrals from all other processors, one list is processed at a time and sent to every processor for the sorting. The master collects the partial lists provided by the processors and puts them together to obtain the integral list that is sent to the processor that originated the label tape. Finally, in the last step the formula tape and the integral list are put together to generate the CI matrix elements; each processor working only on his share of the matrix elements.

The final algorithm which was parallelized in the MRDCI program is the diagonalization of the CI matrices. Single vector diagonalization algorithms generate an improved vector from the product of an initial guess vector and the CI matrix. As the CI matrix is scattered over all the processors (remember that each processor only computed his own part of the CI matrix), each of them receives the whole guess vector, multiplies it by the integrals it has in its file and gives back its incomplete product vector. The master collects these vectors adding them together in one. This is used to generate the improved solution that is iterated again.

Preliminary timing results for the various portions of the MRDCI program are given in Table 1. These results are reported for a computation on the water dimer with 70 orbitals, an active space of 66 orbitals with 1 reference configuration, 14639 states before the CI selection, and 7773 states in the CI diagonalization. In general the performance on the lCAP system is not as efficient as the integrals and SCF programs. The various different parts of the program realize different performance measurements, but overall the speed-up with 4 processors is 2.61, or 63% efficiency. The worst speed-up is observed for the molecular integral transformation, where only a factor of 1.3 with 4 processors is observed. This degradation is due mainly because in the integral transformation each processor must contribute data for the computation of the remaining processors. Therefore, even after the first processor receives his strip of integrals to transform, he must still provide data for the remaining processors and therefore cannot begin the transformation step. This degrades the parallelization in two ways, first no processor can begin the integral transformation before any other processor, and secondly there is a lot of data transfer back and forth among the slave processors. At the time this program was implemented on lCAP there was no mechanism for inter-processor communication except through the master; therefore the amount of communication became a bottleneck. With the recent addition of the shared memory, we expect that the performance of the integral transformation step shall be considerably improved. Work is in progress to implement this portion of the program utilizing the shared memories.

As stated previously, much of our work in chemistry and biophysics also deals with estimation of macroscopic properties. This class of programs utilizes a molecular dynamics approach to simulation.[14] Here the kinetic motion of molecules in bulk liquid or solution is studied as a function of time. The time period for the simulation is divided into many smaller time steps, each step requiring the calculation of the potential energy and inter-molecular forces. In these codes the calculation of the energies and forces constitutes the most time consuming portion of the program. Parallelization is obtained by assigning each processor an equal share of the energies and forces to compute, and then collecting these from each processor to generate the next step. The results for a two body molecular dynamics

Table 1. Preliminary timings for the MRDCI program package on lCAP-1 without shared memory or the fast bus. Results are given in seconds for the water dimer with 70 orbitals (active space = 66 orbitals), 1 reference configuration, 14639 states before selection and 7773 states in the CI diagonalization. The speed-up factors for each portion of the program are given in parenthesis.

	1AP	2AP's	4AP's
Symmetry transf.	1664	1038(1.6)	657(2.5)
SCF	353	202(1.8)	140(2.5)
Molecular transf.	680	528(1.3)	530(1.3)
Conf. selection	58	49(1.2)	53(1.2)
CI matrix	3011	1695(1.8)	887(3.4)
CI diagonalization	1172	643(1.8)	392(3.0)

simulation with 512 water molecules are given in Fig. 4c. This figure indicates that the degradation observed is due only to two effects. Foremost is the overhead associated with the transmission of data by all processors at every time step. Here the effect of the overhead is greater than in the SCF because the time per iteration is much less. In the SCF each iteration is of the order of a few minutes, whereas for the MD each iteration is of the order of seconds. The second effect is the sequential portion of the program, which while only 2% of the program, still yields a maximum speed-up of only 8.95 with 10 processors. The load balancing of this problem is quite good since we have identical molecules and each pair interaction takes equal amounts of time to compute.

Another program which we would like to discuss was originally written by H. Scheraga and co-workers at Cornell,[15] and has now been modified for parallel execution on the lCAP system. It is now currently being used in a collaborative research effort. The program computes the native conformation of a protein by minimizing its empirical conformational energy. The methodology and program have been previously described elsewhere.[15] Here we shall just say that polypeptide chains are built up from smaller segments by minimizing a given set of starting conformations which encompass the multi-dimensional space of the polypeptide. Often thousands or ten of thousands of minimizations must be performed in order to ensure that the multi-dimensional space of the polypeptide is properly analyzed. Each of these minimizations is independent from all remaining minimizations, and therefore this program naturally lends itself to parallel computation.

At this point we would like to stress one point which we have found to be of great importance in migrating codes from sequential to parallel. Previously we explained the obvious method of breaking a loop structure to allow parallel computation. In general, this method is efficient only if each iteration of the loop executes for approximately the same amount of time, or the loop structure is inherently load balanced. For the integral, SCF, MRDCI and molecular dynamics codes this is true. However, for applications where each iteration of a loop does not entail equal amounts of time this procedure is often not very efficient. To achieve load balancing in these cases may require that each processor performs different numbers of iterations of a loop, such that the total time spent computing for each processor is equal. The protein program is one example of such a code. Each of the minimizations is performed as an iterative process running until convergence. Sometimes convergence can be reached in a few iterations, but at other times it is necessary to run for many more iterations.

The protein program, and other programs with these same characteristics are optimally parallelized using what we have termed the "parcel policy". In this method the total amount of work is divided into M parcels where $M >>> N$

processors. Each processor is given an initial parcel to work on; the first processor to finish his assigned work begins working on the next available parcel. This process is repeated until all of the parcels have been assigned. In this manner the time spent in computation is most evenly and dynamically balanced.

The results for the protein program on a rather small example of the glutamine-leucine dipeptide are given in Fig. 4d. This example consisted of 1,000 minimizations of the 17 dimensional space. In this example all three effects, the sequential portion, load balancing, and the overhead, contribute to the overall performance degradation, though the relative amounts are dependent upon the case (number of processors) considered. The sequential portion of the code accounts for almost half of the degradation. The other half is due to either the balancing of the computations, or the overhead involved in the data transmission. The relative weights of these two contributions oscillates from one to the other depending upon the number of processors used. This indicates that the overhead and load balance play almost equal roles in the degradation depending upon the number of parcels and the size of the parcels.

This program has recently been used to predict the low energy conformations of all 400 naturally occurring dipeptides.[16] This information is also currently being used as a database for building a much larger protein, interferon. Interferon is a 155 residue protein believed to be important in combating viral or bacterial infection. Preliminary results for this work are discussed by H. Scheraga in this volume.[17]

To this point we have restricted ourselves to discussing scientific applications in the area of chemistry and biophysics. As stated previously, we are interested in developing a general purpose parallel computer for scientific and engineering applications. With this in mind we now discuss one application in the engineering realm that is of importance in oceanographic studies.

The shallow water equations are a means of studying the dynamics of a two-dimensional, incompressible, barotropic fluid under the hydrostatic approximation.[18] They are based on Newton's Law of Motion and the continuity equation (which ensures conservation of mass in fluids). They may be discretized by the finite difference representation over a grid of the physical system and then solved numerically. Parallelization can then be achieved by dividing the grid equally among the different processors. Each processor computes his own portion of the grid and the final result is obtained by merging the results of all processors to give the total grid.

Usually the evolution of the system over a period of time is studied; this integration is performed by dividing the total time into smaller time steps. At time T(i), results from time T(i-1) are needed. In particular, each processor needs the neighboring boundary values of the grid from its neighboring processor(s). As a result, while the numerical solution of the grid is highly parallelizable, there is a necessity for frequent data transfer among the processors. In particular, in the example studied here the computation time for each time step is of the order of 1.0 second (with 1 processor). Therefore, with 5 processors data transfer must occur on the average every 0.2 seconds. Further details concerning the program and the example studied are available in Ref 18.

The first results observed for a parallel version of this program modelling a small gridded region are given in Fig. 5a. The observed speed-up increases for both 2 and 3 processors, but thereafter increase only slightly, then actually begins to decreases. With 5 processors the maximum speed-up is obtained (2.03), but this is only 40% efficient. From Fig. 5a the reasons for this effect are obvious. The primary factor causing the degradation in this program is the overhead associated with the slave processors; the sequential portion of the code and the load balancing are quite favorable. By transferring data among the slaves through the

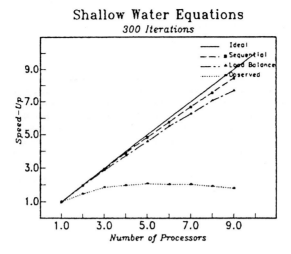

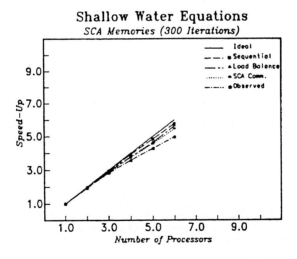

Figure 5.   a. Speed-up curves for 300 iterations of the shallow water equations, discretized by the finite difference method. These results are without the shared memory, and use the master on the IBM host as the

transfer point for all data. b. Speed-up curves for 300 iterations of the shallow water equations discretized by the finite difference method. The communication between FPS-164's that is necessary each iteration is performed through the shared memory.

master program several problems arise. First there is only one master servicing all the slaves.   Therefore each slave effectively becomes dependent upon all the remaining slaves.   Secondly is the data transfer itself. Since in this application program we are not using the SCA shared memories there is no direct communication between APs. Therefore, in order for one AP to transmit data to another AP, the following path must be taken: 1). transmit data from the AP along the IBM channel to the slave, 2). communicate the data from that slave to the master, 3). have the master communicate that data to the second slave, 4). finally the second slave must transfer the data along the IBM channel to the second AP. This rather cumbersome path can be very time consuming; in fact just considering channel speeds the maximum data transfer rate between APs with this method is 1.5 Mbytes/sec. In applications where the the computation time between data transfers is short, this rate becomes an extremely important factor.

In Fig. 5b results for the same program, using the alternative SCA shared memories as communicator paths are shown.   As indicated, there is a marked increase in performance with this modification. In fact, there is almost no degradation due to the data transfer using the SCA memories. Now, almost all the degradation is due to the operating system overhead. The question to be addressed is why is there such a drastic difference between the performance of this code with and without the shared memory.

The most simple answer would be that the shared memory has a significantly higher data transfer rate than the IBM channel.   As previously stated, the maximum transfer rate between APs using the channel is 1.5 Mbytes/sec, while with the SCA memory it is 22 Mbyte/Sec. However, due to overhead, the observed transfer rate for the small data packets of this program with the SCA memory for this program is less than 1.0 Mbyte/Sec, i.e., of the same order of magnitude as the IBM channel. Thus a more detailed examination of the two parallel is warranted.

First, the amount of data transferred in the two programs is drastically different. In the first example the standard FPS software for the Auto Directed Calls (ADC) mode of operation was used; it only allows transfer of arguments or common blocks. The transmission of pieces of arrays or matrices is not possible (unless one modifies the Host AP Subroutine Interface, or HASI, produced by the FPS cross-linker).   Therefore, in the first version of the program each processor had to transfer its entire block of the whole grid rather than the much smaller boundary data needed. Using the SCA memories each block transfer is specified by an address and a length according to the programmers discretion. This makes it possible to transfer only parts of a large grid.   However, again this seems not to be the dominant factor. Results indicate that by writing our own HASI, User Directed Calls (UDC mode), for the FPS data transfer, the observed results would still not be as impressive as those obtained with the SCA memories. This is due in part to the overhead which must be incurred in initiating data transfer from the AP through the host and back up to the AP again. For data lengths of the size required here, most of the time is overhead; by reducing the data size we would not expect to gain much in time. So while for the different size data transfers the rates for the IBM channel and the SCA memory are similar, we could not reduce the time for data transmission with the IBM channel much further by reducing the lengths of the data. Therefore, we expect that the modifications to the HASI would produce almost the same results as with ADC mode. The final answer seems to be that the data transfer in the first case must proceed through the master and each slave must in turn wait for all the rest. By using the SCA memory there is essentially no master in the program. Each slave only waits for those slaves from which it

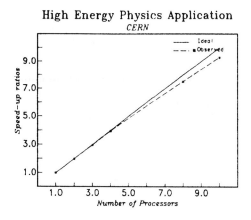

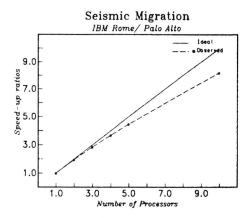

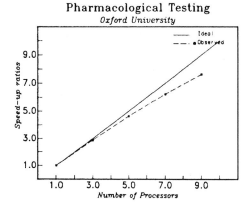

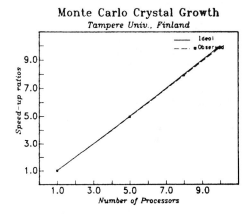

Figure 6.   Speed-up curves for four application programs that have been migrated to parallel via the Visitor's Program. These applications include: a). a high energy physics application from CERN, b). a

seismic data analysis code from IBM Rome/ Palo Alto, c). a pharmacological drug design code from Oxford University, and d). a Monte Carlo code simulating crystal growth from Tampere University, Finland.

requires data, and only at the very end of the computation must they all wait for each other.

For the purpose of brevity we cannot discuss in detail many of the applications that have been acquired and run on our system through the visitors program. This program has been very successful in drawing scientists from both academia and industry to experiment on our parallel computer. We would like, however, to show the performance results of 4 such application programs that have been converted to parallel. These are given in Fig. 6. The first is a high energy physics application from CERN (this application is not vectorizable, but is highly parallelizable), the second is a seismic application from IBM Rome/IBM Palo Alto (originally thought to be very vectorizable but not parallelizable), third is a theoretical chemistry program from Oxford University used in pharmacological testing/drug design and fourth is a Monte Carlo program from Tampere University in Finland which is used to study crystal growth. These four examples are only a few of many different applications that have been successfully parallelized on lCAP. More detailed information on these results, or on the visitors program in general can be obtained by writing the authors directly.

Let us now make a few comments concerning lCAP-2. The results presented thus far are all derived from tests made on lCAP-1. The more powerful former system has recently been made operational and we have begun to perform some of the equivalent tests. The results of these tests, given in Table 2, indicate that the migration of the codes from lCAP-1 to lCAP-2 does not result in any appreciable degradation in parallelism, but results in performance roughly 3 to 4 times greater than that of lCAP-1 without MAX. It is also important to restate that the time necessary to migrate a particular code from lCAP-1 to lCAP-2 has been found to be minimal since the system softwares are as much as possible made to be compatible. Finally, for comparison, Table 2 also gives the elapsed standalone time for the same programs run on one processor of a Cray-XMP. For these applications, lCAP-2 with three FPS-264 processors working in parallel is roughly equivalent to one processor of a Cray-XMP.

Table 2.   Preliminary timings on lCAP-2 with 3 FPS-264's. Results are given in minutes, with the speed-up given in parenthesis.

Time (minutes)

JOB	1AP	2AP's	3AP's	Cray-XMP
Integrals	19.1	9.9	7.3	7.6
(27 atoms)		(1.93)	(2.62)	
SCF	13.6	7.3	5.5	6.8
(27 atoms)		(1.86)	(2.47)	
Molecular	29.9	15.8	11.0	19.8
Dynamics		(1.89)	(2.72)	

VII. STATISTICAL MECHANICAL AND AB INITIO STUDY OF DNA

The computing power offered by the ICAP systems offers us the opportunity to investigate problems in theoretical chemistry and biophysics previously beyond the available computer resources. Two such examples, which has been of interest to us for many years are a study of the hydration pattern and ion positions in DNA via Monte Carlo techniques, and an <u>ab initio</u> quantum mechanical study of the hydrogen bond bridges in DNA base pairs. Both of these problems shall be briefly discussed.

It is well known that water plays an important role in stabilizing the structure and conformation of nucleic acids; it is equally well known that DNA, being a polyelectrolyte, requires the presence of counter-ions for stability.[19]

First we recall that there exists a huge amount of experimental information concerning the structure determination of DNA. However, it has only been recently that X-ray structures of single crystals containing a few water molecules have been obtained,[20] no experimental diffraction pattern for a DNA polymer in solution has been reported. We have performed Monte Carlo simulations at 300K for three full turns of a B-DNA fragment for a variety of relative humidities (starting with one water molecule per nucleotide unit and increasing progressively up to 25, or 1500 water molecules in total) with as many counter-ions as needed to neutralize the phosphate groups. The details of the simulation, as well as the complete data, are collected elsewhere;[21] here we just briefly describe some of the results.

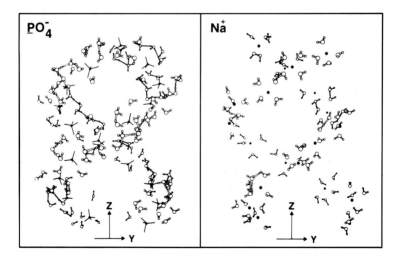

Figure 7.   Patterns of hydration around the phosphates and Na$^+$ counterions for B-DNA at high relative humidities and 300K.

These results are presented from two views; a collective viewpoint and a local viewpoint. From the local viewpoint, the computed orientation of the water molecules hydrating the PO$_4$ groups is with the hydrogen atoms toward PO$_4$ and away from Na$^+$. This is shown in Fig. 7, where the results for the Monte Carlo simulation obtained with 1500 water molecules are given. Here, on the left the hydration of the PO_4 groups are shown, and on the right the hydration of the Na^+ counterions. The PO_4 groups are represented by four bonds from a point, while the Na^+ ions are shown as full dots. Some of the data obtained from these computer simulations have been later confirmed by X-ray data; however, as previously noted, no water-water hydrogen bridge has been experimentally detected up to now, since the X-ray resolution is too low to allow for reliable measurements of the orientation of water molecules. However, the approximate determination of the position for some of the water molecules allowed experimentalists to guess the water molecules orientation, and thus to discuss hydration patterns.

From the collective viewpoint we notice that the water molecules are organized into two predominant patterns formed by filaments of water molecules; one water molecule hydrogen bonded to the next one. These filaments are found at the two grooves; one pattern is nearly a roto-translation along the valley of the minor grooves, and the second pattern parallels the z axis (the long axis of DNA) running across the major groove and connecting PO_4 groups belonging to different strands (see Fig. 8). Again these findings seemed to be confirmed by recent X-ray results. In summary, the connectivity pathways fully envelope DNA and its counterions: the structure of DNA includes these hydrogen bonded structures of water molecules, which are as essential to a proper description of DNA as are the phosphate groups, the base-pairs, or the counter-ions.

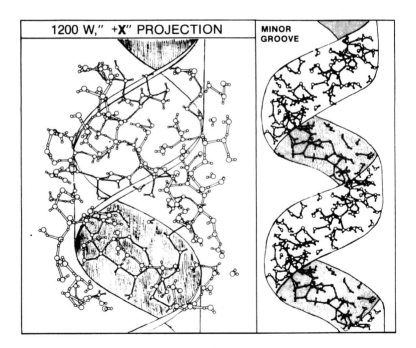

Figure 8.   Patterns of hydration in B-DNA at the major and minor grooves.

Concerning the counterions, in an approximate analysis those hydration sites which are highly stable and have the water's hydrogen atoms pointing toward DNA, are, in principle, good candidates for counterion-binding sites. This is obvious based simply on electrostatic considerations. However, the range of the ion-ion interaction is much greater than that of the water-water interaction. Whereas in water-water interactions the interaction is nearly zero at 9-10 Å, two counterions strongly repel each other at these distances. Ionic hydration cuts down on this repulsion; water molecules hydrating an ion not only add stabilization to the water-ion system, but it also decreases the ion-ion repulsion as a result of the screening effect.

In summary, there are several fundamental aspects in the hydration of DNA. The addition of water to the DNA system decreases the phosphate-phosphate repulsion by screening and stabilizes the entire system. In addition, counterions condense around DNA, further decreasing the phosphate-phosphate repulsion and creating electrostatic stabilization. Water also solvates the counterions, thus decreasing the ion-ion repulsion that would destabilize the system.

A second, yet related study in biophysics concerns the study of proton transfer in the DNA base pair Guanine-Cytosine (GC). In 1963, Lowdin hypothesized that hydrogen bonded protons in DNA base pairs can be transferred from one base to its complementary base by quantum mechanical proton tunnelling.[22] This proton transfer would then lead to a pair of tautomeric bases that could possibly be read incorrectly, resulting in an altered message from the DNA code and leading, possibly, to genetic mutations. Since that time there have been a number of studies aimed at elucidating the shape of the potential energy surface for the hydrogen bond bridges in the DNA base pair Guanine-Cytosine (GC).[23-31] Most of these studies have been undertaken with one major goal in mind; namely to verify the Lowdin hypothesis by searching for a a double potential well on the surface describing the hydrogen bridges. Due to the large number of atoms in the chemical systems of interest, these computer simulations can be extremely complex, especially if one attempts to represent realistically the field of the DNA polymer, for example, in the space in between the bases of a pair. Indeed the application of quantum mechanics to this particular problem can be used as a testing ground for exploring the limits of today's quantum mechanical techniques.

In earlier molecular orbital studies of the ISOLATED GC pair, which assumed separability of the π and σ electrons, both single and double proton transfers were found to be characterized by double potential wells.[23-26] However, more recent semi-empirical investigations on the GC pair[28,29] and ab initio studies of the formic acid dimer[27] (a prototype molecule for double hydrogen bonded systems) have found that single proton transfers are characterized by single well potentials, while double proton transfers are characterized by double well potentials.

While most studies now seem to be in agreement as to the shape of the potential surface for the isolated base pair (single wells for single proton transfers and double wells for double transfers), the applicability of these results to base pairs in DNA is still very unclear. First, in DNA the bases are stacked on top of each other. The effect of the field induced by this stacking is obviously absent in the isolated base pair. Secondly, in DNA the bases are linked by sugar and phosphate groups, and are in an environment characterized by counter-ions and water. These effects cannot be neglected when considering a DNA system, and therefore one should exercise caution when extrapolating results for an isolated base pair towards a base pair in DNA.

Calculations have been performed in order to clarify some of the points left unanswered from previous work. The potential energy surface for the GC pair, including the effects of partial geometry relaxation in the region involved in the proton transfer, the stacking of the base pairs, the sugar and the phosphate groups, and finally, the counter-ions can be nearly routinely calculated. There remain, however, notable approximations, namely the neglect of electron correlation effects, and the truncation of the basis set, which is far from the Hartree-Fock limit. The reader is referred elsewhere for the additional details concerning the computations;[31] here we would like to note that the largest system computed consists of 98 atoms and is one of the largest systems computed with <u>ab initio</u> methods.

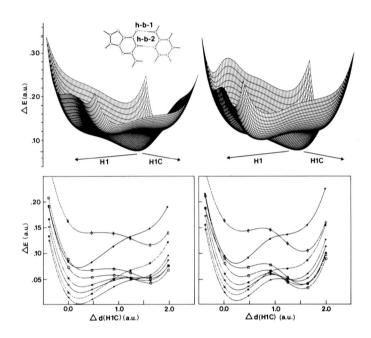

Figure 9.   Three and two dimensional representation of the potential energy surface for the coupled motion of the h-b-1 and h-b-2 hydrogen bonds in the Guanine-Cytosine base pair, with a minimal basis set (left) and with a larger double zeta basis set (right).

The potential energy surfaces for the coupled motion of two protons in the isolated GC system are given in Figs. 9, and 10. In each figure, the top plots indicate three dimensional surfaces for these motions, while the bottom plots are the two dimensional curves from which the three dimensional surfaces are derived. In all diagrams, the energies are represented along the y axis in a.u., and the distances are measured in a.u.; Δ R represents the displacement of the hydrogen atom from the standard N-H bond length. Negative values for Δ R correspond to a shortening of the N-H bond length, and positive values to a lengthening of the N-H bond length. In each figure a label is given for the hydrogen atoms being considered.

Fig. 9 shows the potential energy surfaces for the coupled motion of h-b-1 and h-b-2 (Fig. 9a) yields a double minimum, while the motions of h-b-2 and h-b-3 (Fig. 10a), and h-b-1 and h-b-3 (Fig. 10b) do not give double minima. In addition, curves denoted by a ■ in Figs. 9a, 10a, and 10b (which represent the motion of only a single proton) show that the transfer of a single proton does not occur on a double minimum potential.

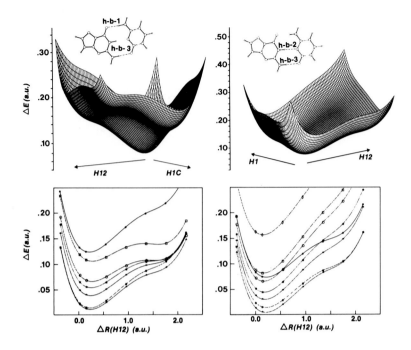

Figure 10.   Three and two dimensional representation of the potential energy
surface for the coupled motion of the h-b-2 and h-b-3 (left), and h-b-1
and h-b-3 (right) hydrogen bonds in the Guanine-Cytosine base pair.

The surface given in Fig. 9a was also calculated using the larger 9/5 basis set, and
is presented in Fig. 9b. Qualitatively, the results are very similar to those obtained
with the minimal basis set. There is most definitely a double minimum for the
coupled motion of h-b-1 and h-b-2. In addition, this second minimum is a bit deeper
than calculated with the smaller basis set, and as a result the corresponding barrier
height is slightly higher.

The effect of the partial geometry optimization has also been considered.
Qualitatively the same result is obtained: there exists a double minimum for the
coupled motion of h-b-1 and h-b-2, but no additional double minima are found.
Quantitatively, the effect of the partial geometry relaxation is only seen in the
region where $\Delta R \cong 1.0$; this effect is to further lower the energy of the tautomeric
form, thereby, increasing the barrier height.

The effect of the field induced by the stacking of the base pairs has been
investigated by performing calculations on a system of three stacked base pairs
with the same relative orientation as in B-DNA (the distance between stacks is 3.34
Å and one base pair is rotated 36 degrees relative to the next

S–P–S–G...C–S–P–S (H1C/H1) 7/3 B.S.

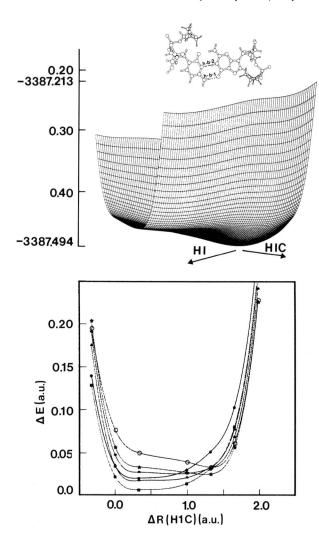

Figure 11.   Three and two dimensional representation of the potential energy
surface for the coupled motion of the h-b-1 and h-b-2 hydrogen bonds in
the Guanine-Cytosine base pair, now including both the sugar and
phosphate groups.

one). In these calculations the hydrogen bond bridges of the middle base pair were examined while fixing the hydrogen atoms in the stacks above and below this pair. The calculated potential energy surfaces are qualitatively similar to the ones obtained for the isolated GC pair. For this reason, we shall not present the corresponding surfaces but simply note that as expected the stacking affects the depth of the minimum corresponding to the tautomeric forms.

In DNA, the base pairs are not only stacked, but linked together by sugar and phosphate groups in the helix. The geometrical nature of the helix necessitates that the base pairs no longer be planar, but rather are twisted in opposite directions along their long axes to create what is referred to as a "propeller twist". This propeller twist ensures maximum overlap of bases along each individual backbone strand. It is found in studies of both fibers and single crystals, though the magnitude and sign of the twist angle is still not completely resolved.[32] When the sugar and phosphate groups are included together with the base pair, we begin to have a helical system large enough to be used as a model for the DNA double helix. However, a GC pair, either alone or stacked, attains its minimum energy conformation in a planar form; in the DNA double helix this is not the case. Therefore, the effect of the sugar and phosphate groups, and the counter-ions can be evaluated by performing another separate and consistent set of calculations using the geometry of the base pair from the DNA helix.

The effect of the attached sugar and phosphate groups and the presence of the K^+ counter-ions on the hydrogen bond potential of the GC base pair is examined by computing potential energy curves and surfaces concomitant to motion of the hydrogen-bonded protons with and without the presence of the sugar and phosphate groups and the counter-ions. The systems considered can be briefly described as sugar(1)-phosphate(1)- sugar(2)-G-C- sugar(3)-phosphate(2) -sugar(4) (GCSP), with the later addition of two K^+ counter-ions (GCSPK). The position of the K^+ counter-ions have been taken from the previously mentioned Monte Carlo studies on DNA hydration.

The potential energy surface and two dimensional curves corresponding to the coupled motion of the h-b-1 and h-b-2 protons are presented for the GCSP and GCSPK systems in Fig. 11. In these surfaces the double well has disappeared and it has been replaced by a single flat well. The presence of the single flat well indicates that the hydrogen bonded protons are not energetically favored to be attached to a particular base, and can rather easily be transferred from one base to the other. Based on this fact, a tentative conclusion is that there seems to be no need to hypothesize proton tunnelling in order to transfer the proton from one base to the other. The use of the partially optimized GC bond lengths does not bring the double well back, nor does the presence of the K^+ counter-ions. counter-ions. These results support the conclusion that the characteristics of the potential energy surface for the hydrogen bonded protons in a GC base pair with a given geometry are largely determined by the interactions with the base pair itself, and that the surrounding environment does not deeply influence directly the behavior of these protons. Of course the environment indirectly influences the computed potential energy surfaces because it is the environment which is responsible for the geometry of the base pair (either planar or non-planar). Further computations are now in progress in order to clarify the above points.

VIII. ON-GOING AND FUTURE EXTENSIONS

From the previous discussions it appears that the lCAP architecture is well adapted to a large and diverse set of scientific/ engineering applications. However, our systems are experimental, and continually evolving in terms of both hardware and software. In this section we wish to present both planned and potential extensions and modifications to lCAP.

First, recall that we plan to expand the 2 shared memories to 5 by the end of 1985 on lCAP-1. This configuration is shown in Fig. 2. Some of the advantages of this particular configuration have already been mentioned: increased fault tolerance, ring computations with 10 processors, aggregate data transfer rates of 110 Mbyte/Sec between APs (5 x 22 Mbytes/Sec), and increased flexibility. This same configuration, with these advantages will also be installed on lCAP-2 by the middle of 1986. This will also have the added merit of maintaining compatibility between our two systems.

The disadvantage of these configurations is that not all of the APs are directly connected to each other. Therefore, in certain instances the transfer of data between two APs must proceed through an intermediate AP/shared memory. Not only does this degrade performance, but seriously complicates programming. To resolve this problem we plan on additionally adding 2 large 512 Mbyte shared memory systems (1 for lCAP-1 and 1 for lCAP-2), each capable of attaching up to 12 FPS-X64's. This shared memory system will have the same hardware and software specifications as their predecessors, and shall be ready by the middle of 1986.

A second but complementary approach to modify our "loosely" coupled array of processors towards a more tightly coupled system is the incorporation of 2 (one for each system) fast common buses that will link all of the FPS-X64's on a particular system into a ring configuration. This bus, designed and tested by FPS, is 32 bits wide and performs at rates of 32 Mbyte/Sec node to node, and 22 Mbyte/Sec from the bus up to the FPS-X64. A prototype model of this bus, and it associated softwares was delivered to us in November 1985. We are in the process of investigating its possible uses. We expect that this bus should complement the use of the shared memories. It should be valuable in applications that require the ability to broadcast small packets of data over the entire ring, or to specific non-nearest neighbor processors on the ring. The principle role of the shared memory could then be for maintaining common data between neighbor processors, or as a fast mass storage device.

With the shared memories and the fast bus, we expect the lCAP architecture to be a general purpose scientific/engineering parallel computer. The general configuration is given in Fig. 12. This configuration exploits both the loosely and tightly coupled approach equally well. Applications that work well with the loosely coupled approach have been well exemplified here, and encompass a large variety of scientific applications. There are also a large set of applications, such as hydro or fluid dynamics that require a more tightly coupled approach. One of these applications, the shallow water equations, already has proven effectiveness with the shared memories. With the additional shared memories and the fast bus we expect that the broad range of fluid dynamics applications shall also perform well. In short, we expect that the lCAP configuration should provide us with a flexible parallel computer capable of solving efficiently extreefly large scale scientific/engineering problems.

In the short foreseeable future we plan to upgrade some of our IBM mainframe host computers. The lCAP-1 4381 and 4341 shall be upgraded to an IBM 3081 and IBM 4381, respectively. In addition, we plan to acquire an IBM 3090 level mainframe which also includes the high performance vector feature.

With the acquisition of the 3090, we plan on developing a flexible, hierarchical, parallel system, lCAP-3. Our proposed configuration for lCAP-3 is given in Fig. 13. As shown, the plan is to use the 3090 as the main processor, and the lCAP-1 and lCAP-2 systems as the support processors. This system should provide a flexible, parallel system with high fault tolerance capable of Gflop computation rate.

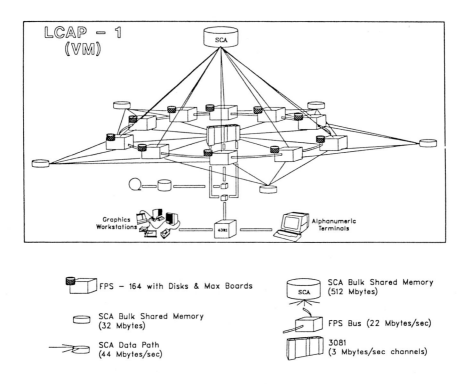

Figure 12.   lCAP configuration (both lCAP-1 and lCAP-2) including both the five
32 Mbyte shared memories, the 512 Mbyte shared memory, and the
FPSBUS. In this configuration only one IBM host is shown.

The 3090 would serve as both the main processor responsible for dispatching jobs to
the support processors, and to execute vector oriented jobs in parallel (the 3090-200
has two tightly coupled processors and the model 400 has 4 processors).
Communication between the lCAP-1 and lCAP-2 support processors could occur
along two distinct paths. The first would be through the 3090 itself, and would be in
the spirit of dispatching separate jobs to different processors. This implies the
ability to properly network our systems. The second path would be through either a
shared memory or through the fast bus. This connection may or may not imply a
FPS-X64 intermediate processor as a "traffic cop"; this perhaps may best be
determined from the results of the modelling. However, this connection would be
made in the spirit of heterogeneous parallelism across systems; and would be used
in passing data required for the computation between systems in a fast
asynchronous manner. This connection could be made through more than one
FPS-X64 or shared memory, or fast bus, or perhaps all of them, thereby increasing
fault tolerance. However, we stress that the configuration for lCAP-3 is not yet
firm. This particular configuration has two especially attractive features for us,
namely the ability to investigate networking our systems, and to study the

characteristics of heterogeneous parallelism. More firm decisions as to the best configuration may be best decided through the results of the modelling previously mentioned.

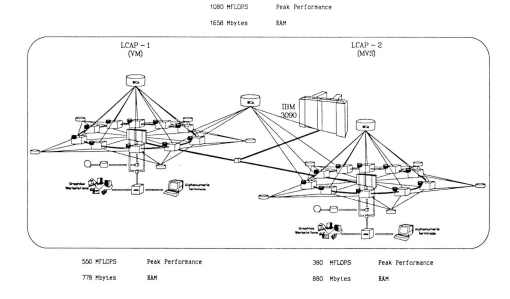

Figure 13.   Potential configuration for lCAP-3.

IX. SUMMARY

The lCAP systems appear to be well adapted to the efficient implementation of a large and diverse collection of scientific and engineering problems. Further ongoing and future extensions to the system would seem to make this parallel processing system a truly general purpose machine. Future experiments will be needed to verify this conclusion.

A superficial examination of the methods employed in making parallel the applications discussed does not reveal a useful abstraction in describing parallel decomposition in general. Each problem or class of problems must find its own mode; this may be the parcel policy of the protein program or the more common loop partitioning scheme. Some problems may involve algorithm changes or even algorithm development, or possibly a restatement of the problem. Clearly, there is a need for an "in depth" analyser for general methods for migrating codes from sequential to parallel. We should however recall that <u>new</u> super-computing tasks can now be written directly for our parallel systems and with time the need of conversion might be less pressing.

The most fundamental conclusion that we wish to emphasis in closing is the following. It is essential in achieving super computing performance in the domain of scientific and engineering problems that computing systems be designed with maximum abilities in all computational modes. That is sequential, vector and parallel computation may be regarded as the three dimensions describing the space of the computations that we wish to execute. Neglecting any component will severely degrade the solution of a large subset of problems of interest today.

ACKNOWLEDGEMENT

We would like to thank Drs. L. Domingo, R. Buenker, and A. Capotondi for permission to publish portions of their work prior to publication.

REFERENCES

1. a) "International Conference on Parallel Processing" August: (1981-84), Bellaire, MI.; (1985), St. Charles, ILL. b) R.W. Hockney, "Parallel Computers: Architecture, Programming and Algorithms", R.W. Hockney, C.R. Jesshope, Eds., Bristol, Adam Hilger, Ltd. (1981). c) "Parallel Processing Systems", J. Evans, Ed., Cambridge, NY, Cambridge Univ. Press (1982). d) Y. Wallach, "Alternating Sequential/Parallel Processing", Lecture Note in Computer Science, 124, Springer Verlag, Berlin, NY (1982). e) G. Corongiu and J.H. Detrich, "Large Scale Scientific Applications in Chemistry and Physics on an Experimental Parallel Computer System", IBM Journal Res. and Dev., 29, 422 (1985).
2. K.R. Wilson, in "Computer Networking and Chemistry", P. Lykos, Ed., ACS Symposium Series, 19, American Chemical Society (1975).
3. Prof. C.C.J. Roothaan "Improved Vector Functions and Linear Algebra for the FPS-164," Proceedings of the 1985 Array Conference, April 14-17, 1985, New Orleans, LA.
4. M.J. Flynn, "Some Computer Organizations and Their Effectiveness" IEEE Trans. Comp. C-21, 948 (1972)
5. Virtual Machine/System Product System Programmer's Guide, Third Edition (Publication No. SC19-6203-2), International Business Machines Corp. (August, 1983).
6. FPS-164 Operating System Manual, vols. 1-3 (Publication No. 860-7491-000B), Floating Point Systems, Inc. (January, 1983).
7. See the IBM Product description for MVS/XA - MVS/SP Version 2 JES2, IBM Product Number 5740-XC6.
8. a) S. Chin, and L. Domingo, "Parallel Computation on the loosely Coupled Array of Processors: Tools and Guidelines", IBM Technical Report KGN-25, September 15, 1985. b) D.L. Meck, "Parallelism in Executing FORTRAN Programs on the 308X: System Considerations and Application Examples", IBM Technical Report POK-38 (April 2, 1984). For another set of FORTRAN-callable communications subroutines to support parallel execution on the IBM 308x under MVS, see IBM program offering 5798-DNL, developed by P.R. Martin; the Program Description Operations Manual for this program offering is IBM publication number SB21-3124 (release date May 4, 1984). See also the IBM Product description of the VS FORTRAN Version 1 Release 4.1 Compiler which contains the description of the IBM Multi-Tasking Facility (MTF). IBM Product Number 5748-F03.
9. S. Chin, L. Domingo, A. Carnevali, R. Caltabiano, and J. Detrich, "Parallel Computation on the loosely Coupled Array of Processors: A Guide to the Pre-Compiler", IBM Technical Report KGN-42, November 25, 1985.
10. G. Amdahl, "The Validity of the Single Processor Approach to Achieving Large Scale Computing Capabilities", AFIPS Conf. Proc., Vol. 30, 1967.
11. a). H. Kobayashi, The Systems Programming Series: Modeling and Analysis: An Introduction to System Performance Evaluation Methodology, Addison-Wesley Publishing Company, Reading, MA (1981). b) C.H. Saur, E.A. McNair, and J.F.

Kurose, "The Research Queueing Package Version 2: Introduction and Examples", IBM Research Report RA 138 (#41126), April 12, 1982. c) E.A. McNair, and C.H. Saur, "The Research Queueing Package: A Primer", IBM Research Report RC 9784 (#43234), Jan. 6, 1983.

12. a). R.S. Mulliken and C.C.J. Roothaan, Proc. Natl. Acad. Sci. USA, **45**, 394 (1949). b) C.C.J. Roothaan, Rev. Mod. Phys., **23**, 69 (1951). c) C.C.J. Roothaan, Rev. Mod. Phys., **32**, 179 (1960). d) G.G. Hall, Proc. R. Soc. Sect. A., **208**, 328 (1951).

13. a). R.J. Beunker, in Proc. of the Workshop on Quantum Chemistry and Molecular Physics in Wollongong, Australia, February, 1980. b) R.J. Beunker in Studies in Physical and Theoretical Chemistry, Volume 21, pp. 17-34 (1982), R. Carbo, Ed., Elsevier Scientific Publishing Company, Amsterdam. This is the Proceedings of an International Conference and Workshop, Barcelona, Spain, September 28- October 3, 1981.

14. a). B.J. Alder, and T.E. Wainwright, J. Chem. Phys., **31**, 459 (1959). b). D.W. Wood in "Water-A Comprehensive Treatise," **vol.6**, pp. 279-409, Felix Franks, Ed., Plenun Press, NY, 1979.

15. a). F.A. Momany, R.F. McGuire, A.W. Burgess, and H.A. Scheraga, J. Phys. Chem., **79**, 2361 (1975). b) G. Nemethy, and H.A. Scheraga, Q. Rev. Biophys., **10**, 239 (1977) c) G. Nemethy, M.S. Pottle, and H.A. Scheraga, J. Phys. Chem., **87**, 1833 (1983). d) H.A. Scheraga, Carlsberg Res. Commun., **49**, 1 (1984).

16. K. Gibson, S. Chin, E. Clementi, and H.A. Scheraga, in preparation.

17. K. Gibson, S. Chin, M. Pincus, E. Clementi, and H.A. Scheraga, "Parallelism in Conformational Energy Calculations of Proteins", in Lecture Notes in Chemistry, M. Dupuis, Ed., (1986).

18. A. Capotondi, S. Chin, V. Sonnad, and E. Clementi, "Parallel Resolution of the Shallow Water Equations Using an Explicit Finite Difference Algorithm", to be published.

19. M. Falk, K.A. Hartman, and R.C. Lord, J. Am. Chem. Soc. **84**, 3843 (1962); ibid **85**, 387 (1963); ibid **85**, 391 (1963). For a review see for example the monographic works by a). G. S. Manning, Quaterly Review of Biophysics II, 2 pp. 179-246 (1978); b). M.T. Record, Jr., C.F. Anderson, and T.M. Lohman, Quaterly Reviews of Biophysics II, 2, pp. 103-178 (1978).

20. G.J. Quigley, A.H.J. Wang, G. Ughelto, G. Van der Marel, J.H. van Boom and A. Rich, Proc. Natl. Acad. Sci., U.S.A., **77**, 7104 (1981); A. Rich, G.J. Quigley and A.H.J. Wang in Biomolecular Stereodynamics Vol. I, pp. 35-52, R.H. Sarma, Ed., Adenine Press, NY (1981); R.F. Dickerson, H.R. Drew, and B. Conner, Biomolecular Stereodynamics Vol. I, pp. 1-34, R.H. Sarma, Ed., Adenine Press, NY, (1981); A. Klug, A. Jack, M.A. Yismanitra, O. Kennard, Z. Shakked, and T.A. Steitz, J. Mol. Biol., **131**, 669 (1979).

21. E. Clementi, and G. Corongiu, Biopolymers, **20**, 551 (1981); ibid, **20**, 2427 (1981); ibid, **21**, 763 (1982); Int. J. Quantum Chem., **22**, 595 (1982); J. of Biological Physics, **11**, 33 (1983); E. Clementi in Structure and Dynamics: Nucleic Acids and Proteins, E. Clementi and R.H. Sarma, Eds., Adenine Press, NY pp. 321-364 (1983); E. Clementi and G. Corongiu in Biomolecular Stereodynamics, R. Sarma, Ed., Adenine Press, NY pp. 209-259 (1981).

22. P.O. Lowdin, Reviews of Modern Physics **35**, 724 (1963)

23. R. Rein, and F.E. Harris, J. Chem. Phys. **41**, 3393 (1964)

24. R. Rein, and F.E. Harris, J. Chem. Phys. **42**, 2177 (1965) **43**, 4415 (1965)

25. R. Rein, and F.E. Harris, J. Chem. Phys. **45**, 1797 (1966)

26. B.R. Parker, and J. Van Everv, Chem. Phys. Lett. **8**, 94 (1971)

27. E. Clementi, J. Mehl, and W. von Niessen, J. Chem. Phys., **54**, 508 (1971)

28. S. Scheiner, and C.W. Kern, Chem. Phys. Lett., **57**, 331 (1978)

29. S. Scheiner, and C.W. Kern, JACS, **101**, 4081 (1979)

30. E. Clementi, G. Corongiu, J. Detrich, S. Chin, and L. Domingo, Int. J. Quantum Chem., **18**, 601 (1984).

31. S. Chin, L. Domingo, G. Corongiu, and E. Clementi, "Hydrogen Bond Bridges in DNA Base Pairs", IBM Technical Report **KGN-21**, February 20, 1985.

32. R.E. Dickerson, M.L. Kopka and H.R. Drew, "Structure and Dynamics: Nucleic Acids and Proteins", E. Clementi and R.H. Sarma, EDS., Adenine Press, N. Y. (1983) pp. 227-257.

CONVERGED CALCULATIONS OF ROTATIONAL EXCITATION AND V-V ENERGY TRANSFER IN THE COLLISION OF TWO MOLECULES

David W. Schwenke and Donald G. Truhlar

Department of Chemistry and Supercomputer Institute
University of Minnesota
Minneapolis, Minnesota 55455, U.S.A.

ABSTRACT

We present the results of large-scale quantum mechanical calculations of state-to-state transition probabilities for the collision of two hydrogen fluoride molecules. We use a potential energy surface obtained by adding a vibrational dependence to the interaction potential of Alexander and DePristo, and we consider zero total angular momentum. We have calculated converged transition probabilities for rotational energy transfer in the rigid rotator approximation and for vibration-to-vibration energy transfer in calculations including full vibration-rotation coupling. The V-V calculations include up to 948 coupled channels. Final production runs were carried out with a highly vectorized code on the University of Minnesota Cyber 205 and Cray-1 computers; earlier test runs were carried out as well on a Cray Research X-MP/48 machine.

1. INTRODUCTION

In molecular quantum mechanics the amount of effort required to treat problems involving systems with two, three, or four atoms increases enormously as each atom is added to the system. The present status in molecular scattering theory is that atom-atom collisions may be treated routinely and atom-diatom collisions may be treated only with great difficulty unless one makes dynamical approximations or artificially restricts or eliminates one or more coordinates.[1] Exact treatments of four-atom systems are generally considered to lie "beyond the state of the art." Encouraged by the great computational enhancements afforded by the class VI computers (i.e., vector pipeline machines like the Cray-1, the Control Data Corporation Cyber 205, and the Cray X-MP series), we have embarked on a quest to obtain converged results for a prototype diatom-collision. Although it is not the simplest such system we chose HF-HF as the prototype for study. The basic reasoning behind this choice of system is as follows. First, the number of internal states involved is much smaller for hydrides than nonhydrides. Second, among collisions of hydrides the HF-HF system has been most widely studied experimentally.[2] In fact the HF-HF system may even be considered the experimental prototype for vibration-to-vibration (V-V) energy transfer, and V-V energy transfer in turn is the dominant energy relaxation mechanism under most conditions where such relaxation is of interest. Furthermore, there have been several studies of the HF-HF potential energy surface,[3] and a knowledge of the potential energy surface is a prerequisite for a dynamics calculation.

We calculate the scattering matrix for the HF-HF collision system by the close coupling method,[1] which for nonreactive scattering essentially involves the solution of N coupled ordinary differential equations, where N is the number of channels. In two earlier reports[4,5] on V-V energy transfer in HF-HF collisions, we discussed calculations including up to 694 channels and here we report 948-channel calculations. The largest number of channels included in previous studies is 510, for which a single column of the scattering matrix was calculated.[6] The method of solution used here and in Refs. 4 and 5 yields all columns. The largest other coupled channel calculation of the whole scattering matrix of which we are aware involves 250 coupled channels.[7] Most close coupling calculations are, however, still restricted to less than 100 channels. Since the number of floating point operations scales as the cube of the number of channels, the present calculations are computationally more intensive than a 100-channel calculation by a factor of $(948/100)^3$ i.e., two and a half orders of magnitude.

We have carried out two kinds of calculations, the first having the diatoms restricted to be rigid, i.e., nonvibrating but free to rotate, and the second with the diatoms also free to vibrate. In both cases we consider only collisions with zero total angular momentum. The major difficulty in these calculations is the representation of the dependence of the wave function on the rotational degrees of freedom. The rigid-rotator calculations, although of considerable interest in their own right,[8] are also designed to help assess the convergence of the rotational basis to provide guidance for the vibrating rotator calculations, whose goal is to converge V-V energy transfer probabilities.

As mentioned above, there have been many studies of the potential energy surface for the HF-HF collisions. In our own previous dynamical calculations of V-V energy transfer in this system[4,5,9] we have used the potentials of Poulsen et al.[3c] and Redmon and Binkley.[31] In the present work we build a new potential by starting with the rigid-rotator potential of Alexander and DePristo[3b] and modifying it to have a dependence on vibrational coordinates. This is accomplished by extending a procedure proposed by Gianturco and co-workers.[3e]

2. THEORY

In this section we summarize the methods we use for the quantum mechanical scattering calculations of the collisions of two identical molecules. A more detailed treatment is given in Ref. 4. The quantum mechanical wave function is written as

$$\psi_{n_0} = \frac{1}{r} \sum_n X_n(x, \hat{r}) f_{nn_0}(r, E) \quad , \tag{1}$$

where $\vec{r} \equiv (r, \hat{r})$, r is the distance between the centers of mass of the two molecules, $\hat{r}$ denotes the orientation of a vector from the center of mass of molecule 1 to the center of mass of molecule 2 in a laboratory-frame coordinate system, x denotes all other internal coordinates of the two-molecule system, E denotes the total energy, X_n is a symmetrized basis function, and f_{nn_0} is a radial translational wave function. The expansion (1) is called the close coupling expansion, X_n is called a channel function, and n and n_0 are channel indices. The second index n_0 on f_{nn_0} denotes which of the radial translational wave functions is given initial-state boundary conditions in the particular solution of (1) under consideration; all other radial translational wave functions have outgoing or exponentially decaying boundary conditions at large r. Those with outgoing boundary conditions are called open channels, and those with exponentially decaying boundary conditions are called closed channels. The equations to be solved for the radial translational functions are obtained by substituting Eq. (1) into the time-independent Schrödinger equation, multiplying in turn by each X_n^* , multiplying by ($-2\mu_{rel}r/\hbar^2$), where μ_{rel} is the reduced mass equal to one half the molecular mass, and integrating over $\underline{x}$ and $\hat{r}$. This yields

$$\left(\frac{d^2}{dr^2} - \frac{l_n(l_n + 1)}{r^2} + k_n^2 \right) f_{nn_0}(r) = \frac{2\mu_{rel}}{\hbar^2} \sum_m V_{nm}(r) \, f_{mn_0(r)} \quad , \tag{2}$$

l_n is the translational orbital angular momentum quantum number for channel n, k_n is the magnitude of the wave vector for channel n, and V_{mn} is a matrix element of the interaction potential V in the basis X_n. The equations (2) are called the close coupling equations. In practice the summations in (1) and (2) are truncated at N terms, but the results become exact only in the limit as $N \to \infty$. The coupled equations (2) are integrated numerically using the R-matrix propagation algorithm with the usual scattering boundary conditions.[4,10,11] The results of the calculation is a matrix of complex scattering matrix elements whose absolute values squared are probabilities for transitions from state n_0 to state m. For the vibrating molecule calculations, we calculated k_n^2 using

$$k_n^2 = 2\mu_{rel} \, (E - \varepsilon_n) \, / \, \hbar^2 \quad , \tag{3}$$

where ε_n is the sum of two diatomic vib-rotational energy levels evaluated from the constants of Webb and Rao.[12] For the rigid rotator calculations, we calculated rotational energy levels by the formula given by DePristo and Alexander[8a] to be consistent with earlier work on this potential.

In order to simplify the calculations we take advantage of the conservation of total angular momentum and the fact that the spatial wave function (1) must be symmetric under the interchange of the coordinates of the two identical molecules. Thus we use basis functions that are eigenfunctions of the total angular momentum and the molecule interchange operator; this block diagonalizes V_{nm} in the total angular momentum quantum number J, its component M on a laboratory-fixed axis, and the interchange symmetry quantum number η and therefore allows the close coupling equations to be solved separately for each J and η (the solutions are independent of M). We are only going to consider initial states with identical quantum numbers for the two molecules; these states have even interchange symmetries, and so they are only coupled to other symmetric basis functions, and we restrict our basis to such symmetric functions. The index n then specifies a set of quantum numbers (v_1, v_2), (j_1, j_2), j_{12}, l, J, M, and η, where v_1 and v_2 are vibrational quantum numbers, j_1 and j_2 are rotational quantum numbers, and j_{12} is the quantum number for the vector sum of the rotational angular momenta of the two molecules. The symmetric basis functions are given by

$$X_n = \left[2(1 + \delta_{v_1 v_2} \delta_{j_1 j_2}) \right]^{-1/2} \left[\phi_\alpha(x, \hat{r}) + (-1)^{j_1 + j_2 + j_{12} + l} \phi_{\bar{\alpha}}(x, \hat{r}) \right] \tag{4}$$

where ϕ_α is a distinguishable-molecule vibrational-rotational-orbital basis function, α is a set of distinguishable-molecule quantum numbers, and $\bar{\alpha}$ denotes the basis function which has v_1 and j_1 interchanged with v_2 and j_2 as compared to the basis function specified by α. Thus n stands for the quantum numbers v_1, j_1, v_2, j_2, j_{12}, l, J, M, and η, where, because it is not possible to distinguish which molecule has which set of quantum numbers, only sets with $v_1 > v_2$ or $v_1 = v_2$ and $j_1 \geq j_2$ are included, in contrast, α stands for the quantum numbers v_1, j_1, v_2, j_2, j_{12}, l, J, and M, where formally the two molecules are distinguished so that molecule i is known to have vibrational and rotational quantum numbers v_i and j_i, and $\bar{\alpha}$ means that the quantum numbers for the two molecules are exchanged. Another consequence of the interchange symmetry is that the physically meaningful transition probabilities are to symmetric final states that can be labelled either by $v_1 > v_2$ with any pair of j_1 and j_2 or by $v_1 = v_2$ with $j_1 \geq j_2$.

When Eq. (4) is employed, the matrix elements of V over symmetric basis functions are given by[4]

$$V_{nn'} = \left[(1 + \delta_{v_1 v_2} \delta_{j_1 j_2}) \, (1 + \delta_{v'_1 v'_2} \delta_{j'_1 j'_2}) \right]^{-1/2} \times (V_{\alpha\alpha'} + (-1)^{j_1 + j_2 + j_{12} + l} V_{\bar{\alpha}\alpha'}) \quad , \tag{5}$$

where $V_{\alpha\alpha'}$ is a potential matrix element calculated using the basis functions for the formally distinguishable molecules. To evaluate $V_{\alpha\alpha'}$ it is convenient to expand the interaction potential function V in terms of angular functions for which the integrals over angles are easily performed:

$$V = \sum_{q_1 q_2 \mu} v_{q_1 q_2 \mu} \, (r, r_1, r_2) \, \gamma_{q_1 q_2 \mu} \, (\hat{r}_1, \hat{r}_2) \tag{6}$$

where

$$\gamma_{q_1 q_2 \mu} = \frac{4\pi}{\left[2(1 + \delta_{\mu 0}) \right]^{1/2}} \left[Y_{q_1 \mu} (\hat{r}_1) \, Y_{q_2 -\mu} (\hat{r}_2) + Y_{q_1 -\mu}(\hat{r}_1) \, Y_{q_2 \mu}(\hat{r}_2) \right] \quad , \tag{7}$$

$\vec{r}_i$ is a vector indicating the bond length and direction of the bond axis of molecule i measured in the frame of reference where the z axis is along $\vec{r}$, and Y_{lm} is a spherical harmonic. Our vibrational-rotational-orbital basis functions take the form

$$\phi_\alpha = (R_1 R_2)^{-1} \chi_{v_1 j_1}(R_1) \chi_{v_2 j_2}(R_2) \sum_{m_1 m_2 m_{12} m_l} (j_1 m_1 j_2 m_2 \mid j_1 j_2 j_{12} m_{12})$$

$$\text{x} \quad (j_{12}m_{12}lm_l \mid j_{12}lJM) \ Y_{j_1m_1} (\hat{R}_1) \ Y_{j_2m_2} (\hat{R}_2) \ Y_{lm_l} (\hat{r}) \quad , \qquad (8)$$

where χ_{vj} is a vibrational wave function, $(\ldots \mid \ldots)$ denotes a Clebsch-Gordon coefficient, and $\vec{R}_i$ is a vector indicating the bond length and direction of molecule i in the frame of reference where the Z axis is along a laboratory-fixed direction. Then the potential matrix elements are given by

$$V_{\alpha\alpha'} = \sum_{q_1q_2\mu} <v_1j_1v_2j_2 \mid v_{q_1q_2\mu} \mid v'_1j'_1v'_2j'_2> \times \ <j_1j_2j_{12}lJM \mid \gamma_{q_1q_2\mu} \mid j'_1j'_2j'_{12}l'JM> \quad (9)$$

where the first factor is a vibrational integral given by

$$<v_1j_1v_2j_2 \mid v_{q_1q_2\mu} \mid v'_1j'_1v'_2j'_2> = \int dR_1 \int dR_2 \ \chi^*_{v_1j_1} (R_1) \ \chi^*_{v_2j_2}(R_2)$$

$$\times \ v_{q_1q_2\mu} \ (r, R_1, R_2) \chi_{v'_1j'_1} (R_1) \ \chi_{v'_2j'_2} (R_2) \qquad (10)$$

and the second factor is an angular integral, which can be easily evaluated in terms of sums and products of vector coupling coefficients.[4] We evaluate Eq. (10) using a recently proposed quadrature scheme[13] using 6 points per dimension. The $\{\chi_{vj}\}$ vibrational wave functions are determined by the linear variational method in a basis of harmonic oscillator functions for the HF diatomic potential given by Murrell and Sorbie[14] which is a fit to RKR data. We do not neglect vibrational-rotational coupling. The functions $v_{q_1q_2\mu}$ are evaluated as needed by numerically performing the three-dimensional integral

$$v_{q_1q_2\mu} = \frac{1}{4\pi} \int_{-1}^{1} d[\cos (\phi_1 - \phi_2)] \left[1 - \cos^2(\phi_1 - \phi_2)\right]^{-1/2}$$

$$\times \int_{1}^{1} d(\cos \theta_1) \int_{-1}^{1} d(\cos \theta_2) \ \gamma^*_{q_1q_2\mu} V \qquad (11)$$

where θ_i and ϕ_i are the inclination and azimuthal angle for $\vec{r}_i$.

The formalism for the rigid rotator calculations is formally identical to that given above with $v_1 = v_2 = 0$ except that integrals over R_1 and R_2 are replaced by functional values at $R_1 = R_e$ and $R_2 = R_e$, where R_e is the equilibrium intermolecular (1.733 a_0) of HF.

The calculations yield transition probabilities $P_{v_1j_1v_2j_2 \to v'_1j'_1v'_2j'_2}$ or, for the rigid rotator case, $P_{j_1j_2 \to j'_1j'_2}$. In the present article we concentrate on a single initial state, $(v_1j_1v_2j_2) = (1010)$ for the vibrating rotator case and $(j_1j_2) = (00)$ for the rigid rotator case. For convenience we sometimes consider probabilities summed over subsets of final rotational states. Thus we define

$$P^{R}_{j'_{\text{sum}}} = \sum_{\substack{j'_1 j'_2 \\ j'_1 + j'_2 = j'_{\text{sum}}}} P_{00 \to j'_1 j'_2} \tag{12}$$

for the rigid rotator case and

$$P^{VV}_{j'_{\text{sum}}} = \sum_{\substack{j'_1 j'_2 \\ j'_1 + j'_2 = j'_{\text{sum}}}} P_{1010 \to 2j'_1 0j'_2} \tag{13}$$

for the vibrating rotator case. The total V-V probability, obtained by summing the previous probability over j'_{sum}, is called P^{VV}.

For both kinds of calculations we consider three values of the initial relative translational energy E_{rel}, which equals $\hbar^2 k^2_{n_0}/(2\mu_{rel})$. In order to utilize the efficiencies possible in R-matrix propagation when results are to be calculated at more than one energy we propagate the solutions through a given step for all three energies before proceeding to the next step.[4]

All transition probabilities presented in this article are well converged with respect to increasing the integration range or decreasing the step size, with the largest source of error being the truncation to finite N. The integration range is from 2 to 150 a_0, using a fixed step size of 0.06 a_0 from 2 to 10.1 a_0 and a variable step size thereafter. The variable-step size algorithm is described in Ref. 4; in the notation of that reference we use EPS = 0.01 and a maximum step size of 3 a_0 is reached by the end of the integration. A total of 297-303 integration steps are taken in each case, depending on the expansion basis.

We note that R-matrix propagation algorithm involves the transformation at every integration step into a basis which is adiabatic with respect to the coordinate r. An expansion of the wave function ψ_{n_0} in terms of the adiabatic basis functions is more rapidly convergent than the expansion of Eq. (1). This is especially true for large r. In the present calculations we use the algorithm described in Ref. 4 with parameter EPSRED = 0.01 to reduce the number of terms in the adiabatic expansion. This reduction has a negligible effect on the accuracy of our calculations. As an example of this reduction, the wave function for the lowest-energy calculation with N = 948 is expanded in terms of 948 adiabatic basis functions for $r < 51$ a_0, but then the expansion is reduced to 893 terms over the range $r = 51$ to 150 a_0

3. POTENTIAL ENERGY FUNCTION

The original interaction potential of Alexander and DePristo[3b] is a fit to the *ab initio* SCF data of Yarkony *et al.*[3a] in a representation involving space-fixed angular functions. They wrote, for both diatoms at their equilibrium separations,

$$V = (4\pi)^{3/2} \sum_{\lambda_1 \lambda_2 \lambda} U^e_{\lambda_1 \lambda_2 \lambda}(r) Y_{\lambda_1 \lambda_2 \lambda}(\hat{r}, \hat{R}_1, \hat{R}_2) \tag{14}$$

where

$$Y_{\lambda_1 \lambda_2 \lambda} = \sum_{m_1 m_2 m} (\lambda_1 m_1 \lambda_2 m_2 / \lambda_1 \lambda_2 \lambda m) \, Y_{\lambda_1 m_1}(\hat{R}_1) \, Y_{\lambda_2 m_2}(\hat{R}_2) \, Y^*_{\lambda m}(\hat{r}) \tag{15}$$

and they truncated the sum in Eq. (14) to 6 terms, namely those with $(\lambda_1 \lambda_2 \lambda)$ equal to (000), (112), (011), (123), (101), and (213). The coefficients $U^e_{\lambda_1 \lambda_2 \lambda}$ for the fifth and sixth terms are equal by symmetry to minus those for the third and fourth; and the first four coefficients are shown in Fig. 1. We note that the factor before the summation in Eq. (14) is chosen so that U^e_{000} is the spherical average of the interaction potential. The potential of Ref. 3b was used unmodified in the rigid rotator calculations.

Note that this potential has only 6 terms in the laboratory-frame expansion (14), but it gives rise to 9 terms in the body-frame expansion (6), namely those with $q_1 + q_2 \leq 3$ except for $(q_1 q_2 \mu)$ equal to (020), (030), (200), and (300).

The calculations of Yarkony *et al.* were single-configuration SCF calculations with a double-zeta-plus-polarization set. To gain some idea of the reasonableness of this data we compare the coefficients of Alexander and DePristo to the analogous ones we computed for the Redmon-Binkley potential. The Redmon-Binkley is a multiparameter fit, including 2-body, 3-body, and 4-body terms, to fourth-order Møller-Plesset perturbation theory calculations with a double-zeta-core, triple-zeta-valence-plus-polarization basis set. The four coefficients retained by Alexander and DePristo, as evaluated from the Redmon-Binkley potential, are shown in Fig. 2. The coefficients from the two potentials show qualitatively similar behavior.

For the vibrating molecule calculations, it was necessary to introduce vibrational dependence into the interaction potential. This was done by extending an approximation of Gianturco *et al.*[3e] for the vibrational dependence of the short-range repulsive interactions and by using accurate R_i- dependent multipole moments to simulate the vibrational dependence of long-range electrostatic interactions.

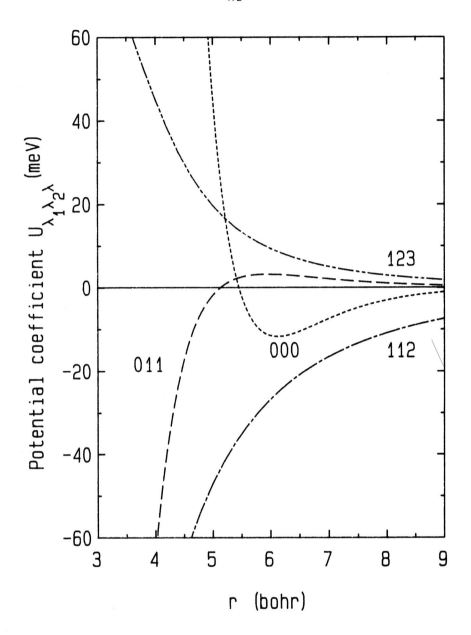

Figure 1. $U_{\lambda_1\lambda_2\lambda}(r, R_1 R_2)$ for the modified Alexander-DePristo potential, with $R_1 = R_2 = R_e$, as functions of the distance between the molecular centers of mass. Short dashed line, U_{000}; long dashed line, U_{011} ; long-short dashed line, U_{112}; long-short-short dashed line, U_{123}.

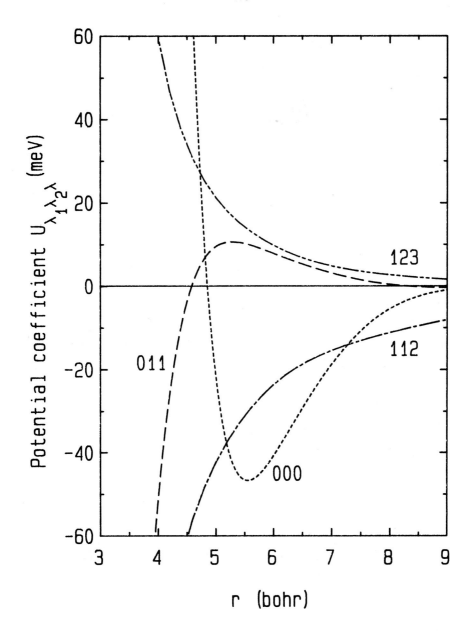

Figure 2.   Same as Fig.1 except for the potential of ref.3*l*

Gianturco et al.[3e] wrote

$$U_{\lambda_1\lambda_2\lambda} = U^e_{\lambda_1\lambda_2\lambda}(r) \, \exp\left[-\alpha_{\lambda_1\lambda_2\lambda}(R_1 + R_2 - 2R_e) \right] \tag{16}$$

with

$$\alpha_{\lambda_1\lambda_2\lambda} = -\left[U^e_{\lambda_1\lambda_2\lambda}(r_0) \right]^{-1} \frac{dU^e_{\lambda_1\lambda_2\lambda}}{dr} \Big|_{r=r_0} \tag{17}$$

for $(\lambda_1\lambda_2\lambda) = (000)$ and we employ this for all $(\lambda_1\lambda_2\lambda)$. Gianturco et al.[3e] suggested that r_0 be the translational classical turning point. We set r_0 equal to 4.1 a_0 because that is approximately the largest distance close to a typical classical translational turning point for which none of the $U^e_{\lambda_1\lambda_2\lambda}$ are passing through zero or near a local extremum. In the original fit[3b] the coefficients U^e_{112} and U^e_{123} take the form

$$U^e_{112} = -c_1 e^{-c_2 r} - c_3 e^{-c_4 r} - (2/15)^{\frac{1}{2}} \mu^2 \, r^{-3} \tag{18}$$

$$U^e_{123} = -(c_5/r - c_6) \, e^{-c_7 r} + (1/7)^{\frac{1}{2}} \mu \, \theta \, r^{-4} \tag{19}$$

where the c_i are constants, and μ and θ are the dipole and quadrupole moments, which were set equal to vibrationally averaged values of 0.716 and 1.93 a.u., respectively. We replaced the multipole moments in these expressions with the linear functions

$$\mu = \mu_e + \beta_{\lambda_1\lambda_2\lambda}(R_i - R_e) \tag{20}$$

$$\theta = \theta_e + \gamma_{\lambda_1\lambda_2\lambda}(R_i - R_e) \tag{21}$$

Table I. Parameters for the interaction potential (in a.u.)

λ_1	λ_2	λ	$\alpha_{\lambda_1\lambda_2\lambda}$	$\beta_{\lambda_1\lambda_2\lambda}$	$\gamma_{\lambda_1\lambda_2\lambda}$
0	0	0	2.38		
0	1	1	2.42		
1	0	1	2.42		
1	1	2	0.738	0.838	
1	2	3	0.850	0.918	3.00
2	1	3	0.850	0.918	3.00
				μe	θe
				0.7066[a]	1.64[b]

[a]   From Ref. 15.

[b]   From Ref. 16.

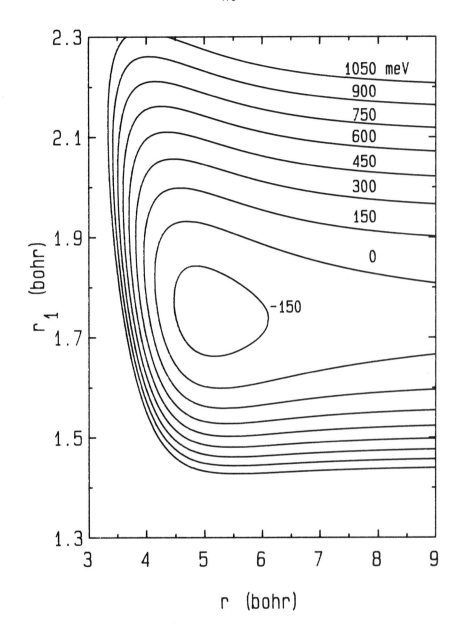

Figure 3.   Total potential based on modified Alexander-DePristo interaction
potential plus Sorbie-Murrell diatom potentials, as a function of r and
r_1 for the collinear configuration with the H of molecule one hydrogen
bonded to the F of molecule two. The contours run from 150 to 1050
meV in steps of 150 meV.

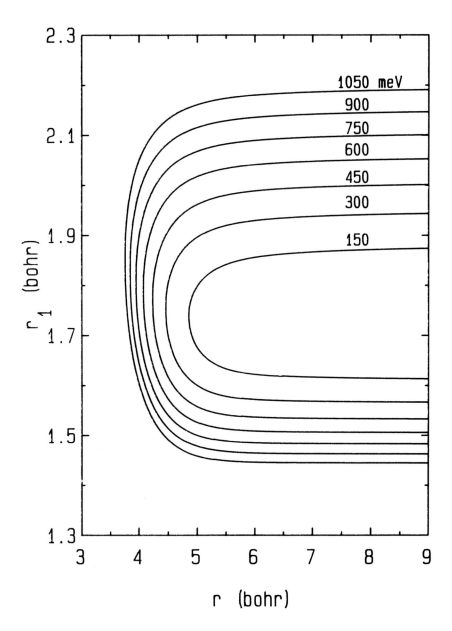

Figure 4.   Same as Fig.3   except for $\theta_1 = 90$ ° and $\theta_2 = 0$ °. The contours run
from 150 to 1050 meV.

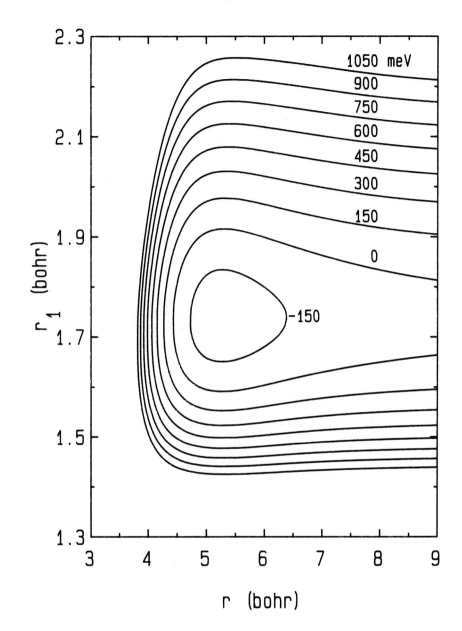

Figure 5.　　Same as Fig.3　except for the potential of ref.31.

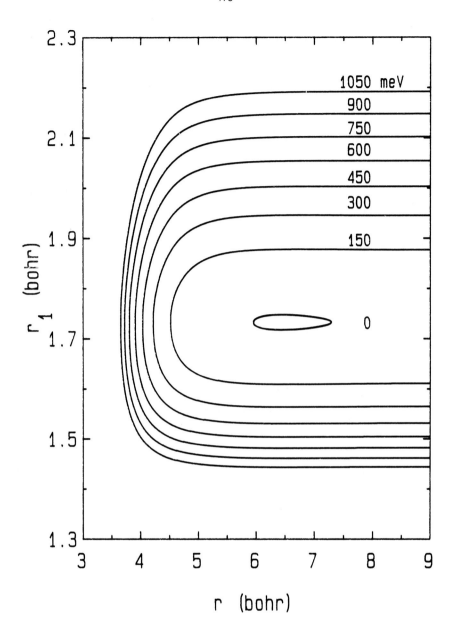

Figure 6.   Same as Fig.4 except for the potential of ref.31 and contour levels run
from 150 to 1050 meV.

where μ_e and θ_e are values corresponding to the equilibrium internuclear separation[15,16] and the $\beta_{\lambda_1\lambda_2\lambda}$ and $\gamma_{\lambda_1\lambda_2\lambda}$ are constants. The constants β_{112}, β_{123}, and γ_{123} were chosen to that $(d/dR_i)\exp(-\alpha_{\lambda_1\lambda_2\lambda}R_i)\,\mu(R_i)$ and $(d/dR_i)\exp(-\alpha_{\lambda_1\lambda_2\lambda}R_i)\,\theta(R_i)$ reproduce the experimental value of $d\mu/dR = 0.3169$ a.u.[15] and the theoretical value of $d\theta/dR_i = 1.601$ a.u.,[17] respectively. The parameters for the modification of the Alexander and DePristo interaction potential are given in Table I.

In order to illustrate the character of the potential energy surfaces resulting from the approximations (16)-(21) we present contour maps of the modified Alexander - DePristo (MAD) potential to see if the interaction potential is physically reasonable. Figure 3 shows a contour map for a linear hydrogen bonded arrangement (F-H...F-H), and Fig. 4 shows a similar map for a configuration with the F of molecule 2 oriented toward the center of mass of molecule 1, $\theta_1 = 90^0$, and $\theta_2 = 0^0$. (Because the center of mass of a monomer is closer to the F than to the bond midpoint, this resembles an "L"). The quantity plotted in the figures is the total potential defined as the sum of the interaction potential and the two diatomic potentials, with the zero of energy at 2HF at R_e. Figures 5 and 6 present comparable maps for the full Redmon-Binkley (RB) potential. These maps illustrate one inherent weakness of the approximation in (16), namely it does not seem to predict the character of the vibrational force equally validly for different orientations of approach. Thus the vibrational force of the MAD potential is opposite to that of the RB potential for the collinear case (Figs. 3 and 5), but the two potentials are very similar for the L case (Figs. 4 and 6). In the real world, perpendicular approaches are more likely than collinear ones, but the extent of rotational participation in V-V energy transfer may depend sensitively on the orientational dependence of the vibrational force.

Another set of attributes of the two potentials that can be compared are the energy and geometry of the van der Waals minima. For the MAD potential the van der Waals dimer is bound by 225 meV, with $r_e = 4.97$ a_0, $R_{1e} = 1.749$ a_0, and $R_{2e} = 1.743$ a_0. For the RB potential the corresponding quantities are 274 meV, 5.04 a_0, 1.742 a_0, and 1.737 a_0. The agreement is reasonable, which gives us further confidence that the MAD potential is physically reasonable.

Although the vibrational forces of the MAD potential are not quantitatively accurate, it was thought to provide a reasonable and portable test potential for which converged V-V energy transfer calculations would be useful as a benchmark, and so we proceeded to carry out such calculations.

4. ROTATIONAL ENERGY TRANSFER IN COLLISIONS OF RIGID ROTATORS

In this section we discuss the results of our rigid rotator calculations. First, however, it is necessary to specify our choice of basis functions. We carried out a series of calculations for $J = 0$ and $\eta = +1$ with basis functions chosen by one of two rules. The first rule was to include all channels coupled by total angular momentum, identical particle interchange symmetry, and parity restrictions which had $j_{sum} \leq j_{sum,max}$, where $j_{sum} = j_1 + j_2$, while the second rule used $\max(j_1,j_2) \leq j_{max}$.

Table II shows the convergence of $P_{j'_{sum}}^R$ for both schemes as a function of the maximum quantum number index and N, the number of channels. The convergence of two of the state-to-state transition probabilities is illustrated in Figs. 7-9. Table II and Figs. 7-9 show that convergence is slow, especially by the standard of previous calculations. [The largest basis set used previously for this potential by Alexander[8b] corresponds to $j_{max} = 3$ (their basis B4). For total angular momentum zero this basis yields n = 20.]

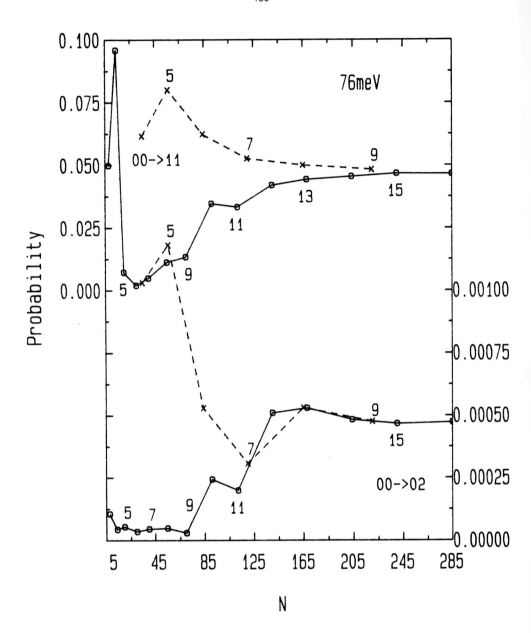

Figure 7.   Convergence of the rotational energy transfer probabilities $P_{00 \to 11}$ and $P_{00 \to 02}$ in rigid rotator calculations for E_{rel} = 76 meV. Solid line and $\square$, $j_{max,sum}$ rule; dashed line and x, j_{max} rule.

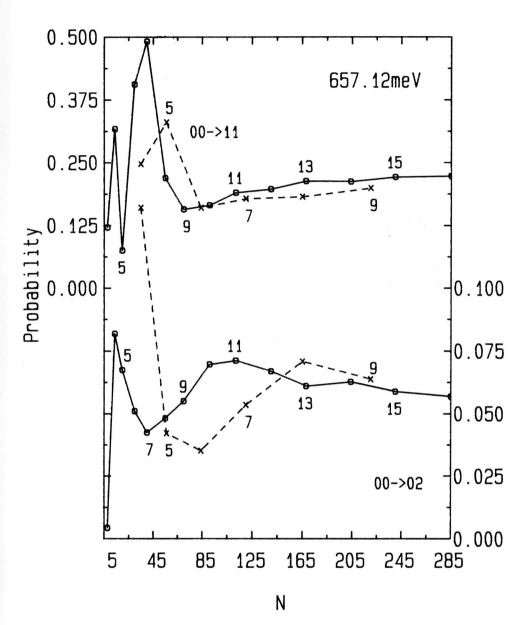

Figure 8.   Same as Fig. 7 except for E_{rel} = 657 meV.

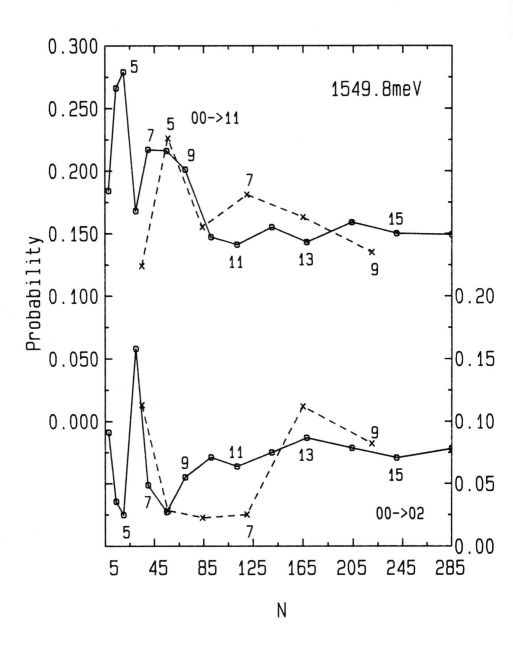

Figure 9.   Same as Fig. 7 except for E_{rel} = 1549.8 meV.

Table II. Rotationally summed probabilities for rigid rotator calculations.

index	N	E_{rel} = 0.076 eV			0.657 eV			1.550 eV		
		P_0^R	P_1^R	P_2^R	P_0^R	P_1^R	P_2^R	P_0^R	P_1^R	P_2^R
$j_{sum,max}$										
3	8	0.948	0.002	0.050	0.819	0.045	0.125	0.634	0.025	0.275
4	14	0.900	0.002	0.096	0.479	0.044	0.399	0.361	0.037	0.302
5	20	0.987	0.001	0.007	0.753	0.038	0.082	0.434	0.105	0.304
6	30	0.991	0.002	0.002	0.358	0.042	0.457	0.050	0.141	0.326
7	40	0.992	0.001	0.005	0.252	0.020	0.534	0.037	0.031	0.266
8	55	0.980	0.002	0.011	0.367	0.043	0.267	0.038	0.019	0.243
9	70	0.977	0.003	0.013	0.473	0.018	0.212	0.104	0.011	0.256
10	91	0.950	0.002	0.035	0.468	0.018	0.234	0.120	0.001	0.218
11	112	0.951	0.003	0.033	0.423	0.024	0.261	0.104	0.004	0.205
12	140	0.940	0.003	0.042	0.430	0.021	0.264	0.106	0.009	0.230
13	168	0.937	0.003	0.045	0.417	0.017	0.274	0.097	0.009	0.230
14	204	0.935	0.004	0.046	0.420	0.019	0.274	0.101	0.007	0.238
15	240	0.934	0.004	0.047	0.414	0.017	0.280	0.106	0.011	0.221
16	285	0.934	0.004	0.047	0.414	0.019	0.280	0.102	0.011	0.227
j_{max}										
4	35	0.924	0.001	0.063	0.372	0.018	0.380	0.075	0.026	0.236
5	56	0.891	0.001	0.081	0.368	0.007	0.373	0.098	0.025	0.254
6	84	0.912	0.003	0.063	0.514	0.008	0.195	0.115	0.023	0.238
7	120	0.925	0.003	0.053	0.468	0.018	0.232	0.139	0.011	0.206
8	165	0.929	0.003	0.050	0.435	0.021	0.252	0.076	0.004	0.276
9	220	0.932	0.004	0.049	0.426	0.020	0.262	0.100	0.025	0.217

Table II shows that the present calculations are converged for both basis-set selection schemes at all three energies, with convergence being achieved typically at $N \cong 200$. At the lowest energy the two basis-set selection schemes converge at about the same rate, but at higher energies the $j_{sum,max}$ rule leads to convergence at a smaller N and hence is more efficient.

Table III gives a set of converged state-to-state transition probabilities for the rigid rotator calculations at each of the three energies. The table shows that the elastic scattering probability decreases from 93% to 10% as the energy is increased. The first-order dipole-dipole transition, (00) → (11), has the largest inelastic transition at each energy; these probabilities are in the range 5-22%. The (00) → (02), (00) → (22), and (00) → (32) transitions have probabilities in the 2-8% range at the two highest energies. None of the other transitions has a probability in excess of 5% at any of the energies.

5. V-V ENERGY TRANSFER IN CALCULATIONS WITH NO COORDINATE OR DIMENSIONAL RESTRICTIONS AND NO DYNAMICAL APPROXIMATIONS

For the vibrating molecule calculations we again considered $J = 0$ and $\eta = +1$, and chose rotational channels by the j_{sum} rule and vibrational channels using an analogous v_{sum} rule which was previously shown to be efficient for reduced dimensionality calculations.[11d] In particular, for each v_{sum}, where $v_{sum} \equiv v_1 + v_2$, we included all channels allowed by total angular momentum, interchange symmetry, and parity restrictions which had $j_{sum} \leq j_{sum,max}$ where $j_{sum,max}$ depends on v_{sum}. We use the same value of $j_{sum,max}$ for $v_{sum} \leq 2$, and smaller values for larger v_{sum}. The compositions of the basis sets are summarized in Table IV. Results were calculated for the process $HF(v_1 = 1, j_1 = 0) + HF(v_2 = 1, j_2 = 0) \rightarrow HF(v'_1 = 2, j'_1) + HF(v'_2 = 0, j'_2)$. Table V summarizes the convergence of the transition probabilities with respect to increasing the size of the basis set. First consider the calculations with $j_{sum,max} = 8$, 9, 10 and 11 for $v_{sum} \leq 2$ and $j_{sum,max} = 6$, 7, 8 and 9 for $v_{sum} = 3$, excluding channels with $v_{sum} > 3$. These calculations involved 400, 530, 694, and 880 channels, respectively.

Table III. Rigid rotator transition probabilities from calculations with $j_{sum,max} = 16.$[a]

$j_1j_2 \rightarrow j_1j_2$	76[b]	657[b]	1550[b]
$0,0 \rightarrow 0,0$	0.934	0.414	0.102
$0,0 \rightarrow 1,0$	0.004	0.019	0.011
$0,0 \rightarrow 1,1$	0.046	0.223	0.149
$0,0 \rightarrow 2,0$	<0.001	0.057	0.08
$0,0 \rightarrow 2,1$	0.004	0.019	0.007
$0,0 \rightarrow 3,0$	0.001	0.002	0.003
$0,0 \rightarrow 2,2$	0.010	0.031	0.076
$0,0 \rightarrow 3,1$	<0.001	0.019	0.003
$0,0 \rightarrow 3,2$	<0.001	0.077	0.022
$0,0 \rightarrow 4,0$	<0.001	0.007	0.005
$0,0 \rightarrow 4,1$	<0.001	0.020	0.012
$0,0 \rightarrow 3,3$	<0.001	0.027	0.020
$0,0 \rightarrow 4,2$	<0.001	0.009	0.03
$0,0 \rightarrow 5,0$	c	0.001	0.005
$0,0 \rightarrow 5,1$	c	0.010	0.008
$0,0 \rightarrow 4,3$	c	0.018	0.04
$0,0 \rightarrow 5,2$	c	0.008	0.04
$0,0 \rightarrow 4,4$	c	0.005	0.050
$0,0 \rightarrow 6,0$	c	<0.001	0.004
$0,0 \rightarrow 5,3$	c	0.004	0.011
$0,0 \rightarrow 6,1$	c	0.001	0.005
$0,0 \rightarrow 6,2$	c	0.001	0.004
$0,0 \rightarrow 5,4$	c	0.006	0.027
$0,0 \rightarrow 6,3$	c	0.002	0.028
$0,0 \rightarrow 7,0$	c	<0.001	0.001
$0,0 \rightarrow 7,1$	c	0.001	0.005
$0,0 \rightarrow 5,5$	c	0.006	0.05
$0,0 \rightarrow 7,2$	c	0.001	0.004
$0,0 \rightarrow 6,4$	c	0.002	0.02
$0,0 \rightarrow 7,3$	c	0.001	0.007
$0,0 \rightarrow 8,0$	c	<0.001	0.001
$0,0 \rightarrow 6,5$	c	0.002	0.019
$0,0 \rightarrow 8,1$	c	<0.001	0.002
$0,0 \rightarrow 7,4$	c	0.001	0.02
$0,0 \rightarrow 8,2$	c	<0.001	0.003
$0,0 \rightarrow 8,3$	c	<0.001	0.003
$0,0 \rightarrow 6,6$	c	0.003	0.04
$0,0 \rightarrow 7,5$	c	<0.001	0.011
$0,0 \rightarrow 9,1$	c	<0.001	0.001
$0,0 \rightarrow 8,4$	c	<0.001	0.003
$0,0 \rightarrow 7,6$	c	0.001	0.008
$0,0 \rightarrow 9,3$	c	<0.001	0.002
$0,0 \rightarrow 8,5$	c	<0.001	0.010
$0,0 \rightarrow 9,4$	c	<0.001	0.001
$0,0 \rightarrow 7,7$	c	0.001	0.02
$0,0 \rightarrow 8,6$	c	<0.001	0.01
$0,0 \rightarrow 9,5$	c	<0.001	0.002
$0,0 \rightarrow 10,3$	c	<0.001	0.001
$0,0 \rightarrow 8,7$	c	<0.001	0.003
$0,0 \rightarrow 10,4$	c	<0.001	0.002
$0,0 \rightarrow 9,6$	c	<0.001	<0.01
$0,0 \rightarrow 10,5$	c	<0.001	<0.01
$0,0 \rightarrow 8,8$	c	<0.001	<0.01
$0,0 \rightarrow 9,7$	c	<0.001	<0.01

[a]   Only transitions with probabilities greater than 0.0005 are included. Transitions are listed in order of increasing excitation energy.

[b] E_{rel} in meV.

[c]   Closed channels.

Comparison of the 880-channel calculations to the 694-channel calculations show convergence of P^{vv} to better than 0.7% at all the energies considered here. In Table V, 13 of the probabilities for the 694-channel and 880-channel calculations agree to within 0.004; for the other two probabilities the discrepancies are 0.006 and 0.011. These calculations demonstrate that the rotational basis of basis set 3 is converged for $v_{sum} \lesssim 3$. It is interesting to note that the convergence is attained at

$j_{sum,max} = 10$ for these V-V calculations, whereas the rigid rotator calculations discussed above required $j_{sum,max}$ values of about 12-14 in order to converge the results. Next we consider calculations that employ the converged rotational basis of basis set 3 for $v_{sum} \leq 3$ and also include channels with $v_{sum} > 3$ in order to converge the vibrational basis. Two additional basis sets were considered, one with $j_{sum,max} = 6$ for $v_{sum} = 4$, and no $v_{sum} \geq 5$ states, which yields 824 channels, and another in which two additional j_{sum} states are included for $v_{sum} = 4$ and 5 ($j_{sum} = 7$ and 8 for $v_{sum} = 4$ and $j_{sum} = 0$ and 1 for $v_{sum} = 5$), yielding our largest basis of 948 channels. Values of $P^{VV}_{j_{sum}}$ and P^{VV} calculated from these additional basis sets are also given in Table V. Comparison of the 824-channel calculations to the 948-channel calculations shows convergence of P^{VV} to better than 1.3% at all energies considered here. If we also consider $P^{VV}_{j_{sum}}$, we see that for 10 of the probabilities in Table V, the 824-channel and 948-channel calculations differ by 0.002 or less and for the other probabilities the discrepancies are between 0.006 and 0.019. This convergence is considered acceptable. Furthermore since the comparison of bases 3 and 4 shows that $j_{sum,max} = 8$ is sufficient to converge the $v_{sum} = 3$ channels, it should also be sufficient for $v_{sum} = 4$.

Table IV. Basis sets for V-V energy transfer calculations.

Basis	$v_{sum} \leq 2$	$v_{sum} = 3$	$v_{sum} = 4$	$v_{sum} = 5$	N
1	8	6	. . .	. . .	400
2	9	7	. . .	. . .	530
3	10	8	. . .	. . .	694
4	11	9	. . .	. . .	880
5	10	8	6	. . .	824
6	10	8	8	1	948

(header: $j_{sum,max}$ spanning the four v_{sum} columns)

The state-to-state transition probabilities $P_{1010 \rightarrow 2j'_1 0j'_2}$ for the four largest basis sets are given in Table VI. As for the results in Table V, if basis 3 agrees with basis 4 and basis 5 with basis 6, then basis 6 should be considered converged. For the most part probabilities with both $j'_1 \leq 1$ and $j'_2 \leq 1$ are well converged.

The most striking aspect of the rotational energy distributions in Table VI is the peaking of the V-V energy transfer cross sections at a final state that is nonresonant by 21 meV and the fact that transition probabilities to other channels with smaller translational energy defects are smaller by factors of thirty or more. It would be interesting to learn the sensitivity of this rotational distribution to the nature of the potential.

6. COMPUTATIONAL CONSIDERATIONS

The close coupling calculations were performed using a code vectorized separately for the Cray-1 and for the Control Data Corporation Cyber 205 pipelined vector

Table V. Partially summed transition probabilities $P_{j'\text{sum}}^{VV}$.

E_{rel} (meV)	open channels	Basis	N	open-channel basis functions	P_0^{VV}	P_1^{VV}	P_2^{VV}	P_3^{VV}	P^{VV}
2.455	1548	1	400	162	0.698	0.045	0.058	0.022	0.823
		2	530	207	0.791	0.032	0.017	0.003	0.843
		3	694	264	0.826	0.031	0.003	0.002	0.863
		4	880	327	0.827	0.031	0.005	0.006	0.868
		5	824	264	0.875	0.034	0.003	0.002	0.914
		6	948	264	0.881	0.035	0.003	0.002	0.921
29	1671	1	400	184	0.873	0.036	0.009	0.004	0.923
		2	530	229	0.904	0.036	0.001	0.005	0.948
		3	694	286	0.906	0.038	0.009	0.005	0.960
		4	880	349	0.902	0.039	0.011	0.008	0.962
		5	824	286	0.925	0.037	0.012	0.005	0.980
		6	948	286	0.924	0.037	0.014	0.005	0.981
76	1888	1	400	239	0.901	0.048	0.013	0.006	0.970
		2	530	284	0.854	0.048	0.045	0.006	0.955
		3	694	341	0.815	0.045	0.065	0.008	0.937
		4	880	404	0.804	0.046	0.068	0.008	0.933
		5	824	341	0.760	0.033	0.085	0.009	0.891
		6	948	341	0.741	0.031	0.093	0.009	0.880

Table VI. State-to-state transition probabilities $P_{1010 \to 2j'_1 0j'_2}$.

j'_1	j'_2	ΔE[a]	Basis	2.455[b]	29[b]	76[b]
0	0	21	3	0.83	0.91	0.81
			4	0.83	0.90	0.80
			5	0.87	0.92	0.76
			6	0.88	0.92	0.74
1	0	17	3	0.026	0.030	0.022
			4	0.024	0.032	0.023
			5	0.028	0.026	0.015
			6	0.028	0.026	0.014
0	1	16	3	0.005	0.009	0.023
			4	0.006	0.006	0.023
			5	0.006	0.011	0.017
			6	0.006	0.011	0.016
1	1	12	3	0.0008	0.008	0.064
			4	0.0007	0.010	0.067
			5	0.0005	0.011	0.084
			6	0.0006	0.0125	0.092
2	0	7	3	0.002	0.0009	0.0006
			4	0.004	0.0007	0.0007
			5	0.002	0.0009	0.0005
			6	0.002	0.0009	0.0006
0	2	6	3	0.0001	0.0002	0.0003
			4	0.0005	0.0004	0.0004
			5	0.0002	0.0002	0.0003
			6	0.0002	0.0002	0.0003

[a] Translational energy defect in meV (21 meV = 172 cm^{-1}).

[b] E_{rel} in meV (76 meV = 613 cm^{-1}).

computers. On both machines we used single precision (64-bit) arithmetic for all calculations. The rigid-rotator production runs were carried out on the Cray-1. The largest rigid-rotator calculation, involving 285 channels, required approximately 1.7 hours of CPU time. Relative to unvectorized calculations on a Digital Equipment Corporation VAX 11/780 with scalar floating point accelerator, we achieved enhancements in computation speed of a factor of about 540 for the whole computation for this calculation. The vibrating rotator production runs were carried out on the two-pipeline Cyber 205, with the largest calculation, with 948 channels, requiring about 17.5 hours of computer time. In this case we estimate a speed enhancement over the VAX of about a factor of 1700, so that the calculation would have required 30,000 hours (1300 days) to execute on the VAX. It is instructive to give more details on how we arrive at the estimated speed enhancement and also to estimate how many millions of floating point operations per second (MFLOPS) we are obtaining. In our initial tests,[4] we compared times on our VAX to the Cyber 205 and the Cray-1 for a test case having $N = 101$. Additional tests of the matrix routines on the VAX indicate that for $N > 50$, the time for a calculation on the VAX scales with N as $N^{3.0}$. Using this factor to scale the execution time for a full calculation with $N = 101$, we estimate for a three-energy calculation taking 297 steps with $N = 948$ a time of 3.0×10^4 hours. We estimate speed in MFLOPS as follows. Most of time (about 95% for the $N = 948$ run and about 92-94% for other N for this potential) used for our calculations is spent on matrix manipulations; in particular for the first energy in each of 297 steps we must diagonalize an NxN matrix, multiply three NxN matrices, and solve two sets of linear equations for N different unknown vectors, while for second and third energies in each of 297 steps we must perform two NxN matrix multiplications and solve a set of linear equations for N different unknown vectors. The number of operations to perform a matrix multiplication is $2N^3$, the number to perform a linear equation solution is $8N^3/3$ ($2N^3/3$ for LU decomposition,[18a] N^2 per vector for forward elimination,[18b] and N^2 per vector for back substitution[18b]), and we will take the number to perform a matrix diagonalization[18c] to be $10N^3$ (which may be a slight, 5-10%, underestimate for our problem), so that the number of operations will be about $28N^3$ per step, which, when combined with our actual execution time for the $N = 948$ run, yields a lower bound for the operation rate of 112 MFLOPS. This compares favorably to the maximum theoretically obtainable value of 200 MFLOPS on a two-pipeline machine running in single precision performing linked triads.[19a]

The speed enhancements of the vector pipelined supercomputers relative to the scalar VAX would increase with increasing N since the number of scalar operations increases as N^3 whereas we found that for the largest N values considered here, the CPU time was increasing proportional to $N^{2.7}$ on the Cray-1 at $N = 200-285$, and proportional to $N^{2.6}$ on the Cyber 205, at $N = 400-948$.

For $N = 285$ we estimate that the ratio of the CPU time for the Cyber 205 to CPU time for the Cray-1 is about 2, which is the same as we observed previously[4] for $N = 101$. We note, however, that the Cray-1 version of our code used here is more efficient than the one described previously[4] because we have implemented the technique of FORTRAN DO-loop unrolling[20,21] for matrix multiplications and linear equation solutions. This increases the rate of execution on the Cray-1 because it decreases the number of unnecessary transfers between main memory and vector registers. Such transfers can become a bottleneck because the Cray-1 has only one data bus between main memory and the vector registers. For the new version of the Cray-1 code, the Cray-1 and Cyber 205 CPU times are almost the same for $N = 101$; however, the Cyber 205 continues to become more efficient as N is further increased whereas the Cray-1 is closer to its asymptotic speed. Further details of our experience with loop unrolling on the Cray-1 are given in an appendix to this chapter. For eigenanalysis steps in the R matrix propagation calculations we use the EISPACK[22] routines of the Cray SCILIB[23] library. On the Cyber 205 we use the

MAGEV library[24] for both eigenanalysis and linear algebra. Loop unrolling is less useful on the Cyber 205 because it is a memory-to-memory machine, i.e., there are no vector registers, and thus the same data transfer bottleneck does not exist.

In separate calculations we estimated that for R matrix propagation calculations, the Cyber 205 and Cray X-MP/48 machine (using a single processor) lead to comparable execution times for $N = 101$. For these tests the linear algebra on the Cray X-MP/48 was carried out by LINPACK[25] routines with the BLAS[26] replaced by their FORTRAN equivalents, as discussed for the Cray-1 in Ref. 4. Since the Cray X-MP/48 has three data buses between main memory and vector registers, we anticipate that the effect of DO-loop unrolling would not be large.

7. CONCLUSIONS

We have shown that it is possible to converge three-dimensional calculations of V-V energy transfer for diatom-diatom collisions, although a very large number of channels are required. Although the present calculations, being restricted to zero total angular momentum, cannot be compared to experiment, they provide benchmarks for testing approximate dynamical theories that can more easily be applied to all total angular momenta.

In future work we plan to study the dependence of the results on the nature of the intermolecular potential.

8. ACKNOWLEDGEMENTS

We are grateful to R. W. Numrich and T. F. Walsh for administrative assistance that facilitated these calculations, to J. J. Dongarra for sharing his improved code for the decomposition step of linear equations solution, and to David Cochrane for comments on the manuscript. This work was supported in part by the National Science Foundation under grant no. CHE83-17944, by the University of Minnesota Supercomputer Institute under grant SCIH001, by the Control Data Corporation, and by Cray Research, Inc.

APPENDIX. Loop Unrolling

As mentioned in Sect. 6, we found significant enhancements in performance on the Cray-1 computer when we incorporated the loop unrolling techniques suggested by Dongarra and Eisenstat.[20,21] Dongarra has also found significant enhancements on a variety of advanced scientific computers.[27] Since these techniques offer significant time savings for widely used matrix algebraic procedures, they should be of general interest, and in this appendix we report on further details of our experience with them. We will compare times for various levels of loop unrolling to times for optimized library routines. On the Cray-1, we will compare to routines from the SCILIB[23] library; in particular, we give reference times for the matrix multiplier MXM, the linear equation solver MINV, which uses Gaussian elimination, and the routines SGEFA and SGESL, which are SCILIB versions of LINPACK[25] codes, which perform the LU decomposition of a matrix and use the LU decomposition of a matrix to solve linear equations, respectively. On the CYBER 205 we will compare to the MAGEV library[24] matrix multiplier MXMPY. The University of Minnesota Supercomputer Institute (UMSI) Cray-1 has COS 1.12 as its operating system and has two versions of the FORTRAN compiler, CFT 1.11 and CFT 1.13. We will discuss execution times for codes compiled by both compilers in what follows, and we will see that the latter compiler often produces faster code;

however, for the production runs discussed in Sect. 6 above we used version 1.11 of the compiler because version 1.13 does not interfere properly with version 1.12 of the operating system in our production context. The UMSI Cyber 205 has the VSOS operating system, and we use the FORTRAN 2.1.6 compiler for the timings in this section.

It should be noted that the Cray-1 is able to hold a maximum of 64 elements in one vector register. Thus all vector operations occur with vectors of length 64 or less. When FORTRAN programs have vectors longer than 64 elements, the compiler automatically breaks the longer vector into shorter vectors containing no more than 64 elements, and it operates on each of the shorter vectors separately. The inclusion of vector lengths greater than 64 in the discussion below would complicate the presentation without adding any important new ideas. Thus, we will discuss the operation of the Cray computers as if the vector registers were infinitely long or all vectors were of length 64 or less, but we give timing examples involving longer vectors.

A. Matrix multiplication

The basic idea of loop unrolling can be illustrated clearly for the operation of matrix multiplication. Consider the following FORTRAN code which multiplies an NxM matrix A times an MxP matrix B to produce an NxP matrix C:

```
      DO 1 I1 = 1,P
      DO 1 I2 = 1,N
1        C(I2,I1) = 0
      DO 2 J = 1,P
      DO 2 K = 1,M
      DO 2 I = 1,N
2        C(I,J) = C(I,J) + A(I,K)*B(K,J).
```
(A1)

The Cray-1 compiler will vectorize the innermost DO-loop over the I variable by giving the instructions: (i) load the vector with N elements starting with A(1,K) from main memory into vector register V_0, (ii) multiply the elements of the vector register V_0 times the scalar B(K,J) and store the results into vector register V_1, (iii) load the vector with N elements starting with C(1,J) from main memory into vector register V_2, (iv) add the elements of the vector register V_2 to the elements of the vector register V_1 and store the result in vector register V_3, and (v) store the elements of the vector register V_3 back into main memory. The Cray-1 is able to "chain" many of these operations together so that they form one continuous pipeline.[19b] For two operations to chain, an instruction must be issued at the "chain time slot," which is the functional unit time plus two clock periods. When this is done, the result from one functional unit is immediately sent to another functional unit as an operand. Chaining is essentially equivalent to connecting two functional units together as one continuous unit. For algorithm (A1), it is possible for step (i) to be chained with step (ii) and step (iii) to be chained with step (iv). The CPU time required to complete a vectorized operation on a vector of length N is $N\tau + T_{overhead}$, where τ is a cycle time and $T_{overhead}$ is an overhead time which is independent of N. For two vectorized operations that are chained, the CPU time is not $2(N\tau + T_{overhead})$ but rather $N\tau + T'_{overhead}$. Normally τ is the clock period, but if memory bank conflicts occur τ will be larger. Thus if N is large enough so that $T'_{overhead}$ can be neglected and no memory bank conflicts occur, the time for one pass through steps (i)-(v) of algorithm (A1) will be about 3N clock periods (whereas without chaining it would be about 5N clock periods). Unfortunately, however, the

FORTRAN compiler is not always able to properly recognize potential chaining opportunities. It appears that the CFT 1.11 compiler is not able to chain steps (iii) and (iv) whereas the CFT 1.13 compiler schedules instructions so that steps (iii) and (iv) do chain. It turns out, for this particular case, that comparable savings can be achieved by placing parentheses around $C(I,J)$ so that it is loaded first and so that the loading of $A(I,K)$ is chained to both the multiplication and the addition. However, even more savings can be achieved by loop unrolling as discussed next.

The key to further efficiency in accomplishing the objectives of algorithm (A1) is to observe that the vector which is stored back into main memory in step (v) is the same vector which will be loaded from memory in step (ii) of the next pass (with K incremented) through steps (i)-(v). Ideally we would like to have the compiler eliminate this unnecessary step and maximize chaining by having it generate the instructions: (i′) load the vector with N elements starting with $C(1,J)$ from main memory into vector register V_2; (ii′) load the vector with N elements starting with $A(1,K)$ from main memory into vector register V_0: (iii′) multiply the elements of the vector register V_0 times the scalar $B(K,J)$ and store the results into vector register V_1; (iv′) add the elements of vector register V_1 to the elements in vector register V_α and store the results into vector register V_β; then repeat steps (ii′)-(iv′) M times, successively increasing K so that $B(K,J)$ is different each time and using $\alpha=2$, $\beta=3$ for even K and $\alpha=3$, $\beta=2$ for odd K; and (v′) store the elements of vector register V_β into main memory. In this scheme steps (ii′)-(iv′) can all be chained together so that the time is about N clock periods, or about 1/3 of the previous example. One way one could attempt to cause the FORTRAN compiler to recognize this additional chaining possibility would be to replace the J, K, and I DO-loops of algorithm (A1) by

$$\text{DO 2 J}=1,\text{P} \qquad\qquad (A2)$$
$$\text{DO 2 I}=1,\text{N}$$
$$2 \quad C(I,J)=(...(C(I,J))+A(I,1)*B(1,J))+...+A(I,M)*B(M,J).$$

Table VII. Times in central processor seconds to multiply two square matrices of order 300.

	execution time (CPU seconds)	
unrolling depth	Cray-1	Cyber 205
1	1.63 (1.31)[a]	0.51
2	1.03 (0.87)	0.51
4	0.75 (0.71)	0.51
8	0.63 (0.56)	0.52
16	0.58 (0.52)	0.51
32	0.55 (0.53)	0.50
MXM	0.37 (0.37)	...
MXMPY	...	0.43

[a] Times in parenthesis are for the CFT 1.13 compiler, while the other Cray-1 times are for the CFT 1.11 compiler.

This approach suffers from the drawback that it is necessary to have a separate program for each value of M. A more general yet still partially optimized code would be:

```
DO 2 J = 1,P                                                    (A3)
DO 2 K = 1,M,NU
DO 2 I = 1,N
2    C(I,J) = (...(C(I,J)) + A(I,K)*B(K,J)) + ... + A(I,K + NU-1)*B(K + NU-1,J)
```

where it is assumed here that M is an integral multiple of NU. The procedure employed in algorithm (A3) is called unrolling the K loop to a depth of NU. Table VII contains some CPU times for various depths of unrolling the K loop in the example above for the two Cray-1 compilers mentioned above. It also contains CPU times for performing the same matrix multiplications with subprogram MXM of the Cray SCILIB library. (The last column of this table will be explained in the next paragraph.) For the case in the table the largest speedup obtained with algorithm (A3) and for the CFT 1.11 compiler is observed for unrolling to a depth of 32, which is the largest amount of unrolling we considered, while for CFT 1.13 the largest speedup occurs when unrolling to a depth of 16. It can be seen from the table that even for CFT 1.11 unrolling to depths greater than 32 will not significantly increase the speedup factor. The code produced by CFT 1.13 runs consistently 4-20% times faster than that produced by CFT 1.11. The speedup factor of about three for the programs with loops unrolled as compared to the original program is in accordance with the analysis above. It is interesting to note that the MXM routine is about 1.4 times faster than the fastest FORTRAN program. It is, however, unfortunate that MXM is coded in a restrictive manner that requires that the orders, i.e., N, M, and P, of the matrices involved must correspond to their first dimension. This makes MXM unusable for our application because during the course of our calculation we change the order of the matrices which we need to multiply. This arises because the adiabatic basis set size is reduced at large r as discussed at the end of Sect. 2. Another reason we do not use MXM is that our application requires multiplications of the form

$$\underset{\sim}{C} = \underset{\sim}{A}^T \underset{\sim}{B} \tag{22}$$

with the elements of $\underset{\sim}{C}$ stored in the same space as the elements of $\underset{\sim}{B}$. With our own FORTRAN code we can easily implement multiplications of the form of (22) without explicitly constructing $\underset{\sim}{A}^T$. In this manner we avoid wasting the time necessary to form $\underset{\sim}{A}^T$ as well as the extra memory required to store $\underset{\sim}{A}^T$. For these two reasons we do not use MXM, and the DO-loop unrolling procedures described above are indeed very helpful. In particular the version of our code which we used to perform the rigid rotator calculations used matrix multiplication routines unrolled to a depth of 8.

Table VII also shows CPU times for algorithm (A3) as run on the two-pipeline Cyber 205 and compares them to times obtained with subprogram MXMPY from the Cyber 205 MAGEV library. On the Cyber 205 the effect of DO-loop unrolling is much smaller than one the Cray-1; the difference in performance with various values of NU is on the order of two percent, which is understandable because the memory-to-register transfers discussed above are not required on this machine. (The Cyber 205 feeds the vector pipeline directly from memory without requiring a prior transfer into vector registers.)

B. Solution of linear equations

We now turn to the problem of solving linear equations. The problem is written in the form

$$A\vec{x} = \vec{b} \tag{23}$$

where A is a nonsingular, nonsymmetric NxN matrix and $\vec{b}$ and $\vec{x}$ are column vectors; A and $\vec{b}$ are known, and the problem is to solve for $\vec{x}$. The usual method involves first decomposing A into its so called LU form, i.e., determining matrices L and U such that

$$U = L^{-1}A \tag{24}$$

where U has zeros below its diagonal (see e.g., Ref. 18d or 28 for explicit definitions of L and U). Usually the matrices L and L^{-1} are not explicitly formed, however information is retained to allow the easy determination of products of the form of Eq. (24). The solution continues by forming $L^{-1}\vec{b}$, which is called the forward elimination step, followed by

$$\vec{x} = U^{-1}L^{-1}\vec{b} \ . \tag{25}$$

which is called the back substitution step. The number of arithmetic operations to perform the LU decomposition is $0(N^3)$ and the number of arithmetic operations to compute $\vec{x}$ (the forward elimination and back substitution steps) is $0(N^2)$. Thus in cases where there is only one vector $\vec{b}$ most of the time will be spent in the decomposition step, while in cases where there are N vectors $\vec{b}$, about equal time will be spent on the decomposition and on the forward elimination and back substitution steps.

1. Decomposition step

Dongarra[20] has pointed out that it is possible to re-organize the usual sequence of operations in the decomposition step in order to cast it into a form in which FORTRAN DO-loop unrolling as discussed above can be used to improve the performance. An example of a code re-organized this way is (this code is base on one given to us by Dongarra, but differs in that subroutine calls to matrix kernels are replaced by in-line code)

```
          DO 1 J=1,N                                      (A4)
            IF (J.EQ.1) GO TO 2
            DO 3 K=1, J-1
            DO 3 L=1, N-J+1
        3      A(J+L-1,J)+A(J+L-1,J)+A(K,J)*A(J+L-1,K)
        2   T=ABS(A(J,J))
            K=J
            IF (J.EQ.N) GO TO 4
            DO 5 I=J+1,N
              IF (ABS(A(I,J)).LE.T) GO TO 5
              T=ABS(A(I,J))
              K=I
```

```
      5     CONTINUE
      4     IPVT(J) = K
            IF (T.EQ.0) STOP
            IF (K.EQ.J) GO TO 6
            DO 7 M = 1,N
              T = A(J,M)
              A(J,M) = A(K,M)
      7       A(K,M) = T
      6     A(J,J) = 1./A(J,J)
            IF (J.EQ.N) GO TO 1
            IF (J.EQ.1) GO TO 8
            DO 9 K1 = 1,J-1
            DO 9 L1 = 1, N-J
      9       A(J,J + L1) = A(J,J + L1) + A(J,K1)*A(K1,J + L1)
      8     T = -A(J,J)
            DO 10 I1 = J + 1,N
     10       A(J,I1) = T*A(J,I1)
      1     CONTINUE
```

In this example the K loop and the K1 loop would be unrolled, i.e., the first loop would become, assuming NU divides J-1.

$$\text{DO 3 K} = 1,\text{J-1,NU} \tag{A5}$$
```
            DO 3 L = 1,N-J + 1
      3       A(J + L-1,J) = (...(A(J + L-1,J)) + A(K,J)*A(J + L-1,K)) + ...
      $       + A(K + NU-1,J)*A(J + L-1,K + NU-1)
```

and similarly for the K1 loop. Of course in an actual code provision would have to be made for the case when NU is not an integral divisor of J-1. It should be noted that the L1 loop causes the second index of the array A to vary fastest. This causes

Table VIII. Times in central processor seconds on the Cray-1 to solve linear equations of order 300 in which there are 300 right hand side vectors.

unrolling depth	time for decomposition step	time for forward elimination and back substitution steps	total time
1	0.62 (0.56)a	1.38 (1.30)	2.00 (1.87)
2	0.46 (0.39)	1.05 (0.88)	1.51 (1.27)
4	0.35 (0.33)	0.77 (0.72)	1.11 (1.05)
8	0.30 (0.27)	0.64 (0.57)	0.94 (0.84)
16	0.28 (0.25)	0.57 (0.51)	0.85 (0.75)
32	0.26 (0.24)	0.54 (0.53)	0.80 (0.78)
LINPACKb	0.52 (0.49)	1.58 (1.49)	2.10 (1.99)
MINVc	. . .	. . .	1.78 (1.78)

[a] Times in parentheses are for the CFT 1.13 compiler while the other times are for the CFT 1.11 compiler.

[b] Using SCILIB routines SGEFA for decomposition step and SGESL for substitution step.

[c] SCILIB routine.

the memory to be accessed in a nonsequential manner. This can be a problem on the Cray-1 if the first dimension of the array A is an integral multiple of half the number of memory banks. If that occurs, memory bank conflicts will arise which will greatly slow down the calculation. Some times for the decomposition step for various levels of unrolling are given in Table VIII along with the time for the Cray-1 SCILIB routine SGEFA. The maximum speedup, a factor of about 2.3 for both CFT 1.11 and CFT 1.13 compilers, occurs for unrolling to a depth of 32, which is the maximum unrolling depth we considered. It can be seen from the table that unrolling to depths greater than 32 will not significantly increase the speedup factor. The CFT 1.13 compiler produces code which runs 6-15% faster than the CFT 1.11 compiler.

2. Forward elimination and back substitution step

Both the forward elimination and back substitution steps can also benefit from loop unrolling.[21] For a single vector $\vec{b}$ the code to determine $\vec{x}$ from the decomposition performed above (the values of $\vec{b}$ in the array B are overwritten with the values of $\vec{x}$) can be written (this code was given to us by Dongarra)

```
        DO 1 K=1, N                                      (A6)
        L=IPVT(K)
        T=B(L)
        B(L)=B(K)
 1      B(K)=T
        DO 2 K1A=1,N-1
        T=B(K1A)*A(K1A,K1A)
            DO 3 I=K1A+1,N
 3          B(I)=B(I)-A(I,K1A)*T
 2      B(K1A)=T
        B(N)=B(N)*A(N,N)
        DO 4 K2A=N,2,-1
        T=B(K2A)
        DO 4 I1=1,K2A-1
 4      B(I1)=B(I1)+A(I1,K2A)*T
```

The K1A loop and K2A loops can be unrolled, e.g., unrolling the first loop to a depth of two yields

```
        NREM=MOD(N,2)                                    (A7)
        M=N-NREM
            DO 2 K1A=1,M,2
            T1=B(K1A)*A(K1A,K1A)
            B(K1A)=T1
            T2=(B(K1A+1)-A(K1A+1,K1A)*T1)*A(K1A+1,K1A+1)
            B(K1A+1)=T2
                DO 3 I=K1A+2,N
 3              B(I)=((B(I))-A(I,K1A)*T1)-A(I,K1A+1)*T2
 2          CONTINUE
        IF (NREM.EQ.0) GO TO 5
        B(N)=B(N)*A(N,N)
 5      CONTINUE
```

A similar procedure can be used to unroll the K2A loop in (A6). However, if the number, NRHS, of vectors $\vec{b}$ is large, it is more efficient to re-organize the substitution steps in (A6) to yield

```
          DO 1 K = 1,N                                          (A8)
          L = IPVT(K)
          IF (L.EQ.K) GO TO 1
               DO 11 J = 1, NRHS
               T = B(L,J)
               B(L,J) = B(K,J)
   11          B(K,J) = T
    1     CONTINUE
          DO 3 I = 1,N
               DO 2 K1B = 1,I-1
               DO 2 J1 = 1,NRHS
    2          B(I,J1) = B(I,J1)-A(I,K1B)*B(K1B,J1)
          DO 3 J2 = 1,NRHS
    3     B(I,J2) = B(I,J2)*A(I,I)
          DO 4 I1 = N,1,-1
          DO 4 K2B = I1 + 1,N
          DO 4 J3 = 1,NRHS
    4     B(I1,J3) = B(I1,J3) + A(I1,K2B)*B(K2B,J3)
```

In algorithm (A8) the K1B loop and the K2B loop are the ones to be unrolled. This version is faster than algorithm (A7) because the vectorized loops have constant lengths, i.e., vectors of length NRHS are always used, whereas in algorithm (A7) the vector lengths range from 1 to N. It should be noted that the second index of the array B is varied fastest in the J, J1, J2, and J3 loops; since this constitutes nonsequential access from memory, memory bank conflicts may pose a problem. Table VIII shows times for the forward elimination and back substitution steps for algorithm (A8) with NRHS = N using various levels of loop unrolling along with the time for the Cray-1 SCLIB routine SGESL. For the CFT 1.11 compiler, unrolling to a depth of 32, which is the maximum considered, gives the largest speedup — a factor of about 2.6, while for the CFT 1.13 compiler the largest speedup is a factor of 2.4 and occurs when unrolling to a depth of 16. As before, we see from the table that further unrolling will not produce significant additional speedups for the CFT 1.11 compiler. The CFT 1.13 compiler produces code which runs 2-16% faster than the CFT 1.11 compiler for a given unrolling depth. Overall for both decomposition and for the forward elimination and back substitution steps we obtain a factor of about 2.6 over using the SCILIB routines SGEFA and SGESL and a factor of about 2.2 over using the SCILIB routine MINV.

In the code used in the rigid rotator calculations reported here, we used algorithm (A4) and (A6) unrolled to a depth of 8. Additional performance enhancements of our code would be obtained by using algorithm (A4) and (A8) unrolled to a depth of 32.

It should be noted that since both algorithm (A4) and (A8) involve varying the second index of an array fastest, these algorithms are not well suited for the Cyber 205.

We have found that the larger the amount of loop unrolling, the better the performance on the Cray-1. In the standard linear algebraic examples considered here, the algebraic expressions to be evaluated in the innermost loop are relatively simple, and we have seen that unrolled loops are evaluated very efficiently on the Cray-1. Although this need not be the case for more complicated examples, in our application a large fraction of the CPU time is spent on these standard linear algebraic steps. Possible difficulties in extending the techniques presented here to more complicated expressions is that the number of instructions may eventually exceed the size of the instruction buffer, causing chaining to be inhibited, and the compiler may not be able to determine the optimum utilization of the vector registers and so be forced to save temporary data in main memory.

References

1. R.B. Bernstein, "Atom-Molecule Collision Theory," Plenum, New York, 1979.
2. S.R. Leone, J. Phys. Chem. Ref. Data 11, 953 (1982).
3. An incomplete survey of HF-HF potential energy surfaces and related calculations includes (a) D.R. Yarkony, S.V. O'Neil, H.F. Schaefer III, C.P. Baskin, and C.F. Bender, J. Chem. Phys. **60**, 855 (1974); (b) M.H. Alexander and A.E. DePristo, J. Chem. Phys. **65**, 5009 (1976); (c) L.L. Poulsen, G.D. Billing, and J.I. Steinfeld, J. Chem. Phys. **68**, 5121 (1978); (d) M.L. Klein, I.R. McDonald, and S.F. O'Shea, J. Chem. Phys. **69**, 63 (1978); (e) F.A. Gianturco, U.T. Lamanna, and F. Battiglia, Int. J. Quantum Chem. 19, 217 (1981); (f) R.L. Redington, J. Chem. Phys. **75**, 4417 (1981); (g) A.E. Barton and B.J. Howard, Faraday Disc. Chem. Soc. **73**, 45 (1982); (h) J.T. Brobjer and J.N. Murrell, Mol. Phys. **50**, 885 (1983); (i) M.E. Cournoyer and W.L. Jorgensen, Mol. Phys. **51**, 119 (1984); (j) D.W. Michael, C.E. Dykstra, and J.M. Lisy, J. Chem Phys. **81**, 5998 (1984); (k) D.W. Schwenke and D.G. Truhlar, J. Chem. Phys. **82**, 2418 (1985); and (l) M.J. Redmon and J.S. Binkley, unpublished.
4. D.W. Schwenke and D.G. Truhlar, in "Supercomputer Applications," edited by R.W. Numrich, Plenum, New York, 1985, p. 215.
5. D.W. Schwenke and D.G. Truhlar, paper presented at the 1985 Cray Science and Engineering Symposium, Bloomington, Minnesota, April 14-17, 1985, University of Minnesota Supercomputer Institute report UMSI85/5.
6. L.D. Thomas, J. Chem. Phys. **76**, 4925 (1982).
7. J.F. McNutt and M.J. Redmon, unpublished.
8. (a) A.E. DePristo and M.H. Alexander, J. Chem. Phys. **66**, 1334 (1977); (b) M.H. Alexander, J. Chem. Phys. **73**, 5135 (1980); (c) H.K. Haugen, W.H. Pence, and S.R. Leone, J. Chem. Phys. **80**, 1839 (1984); (d) P.H. Vohralik and R.E. Miller, J. Chem. Phys. **83**, 1609 (1983).
9. D.W. Schwenke, D. Thirumalai, D.G. Truhlar, and M.E. Coltrin, J. Chem. Phys. **78**, 3078 (1983).
10. J.C. Light and R.B. Walker, J. Chem. Phys. **65**, 4272 (1976).
11. (a) N.A. Mullaney and D.G. Truhlar, Chem. Phys. **39**, 91 (1979); (b) D.G. Truhlar, N.M. Harvey, K. Onda, and M.A. Brandt, in "Algorithms and Computer Codes for Atomic and Molecular Quantum Scattering Theory," Vol. 1, edited by L. Thomas, National Resource for Computation in Chemistry, Lawrence Berkeley Laboratory, Berkeley, CA, 1979, p. 220; (c) N.M. Harvey and D.G. Truhlar, Chem. Phys. Lett. **74**, 252 (1980); (d) D. Thirumalai and D.G. Truhlar, J. Chem. Phys. **76**, 5287 (1982).
12. D.U. Webb and K.N. Rao, J. Mol. Spectry. **28**, 121 (1968).
13. D.W. Schwenke and D.G. Truhlar, Comp. Phys. Comm. **34**, 57 (1984).
14. J.N. Murrell and K.S. Sorbie, J. Chem. Soc. Faraday Trans. II **70**, 1552 (1974).
15. R.N. Siko and T.A. Cool, J. Chem. Phys. **65**, 117 (1976).
16. K.D. Huber and G. Herzberg, "Constants for Diatomic Molecules" Van Nostrand, Princeton, New York, 1979.
17. D. Maillard and B. Silvi, Mol. Phys. **40**, 933 (1980).
18. G.H. Golub and C.F. Van Loan, "Matrix Computations," Johns Hopkins University Press, Baltimore, Maryland, 1983, (a) p. 69, (b) p. 53, (c) p. 282, (d), chapter 4. Note that the authors count operations in terms of a unit that corresponds to a floating point add plus a floating point multiply. We have converted the operation count to the more usual definition by which each add or multiply counts as an operation.
19. R.W. Hockney and C.F. Jesshope, "Parallel Computers," Adam Hilger Ltd., Bristol, England, 1981, (a) p. 124, (b) p.80.
20. J.J. Dongarra and S.C. Eisenstat, ACM Trans. Math. Software **10**, 221 (1984).
21. J.J. Dongarra, L. Kaufman, and S. Hammarling, Argonne National Laboratory, Mathematics and Computer Science Division, Technical Memorandum No. 46, Jan., 1985, unpublished.
22. B.T. Smith, J.M. Boyle, J.J. Dongarra, B.S. Garbow, Y. Ikebe, V.C. Klema, and C.B. Moler, "Matrix Eigensystem Routines-EISPACK Guide," Springer-Verlag, New York, 1976.

23. Cray-1 and Cray X-MP Computer Systems Library Reference Manual SR-0014, Cray Research, Inc., Mendota Heights, Minnesota, 1984.
24. Mathematical-Geophysical Vector Library for the Cyber 205, Control Data Corporation, unpublished.
25. J.J. Dongarra, C.B. Moler, J.R. Bunch, and G.W. Stewart, "LINPACK User' Guide," Society for Industrial and Applied Mathematics, Philadelphia, 1979.
26. C. Lawson, R. Hanson, D. Kincaid, and F. Krogh, ACM Trans. Math. Software **5**, 308 (1979).
27. J.J. Dongarra, Argonne National Laboratory, Mathematics and Computer Science Division, Technical Memorandum No. 23, revised version of Dec., 1984.
28. L.W. Johnson and R.D. Riess, "Numerical Analysis," 2nd ed., Addison-Wesley, Reading, Massachusetts, 1982, chapter 2.

PARALLELISM IN CONFORMATIONAL ENERGY CALCULATIONS ON PROTEINS*

K.D. Gibson,[1] S. Chin,[2] M.R. Pincus,[3] E. Clementi,[2] and H.A. Scheraga[1]

[1]Baker Laboratory of Chemistry, Cornell University, Ithaca, NY 14853-1301

[2]IBM Corporation, Dept. 48B/MS 428, Kingston, NY 12401

[3]Department of Pathology, New York University Medical Center, New York, NY 10016

---

*The work carried out at Cornell University was supported by the National Institutes of Health (GM-14312), the National Science Foundation (DMB84-01811) and Hoffman-La Roche, Inc. The collaborative effort was supported by the National Foundation for Cancer Research.

ABSTRACT

Attempts are being made to identify the native conformation of a protein by minimizing its (empirical) conformational energy. To circumvent the multiple-minima problem, a polypeptide chain is built up from low-energy conformations of shorter segments. After the minimum-energy conformations have been located, their energies are recalculated with the inclusion of a term for the free energy of solvation; this leads to a re-ordering of the energies of each segment, and the low-energy conformations of the re-ordered sets are used to build up larger segments. The methodology has been applied successfully to various model peptide systems, and is used here to begin the computation of the 156-residue protein, human leukocyte interferon. In spite of the large number of low-energy conformations of small segments that must be stored, and the large amount of time required to minimize the energy of longer and longer segments, the computations are feasible because they can be parallelized. The procedure is readily adapted for parallel processing by several array processors operating in a quasi-independent manner, and such computations are being carried out on a system of "loosely coupled array processors". Preliminary results are reported for three segments of interferon containing 29, 9, and 8 residues, respectively. The low-energy conformations of the 29-residue segment from position 122 to 150 all take the form of two roughly anti-parallel α-helices joined by a hinge containing Cys[139]. The 9-residue segment from position 60 to 68 adopts low-energy conformations that are almost totally α-helical, suggesting strongly that it forms part of an α-helix in the native protein. The low-energy conformations of the 8-residue fragment from position 100 to 107 are all irregular with an apparent chain reversal, suggesting that this part of the molecule links two α-helices in the native protein. The results show that parallel processing constitutes a powerful approach to the massive computational effort involved in predicting the native conformations of globular proteins from thermodynamic principles.

INTRODUCTION

On the basis of experiments of Anfinsen,[1] it is now well-established that the three-dimensional structure of a globular protein is determined by its amino acid sequence. Presumably, the native conformation "is the one in which the Gibbs free energy of the *whole system* is a minimum with respect to all degrees of freedom, i.e. this conformation is determined by the various interatomic interactions and hence by the amino acid sequence, in a *given environment*".[2] Much effort is therefore being devoted to the identification of the native protein by computational methods. While entropic effects can be included in the computations,[3] most of the initial effort concerns the minimization of an empirical potential energy function that takes into account all interatomic interactions and also the effects of the solvent, viz. water.

Even though efficient algorithms are currently available for minimizing a function of many variables,[4,5] the problem is made very difficult by the existence of many minima in the multi-dimensional conformational energy space. Several approaches are being explored to surmount the multiple-minima problem; one of them, the build-up procedure[6] is considered here. In this procedure, a polypeptide chain is built up by stepwise ligation of shorter segments, using only low-energy conformations of these segments as input for the next round of computations. The rationale for this approach is that interactions between neighboring and near-neighboring amino acid residues dominate the overall energy of any minimum-energy conformation;[6] hence it is reasonable to regard all minimum-energy conformations of the whole chain, in which a short segment would have a high energy, as making a negligible contribution to the native state. In this paper, we shall illustrate some of the results that have been obtained with this methodology.

Even with the successful application of this methodology to small polypeptides, the computational problems become insurmountable with conventional computers, when applied to larger polypeptides, but are tractable if use is made of parallelism in the computations. We shall therefore present a discussion of our current efforts to introduce parallelism into the computations, and a description of our preliminary results with human leukocyte interferon, a 156-residue protein.

METHODOLOGY

Our current methodology is described briefly here; more extensive discussion can be found elsewhere.[7,8] The conformational energy of a polypeptide is computed with the following expression which forms the basis of the ECEPP/2 (Empirical Conformational Energy Program for Peptides) algorithm.[9,10]

$$U = \sum_{i \neq j} \varepsilon_{ij} \left[\left(\frac{r^o_{ij}}{r_{ij}} \right)^{12} - 2 \left(\frac{r^o_{ij}}{r_{ij}} \right)^M \right] + \sum_{i \neq j} \frac{q_i q_j}{D r_{ij}} + \sum_k \frac{A_k}{2} (1 \pm \cos n\theta_k) \qquad (1)$$

where ε_{ij} and r_{ij}^e are the potential well depth and position of the minimum of the pair-interaction energy (for nonbonded energy, M = 6; for hydrogen-bonding energy, M = 10), q is the partial atomic charge, D is the dielectric constant, r_{ij} is the distance between the two interacting atoms, A_k is the barrier height for rotation around the kth bond, θ_k is the dihedral angle, and n is the n-fold degeneracy of the torsional potential. Bond lengths and bond angles are kept fixed while the energy is minimized; the dihedral angles are the independent variables. In the work reported here, the secant-type routine for unconstrained minimization (SUMSL) of Gay[5] has been used for locating minimum-energy conformations from selected starting points, as outlined in the next paragraph. When all minimum-energy conformations have been located, their energies are recalculated with the inclusion of a term for solvation free energy based on a shell model.[11,12]

These potential functions have been used to compute the single-residue minima for all of the naturally occurring amino acids.[13-15] Then a library of selected dipeptides was created by the build-up procedure; i.e. all possible combinations of minimum-energy single-residue conformations were used as starting points to obtain the low-energy conformations of dipeptides.[16] More recently, with the availability of the system described below, we have completed this library which now contains the low-energy conformations of all 400 possible dipeptides.[17]

This build-up process can be continued to build up larger structures from the low-energy conformations of smaller ones. For this purpose, a variety of procedures have been used to reduce the library of stored conformations to a manageable size. For example, many minima for oligopeptide segments have essentially identical backbone conformations but different rotational isomeric states for their side chains. Thus, only conformations with uniquely different backbone dihedral angles need be considered in the chain build-up procedure. Combination of the unique or non-degenerate minima for neighboring peptide segments in a polypeptide chain allows computation of the allowed structures for long polypeptide sequences. This method of combining non-degenerate minima[18,19] has been applied to a number of polypeptides and proteins known to require membranes or nonpolar environments to fold correctly. In order to adapt this method for the computation of protein structures in water, we have modified the procedure to allow for a more complete sampling of the possible rotational isomeric states of the side chains, and to include the effect of solvation; we apply the modified procedure here to interferon.

EARLIER APPLICATIONS

This methodology was applied previously to various model peptide systems to gain an understanding as to how the conformational energy determines the observed behavior. It was thus possible to identify, and verify by experiment, the handedness (twist) of various α-helices[20-24] and β-sheets,[25-28] the favorable arrangements of a α-helices[29-31] and collagen triple helices[32,33] with respect to each other, the interactions between α-helices and β-sheets,[34] and the transition curves for the interconversion of the cis and trans forms of polyproline,[35] of the α- and ω-helical forms of a crystalline homopolyamino acid,[36] and of the α-helical and coil forms of various homopolyamino acids.[37-39]

For polypeptides and proteins, where the regularity of model systems does not exist, the multiple-minima problem has been overcome for open-chain and cyclic peptides and fibrous proteins by the build-up procedure.[6] For example, the computed structures of gramicidin S,[40,41] the membrane-bound portion of melittin,[19] and the collagen model, poly(Gly-Pro-Pro),[42] have all been verified by experiment.[43-46] Recently, we have shown that the build-up procedure[47] and an importance-sampling Monte Carlo method[48] both lead to the same structure for enkephalin.

LARGER PEPTIDES

As the build-up procedure is applied to polypeptides of increasingly longer chain lengths, two problems arise. First, with the increasing number of independent variables, each minimization needs more iterations to converge (approximately in proportion to the number of variables) and takes more time per iteration (in proportion to the square of the number of variables). By coding part of the computation for a Floating Point Systems AP120B array processor, it has been possible to speed up the calculations significantly,[49] to the point at which minimum-energy conformations for hexapeptides and octapeptides are calculated routinely. However, peptides larger than fifteen or twenty residues still require an inordinate amount of time to build up on this system.

The second problem that comes with increasing peptide size arises from the need to retain more and more minima at each step, in order to be sure that the native conformation is not discarded. Since there are, in principle, of the order of 10^N possible minimum-energy conformations, where N is the number of amino acid residues in the peptide, it is necessary to discard the higher-energy conformations at every stage of the build-up. Our practice has been to determine a cutoff energy at each stage and keep only those minimum-energy conformations whose energies are within this cutoff energy of the lowest energy found for that stage. In the original use of the method, the cutoff energy was chosen to be 3.0 kcal/mol, based on considerations of statistical weights of conformers of amino acids and small peptides.[19] For a peptide in a nonpolar environment, such as a membrane-bound protein, the relative stability of two conformers should be approximated quite well by calculating statistical weights directly from the ECEPP potential; hence application of a cutoff energy of 3.0 kcal/mol to the ECEPP energies of the minimum-energy conformations of such peptides is not likely to eliminate any conformation that might ultimately contribute to the native conformation of the whole peptide or protein. However, with a peptide in aqueous solution, the situation is different. The cutoff energy must now be applied to the minimum-energy conformations of the solvated peptide if the calculations are to reflect the aqueous environment. Our practice is to perform all minimizations with the ECEPP potential, as though the peptide were in a medium of low dielectric constant, and without any cutoff energy; then, when all minimum-energy conformations have been generated, their energies are recalculated as though they were in aqueous solution, without changing the dihedral angles, and only those conformations whose energies with solvation are within 3.0 kcal/mole of the lowest energy are retained for future use. This procedure is based on the observation that the bulk of the water of hydration in protein crystals is unstructured, and does not appear to interact in a specific manner with the individual atoms of the protein other than to partially solvate exposed polar atoms and groups on the surface of the protein.[50] Hence, the actual dihedral angles of a minimum-energy conformation should not be greatly affected by the presence or absence of solvent, although of course the relative energies of different conformations might be very much altered because of unequal energies of hydration of solvent-exposed polar groups. In agreement with this, it may be noted that the dihedral angles of the solvated minimum-energy conformations of terminally blocked alanine[11] and other amino acids[12] differed by at most 5° from those of the unsolvated minimum-energy conformations,[14] although the relative energies and ordering of the conformations diverged sharply.

Usually only non-degenerate minimum-energy conformations are kept at each step; nevertheless, the build-up procedure can sometimes result in the retention of several hundred conformations of comparable energies (with solvation), because inclusion of solvation energy tends to level out differences in the ECEPP energy between the various local minima. [This trend was noted previously with terminally blocked dipeptides.[51]] Each of these conformations must then be combined with tens or even hundreds of conformations of the next peptide to which it is to be joined.

PARALLELIZATION

The two problems of more time per minimization and more minimizations at each step lead to a huge increase in the amount of computation as the length of the peptide increases. It becomes imperative to reduce the time spent in computation by some means, and parallelizing the computations is an attractive solution. The present problem is rather easily adapted for parallel processing. If there are n_1 minimum-energy conformations of peptide fragment 1 and n_2 minimum-energy conformations of fragment 2, we have to perform $n_1 n_2$ minimizations in all. With m processors running in parallel, we can split the minimizations into packages of $n_1 n_2/m$ and send one package to each processor. In practice, we do this in one of two ways. (1) The kth processor receives the kth input minimum-energy conformation of fragment 1, together with the $(k + m)^{th}$, $(k + 2m)^{th}$, etc., and joins each of these to each of the n_2 input minimum-energy conformations of fragment 2. (2) Since n_1 is usually at least ten times as large as m, we can send packages of up to 10 minimum-energy conformations of the first fragment to each processor and allow it to combine these with each of the n_2 conformations of the second fragment; then, when any processor finishes its allotted minimizations, it can receive another package of conformations of fragment 1 for the next task. At the end, all computed minimum-energy conformations of the whole peptide are returned from all the processors and combined, checked for redundancy and sorted by energy.

Parallel computations have been carried out on the system of "loosely coupled array processors" (LCAP), consisting of an IBM 4381 computer, two IBM 4341 computers and ten Floating Point Systems FPS-164 array processors.[52,53] The programs are coded so that, at least for peptides up to about 150 residues in length, all the information needed for a package of minimizations is transferred into the memory of one FPS-164 array processor. This cuts out most of the exchange of data between the IBM host computer and the array processors. Benchmark results for a small computation are shown in Figure 1. This figure shows the speed-up ratios for computing 1,000 minimum-energy conformations of the terminally-blocked dipeptide

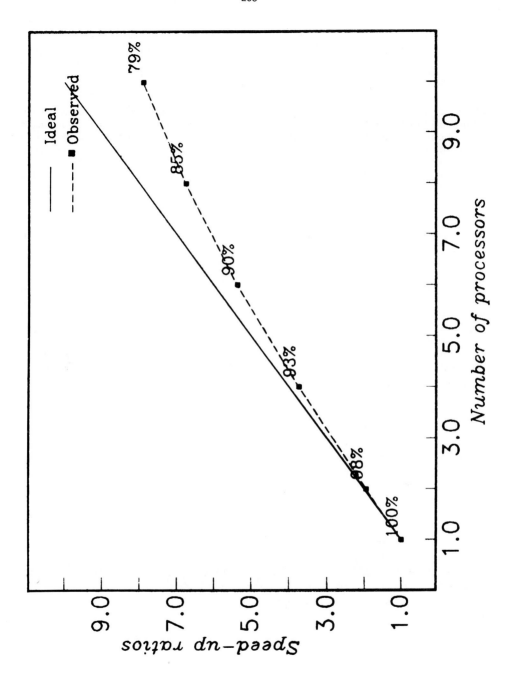

Figure 1. Speed-up ratios for parallel processing by parcelled assignments. One thousand minimizations of the blocked dipeptide N-acetyl-Gln-Leu-N′-methyl amide were carried out on 1, 2, 4, 6, 8 or

10 FPS-164 array processors run in parallel. Speed-up ratios were calculated from the elapsed times and compared with the ideal ratios.

N-acetyl-Gln-Leu-N'-methyl amide, with 17 variable dihedral angles, using 1, 2, 4, 6, 8 or 10 array processors in parallel. The observed speed-up ratio is compared with the theoretical speed-up ratio for the same number of array processors. The efficiency began to drop off when 4 processors were run in parallel, but even with 10 processors the efficiency was still nearly 80% of theory. With a peptide as small as this one, the time taken to pass data into and out of the host computer and to compare and sort all the conformations at the end of the calculations is a significant fraction of the whole elapsed time. Since these two processes are not parallelized, it was expected that the benchmark results would not agree perfectly with theory, and the results in Figure 1 are satisfactory.

COMPUTATIONS WITH INTERFERON

Human leukocyte interferons (IFN-α) are antiviral proteins of 155 or 156 amino acid residues, which are synthesized and secreted by leukocytes in response to several types of viral or bacterial infection.[54,55] At least 15 different IFN-α's have been identified and sequenced, either directly[56] or through their genes or cDNAs.[57-59] The genes code for 165 or 166 residues, but the biologically active proteins isolated from cultures of leukocytes lack the ten C-terminal amino acids.[56] There is a high degree of homology between these sequences, and about half of the amino acids are completely conserved throughout the family of IFN-α's. There is also a 33% homology between the IFN-α's and IFN-β, the single known human fibroblast interferon.

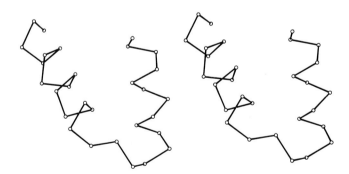

Figure 2.   Stereo drawing of the lowest-energy conformation of the blocked 29-residue peptide of IFN-αA (122-150). Only the α-carbon atoms are shown. The longer helix is at the N-terminus of the peptide.

Although IFN-αA has been crystallized,[60] there has been no report of an X-ray diffraction study to date. Attempts have been made to determine regular backbone structure by circular dichroism and Raman spectroscopy,[61-63] with general agreement that the protein is rich in α-helix. The most recent estimate for IFN-αA by circular dichroism assigns 59% of the residues to α-helix, 16% to antiparallel β-sheet, 18% to β-turns and 13% to other conformations, with no parallel β-sheet.[63] The results of Raman spectroscopy suggest 71 to 75% α-helix and 7 to 10% β-sheet.[64] Several attempts to predict regular backbone structure, using methods based on statistical analysis of known protein conformations, have yielded estimates of α-helix varying between 40% and 75%, with less than 20% β-sheet.[63-67] Sternberg and Cohen[66] carried their prediction further, by identifying four sequences ranging from 11 to 25 residues in length as being most probably α-helical and proposing two arrangements for packing these together in space. An even more detailed prediction has recently been put forward by Ptitsyn et al.,[67] who propose that the first 110 residues of IFN-α and IFN-β comprise one domain consisting of three large contiguous α-helices, with a short α-helix connecting the first two large helices in IFN-αA and IFN-β; the C-terminal portion of the molecule is predicted to fold either as three roughly equal close-packed helices or, more probably, as two α-helices lying against a three-stranded β-sheet. The general consensus from all these experimental and predictive investigations is that the protein is rich in α-helix with at most a small amount of β-sheet, and can probably be placed in Richardson's category I of proteins with antiparallel α-domains.[50]

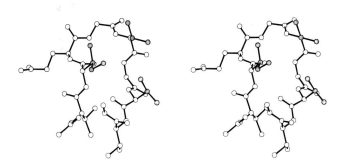

Figure 3. Stereo drawing of the lowest-energy conformation of the blocked octapeptide Val[100]-Ile-Gln-Gly-Val-Gly-Val-Thr [107]. Only heavy atoms are shown. The close proximity of the side chains of Ile[101], Val[104] and Val[106] (shaded in the figure) should be noted.

In our work, we have so far concentrated on three regions of the sequence of IFN-αA. This particular IFN differs from most other IFN-α's in having 155 rather than 156 residues, the missing residue being at position 44.[56-59] The first region that we chose lies toward the C-terminus, from residues 122 to 150. This region contains the sequence Cys-Ala-Trp-Glu-Val-Val-Arg-Ala-Glu (residues 139 to 147), which is conserved in all human and murine IFN-α genes,[58,68] and the half-cystine residue at position 139 is involved in the only disulfide bridge needed for activity.[62,69] For these reasons, we judged this part of the molecule to be functionally important [however, recent work by Valenzuela et al.[70] raises strong doubts about the functional significance of sequence conservation in the IFN-α family].

Table 1. Protocol for the build-up procedure

Peptide No. 1, residues 122 to 150

N-terminal portion[a]	C-terminal portion[a]	Total Number of starting points
AW	E	1,224
AWE	VV	1,887
AWEVV	R	1,545
E	IM	2,006
EIM	R	2,505
A	EIMR	1,773
AWEVVR	AEIMR	24,948
SPC	AWEVVRAEIMR	29,072
YF	Q	1,394
YFQ	R	14,460
LY	L	594
IT	LYL	26,474
ITLYL	K	6,540
YFQR	ITLYLK	67,032
K	YFQRITLYLK	5,040
KYFQRITLYLK	E	7,769
KYFQRITLYLKE	K	6,150
KYFQRITLYLKEK	K	2,550
KYFQRITLYLKEKK	Y	630
KYFQRITLYLKEKKY	SPCAWEVVRAEIMR	8,000

Peptide No. 2, residues 60 to 68

N-terminal portion[a]	C-terminal portion[a]	Total Number of starting points
IQ	QI	9,942
M	IQQI	2,353
MIQQI	F	4,410
MIQQIF	N	5,967
MIQQIFN	L	2,706
MIQQIFNL	F	5,782

Peptide No. 3, residues 100-107

N-terminal portion[a]	C-terminal portion[a]	Total Number of starting points
VI	Q	264
VIQ	GV	5,434
VIQGV	GV	6,422
VIQGVGV	T	3,663

[a]One-letter amino acid code[72]

The second region lies between residues 60 and 68, and includes Gln[62], which is conserved throughout all known human, murine and bovine IFN-α's and IFN-β's.[70] Application of an algorithm for predicting nucleation sites for protein folding[71] suggested that the sequence Met-Ile-Gln-Gln-Ile-Phe, residues 60 to 65, would be a nucleation site for IFN-αA; hence it seemed logical to include this part of the molecule in our conformational search.

Table 2. Low-Energy Minima of the 29-residue peptide, 122-150

Residue No. in sequence	Type[a]	Backbone Conformation of Minimum No.:				
		1	2	3	4	5
122	K	A	A	A	A	A
	Y	A	A	A	A	A
	F	A	A	A	A	A
125	Q	A	A	A	A	A
	R	A	A	A	A	A
	I	A	A	A	A	A
	T	A	A	A	A	A
	L	A	A	A	A	A
130	Y	A	A	B	A	B
	L	A	A	G	A	G
	K	A	A	A	A	A
	E	A	E	A	A	A
	K	A	C	E	A	A*
135	K	A	E	C	A	A
	Y	E	D	A	B	G
	S	D	F	D	D	E
	P	A	A	A	C	A
	C	F	A	F	C	C
140	A	A	A	A	A	A
	W	A	A	A	A	A
	E	A	A	A	A	A
	V	A	A	A	A	A
	V	A	A	A	A	A
145	R	A	A	A	A	A
	A	A	A	A	A	C
	E	A	A	A	A	A
	I	A	A	D	A	A
	M	A	A	G	A	A
150	R	B	D	A	B	A
Energy (with solvation)[b]		0.0	1.79	1.80	1.84	2.50

[a]One-letter amino-acid code[72]

[b]In kcal/mol, relative to minimum No. 1

The third region runs between residues 100 and 107. In contrast with the other two regions, this one is not conserved among IFN-α's but is quite variable.[59] In human IFN-αA, the amino acid sequence is Val-Ile-Gln-Gly-Val-Gly-Val-Thr. The high proportion of branched-chain amino acids and the presence of two glycine residues suggested to us that the conformation of this part of the molecule might include some form of chain reversal, such as a β-bend.

Our protocol for building up the three fragments is shown in Table 1. The guiding principle was to build nuclei of nonpolar residues wherever possible, since these have fewer single-residue minima[14] and thus require less computation than oligopeptides containing polar residues.[19] Residues with polar side chains have usually been added individually to the ends of longer chains. As far as possible, ligation of two large peptides has been avoided, since this is time-consuming. However, in the last step of construction of the 29-residue peptide,it was necessary

Table 3. Low-Energy Minima of the Nonapeptide Segment 60-68

Residue:

No. in sequence	60	61	62	63	64	65	66	67	68
Type[a]	M	I	Q	Q	I	F	N	L	F

Minimum No.				Backbone Configuration						Energy (with solvation)[b]
1	A	A	A	A	A	A	A	A	C	0.0
2	A	A	A	A	A	A	A	A	E	0.62
3	A	A	A	A	A	A	B	G	A	0.95
4	A	A	A	A	A	A	A	E	A	1.19
5	A	A	A	A	A	A	A	A	A	1.21
6	A	A	A	A	A	A	C	G	G	1.50
7	A	A	A	A	A	A	A	A	D	1.59
8	A	A	A	A	A	A	A	E	C	1.69
9	A	A	A	A	A	A	D	G	A	1.72
10	F	A	A	A	A	A	A	A	D	2.09
11	F	A	A	A	A	A	A	A	A	2.10
12	A	A	A	A	A	A	A	A	G	2.12
13	C	A	A	A	A	A	A	A	A	2.14
14	C	A	A	A	A	A	A	A	D	2.20
15	E	A	A	A	A	A	A	A	A	2.37
16	A	A	A	A	A	A	A	D	A	2.49
17	A	A	A	A	A	A	A	D	C	2.62

[a]One-letter amino-acid code[72]

[b]In kcal/mol, relative to minimum No. 1

to join a 15-residue peptide to a 14-residue peptide, a process that required a few hundred array-processor hours.

Our results to date are presented in Tables 2 to 4 and Figures 2 and 3. The predicted conformations must still be regarded as preliminary because we have not yet sampled enough of conformational space to be sure that other, more stable, conformations do not exist. Nevertheless, certain trends have appeared, which we believe will be maintained without extensive change. Firstly, the lowest-energy conformations of the 29-residue peptide near the C-terminus all take the form of two α-helices joined by a hinge consisting of between four and ten residues around the tripeptide Ser-Pro-Cys, with the helices packed together in a more or less antiparallel manner. The conformations of the five minima whose energies with solvation lie within 3.0 kcal/mol of the lowest energy are presented in Table 2, in the notation of Zimmerman et al.,[14] in which the φ-ψ map of each residue is divided into 16 regions denoted by a letter of the alphabet; these regions are shown in Figure 4. The conformation with the lowest energy with solvation also had the lowest energy without solvation, this being 6.8 kcal/mol lower than the energy without solvation of the next conformation. Thus, the lowest-energy conformation in Table 2 is the most stable one by either criterion that we have identified to date for the portion of the protein surrounding Cys[139]. A stereo drawing of the conformation is shown in Figure 2. An α-helix from residue 122 to residue 135 is joined by a four-residue hinge to another α-helix from residue 140 to residue 149, with the centers of the two helices at a distance of 24 Å and the helix axes making a dihedral angle of − 130°. In terms of helix packing, this is one of the stable arrangements identified by Chou et al.[30] The two helical regions in Figure 2

Table 4. Low-Energy Minima of the Octapeptide Segment 100-107

Residue:

No. in sequence	100	101	102	103	104	105	106	107
Type[a]	V	I	Q	G	V	G	V	T

Minimum No.				Backbone Configuration					Energy (with solvation)[b]
1	C	C	C	D*	A	A*	E	C	0.0
2	E	A	C	D*	A	D*	E	C	0.55
3	A	A	A	B*	A	A	A	A	0.56
4	E	A	C	D*	A	B*	E	F	1.13
5	E	A	C	D*	A	B*	E	A	1.46
6	E	C	C	D*	A	A*	E	F	1.52
7	E	A	C	D*	A	B*	E	E	1.54
8	C	A	C	D*	A	A	A	A	1.55
9	E	C	C	D*	A	A*	E	C	1.56
10	C	A	C	D*	A	A	A	C	1.66
11	A	A	A	B*	A	A	A	G	1.80
12	E	A	C	D*	A	A	A	A	1.89
13	F	C	C	D*	A	A*	E	A	2.08
14	E	A	C	D*	A	D*	E	A	2.09
15	C	D	C	A*	E	C*	A	A*	2.32
16	F	C	C	D*	A	A*	E	C	2.34
17	E	A	C	D*	A	A*	E	C	2.63
18	F	C	C	D*	A	A*	E	F	2.72
19	E	A	C	D*	A	A	E	A	2.76
20	E	A	C	D*	A	A	E	F	2.82
21	C	A	E	C*	D	E	A	A	2.83
22	E	A	C	D*	A	A	A	C	2.87
23	C	A	C	D*	A	A	E	A	2.96
24	E	F	A	A	A	A	A	A	2.96
25	A	C	A*	A*	E	C	A	A	2.97

[a]One-letter amino-acid code[72]

[b]In kcal/mol, relative to minimum No. 1

correspond quite well with portions of the E and E′ helices in the less probable of the two conformations proposed by Ptitsyn et al.[67] for the C-terminal domain of IFN-α. However, in our structure the helices pack with a left-handed twist rather than the right-handed twist suggested by Ptitsyn et al.

Turning to the second region that we have examined in IFN-αA, the seventeen minima of Met-Ile-Gln-Gln-Ile-Phe-Asn-Leu-Phe whose energies with solvation are within the cutoff are all α-helical except possibly at the ends (Table 3). We believe that this part of the IFN-αA molecule will turn out to be α-helical in the whole protein. Since this region is either fully conserved in IFN-α's and IFN-β, or at most subject to conservative amino acid substitutions,[59] we predict that all IFN-α's and IFN-β will have an α-helix in this portion of the molecule. It is interesting to note that three commonly used predictive algorithms that are based on analysis of known protein conformations indicated that this region of IFN might be α-helical;[66] so also does the predictive scheme of Ptitsyn and his colleagues.[67]

210

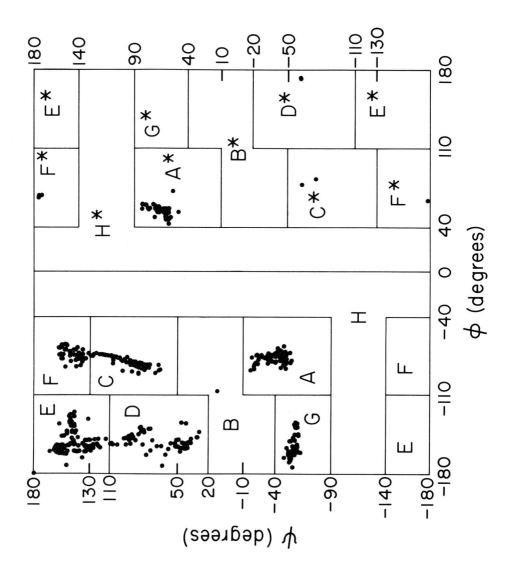

Figure 4.   φ-ψ map showing the regions defining the conformational letter code. The closed circles represent 969 low-energy minima (ΔE ≤ 5 kcal/mol) for the 20 naturally occurring amino acid residues (Reference 14).

The third region that we have examined in IFN-αA, surrounding residue 105, appears not to have a regular conformation according to our calculations. A total of twenty-five conformations have been identified whose energies with solvation lie within 3.0 kcal/mol of the lowest energy (Table 4). In all but one of these, Gly^{103} is in a conformation in the right half of the φ-ψ map of Fig. 4 (starred in the notation of ref. 14). In the lowest-energy minima, this glycine residue consistently adopts a starred conformation. The other glycine residue, at position 105, takes on a starred conformation in 52% of all cases. The four lowest-energy conformations all have a compact form, with two or three of the four nonpolar side chains packed together in a single hydrophobic globule. A stereo drawing of the lowest-energy conformation, in which two Val side chains and the Ile side chain have come together to form an "oil droplet", is shown in Fig. 3. Generalizing these observations to other IFN-α's, we propose that the portion of the molecule between residues 100 and 110 links two α-helices together, either through a β-turn or as an unclassified "irregular" structure.

CONCLUDING REMARKS

Our experience and that of other workers has shown that the computational effort involved in predicting the native conformations of proteins from their amino acid sequences is immense. We have outlined here a promising path for future development by showing that parallelizing the build-up procedure on a well-designed computer system, with a proven semi-empirical energy formulation supported by an efficient technique for energy minimization, can surmount some of the computational hurdles and perhaps accelerate the resolution of this challenging biological problem.

ACKNOWLEDGEMENTS

We thank George Nemethy for help with the determination of helix packing, and for comments on this manuscript, Shirley Rumsey for help with the stereo drawings, and the members of the IBM LCAP Laboratory for their continued cooperation.

REFERENCES

1. C.B. Anfinsen, Science **181**, 223 (1973).
2. C.B. Anfinsen and H.A. Scheraga, Adv. Protein Chem., **29**, 205 (1975).
3. H.A. Scheraga, Chem. Revs., **71**, 195 (1971).
4. J.E. Dennis and H.H.W. Mei, Technical Report No. 75-246, Department of Computer Science, Cornell University, Ithaca, New York (1975).
5. D.M. Gay, ACM Trans. Math. Software, **9**, 503 (1983).
6. H.A. Scheraga, Biopolymers, **22**, 1 (1983).
7. G. Némethy and H.A. Scheraga, Q. Rev. Biophys., **10**, 239 (1977).
8. H.A. Scheraga, Carlsberg Res. Commun., **49**, 1 (1984).
9. F.A. Momany, R.F. McGuire, A.W. Burgess and H.A. Scheraga, J. Phys. Chem., **79**, 2361 (1975).
10. G. Némethy, M.S. Pottle and H.A. Scheraga, J. Phys. Chem., **87**, 1883 (1983).
11. G. Némethy, Z.I. Hodes and H.A. Scheraga, Proc. Natl. Acad. Sci., U.S., **75**, 5760 (1978).
12. Z.I. Hodes, G. Némethy and H.A. Scheraga, Biopolymers, **18**, 1565 (1979).
13. P.N. Lewis, F.A. Momany and H.A. Scheraga, Israel J. Chem., **11**, 121 (1973).
14. S.S. Zimmeraman, M.S. Pottle, G. Némethy and H.A. Scheraga, Macromolecules, **10**, 1 (1977).
15. M. Vasquez, G. Némethy and H.A. Scheraga, Macromolecules, **16**, 1043 (1983).

16. S.S. Zimmerman and H.A. Scheraga, Biopolymers, **16**, 811 (1977); **17**, 1849, 1871, 1885 (1978).
17. K.D. Gibson, S. Chin, M.R. Pincus, E. Clementi and H.A. Scheraga, in preparation.
18. M.R. Pincus and R.D. Klausner, Proc. Natl. Acad. Sci., U.S., **79**, 3413 (1982).
19. M.R. Pincus, R.D. Klausner and H.A. Scheraga, Proc. Natl. Acad. Sci., U.S., **79**, 5107 (1982).
20. R.A. Scott and H.A. Scheraga, J. Chem. Phys., **45**, 2091 (1966).
21. T. Ooi, R.A. Scott, G. Vanderkooi and H.A. Scheraga, J. Chem. Phys., **46**, 4410 (1967).
22. J.F. Yan, G. Vanderkooi and H.A. Scheraga, J. Chem. Phys., **49**, 2713 (1968).
23. J.F. Yan, F.A. Momany and H.A. Scheraga, J. Am. Chem. Soc., **92**, 1109 (1970).
24. E.H. Erenrich, R.H. Andreatta and H.A. Scheraga, J. Am. Chem. Soc., **92**, 1116 (1970).
25. K.C. Chou, M. Pottle, G. Némethy, Y. Ueda and H.A. Scheraga, J. Mol. Biol., **162**, 89 (1982).
26. K.C. Chou and H.A. Scheraga, Proc. Natl. Acad. Sci., U.S., **79**, 7047, (1982).
27. K.C. Chou, G. Némethy and H.A. Scheraga, J. Mol. Biol., **168**, 389, (1983).
28. K.C. Chou, G. Némethy and H.A. Scheraga, Biochemistry, **22**, 6213 (1983).
29. K.C. Chou, G. Némethy and H.A. Scheraga, J. Phys. Chem., **87**, 2869 (1983).
30. K.C. Chou, G. Némethy and H.A. Scheraga, J. Am. Chem. Soc. **106**, 3161 (1984).
31. M. Gerritsen, K.C. Chou, G. Némethy and H.A. Scheraga, Biopolymers, **24**, 2177 (1985).
32. G. Némethy, Biopolymers, **22**, 33 (1983).
33. G. Némethy and H.A. Scheraga, Biopolymers, **23**, 2781 (1984).
34. K.C. Chou, G. Némethy, S. Rumsey, R.W. Tuttle and H.A. Scheraga, J. Mol. Biol., **186**, 591 (1985).
35. S. Tanaka and H.A. Scheraga, Macromolecules, **8**, 516 (1975).
36. Y.C. Fu, R.F. McGuire and H.A. Scheraga, Macromolecules, **7**, 468 (1974).
37. M. Gō, N. Gō, and H.A. Scheraga, J. Chem. Phys. **54**, 4489 (1971).
38. M. Gō, F.T. Hesselink, N. Gō and H.A. Scheraga, Macromolecules, **7**, 459 (1974).
39. M. Gō, and H.A. Scheraga, Biopolymers, **23**, 1961 (1984).
40. M. Dygert, N. Gō and H.A. Scheraga, Macromolecules, **8**, 750 (1975).
41. G. Némethy and H.A. Scheraga, Biochem. Biophys. Res. Commun., **118**, 643 (1984).
42. M.H. Miller and H.A. Scheraga, J. Polymer Sci., Polymer Symposia No. **54**, 171 (1976).
43. S.E. Hull, R. Karlsson, P. Main, M.M. Woolfson and E.J. Dodson, Nature, **275**, 206 (1978).
44. V.M. Naik, S. Krimm, J.B. Denton, G. Némethy and H.A. Scheraga, Int. J. Peptide Protein Res., **24**, 613 (1984).
45. T.C. Terwilliger, L. Weissman and D. Eisenberg, Biophys. J., **37**, 353 (1982).
46. K. Okuyama, N. Tanaka, T. Ashida and M. Kakudo, Bull. Chem. Soc. Japan, **49**, 1805 (1976).
47. M. Vasquez and H.A. Scheraga, Biopolymers, **24**, 1437 (1985).
48. G.H. Paine and H.A. Scheraga, Biopolymers, **24**, 1391 (1985).
49. C. Pottle, M.S. Pottle, R.W. Tuttle, R.J. Kinch and H.A. Scheraga, J. Computational Chem., **1**, 46 (1980).
50. J.S. Richardson, Adv. Protein Chem., **34**, 167 (1981).
51. Z.I. Hodes, G. Némethy and H.A. Scheraga, Biopolymers, **18**, 1611 (1979).
52. E. Clementi, G. Corongiu, J.H. Detrich, H. Khanmohammadbaigi, S. Chin, L. Domingo, A. Laaksonen and H.L. Nguyen, in Structure and Motion: Membranes, Nucleic Acids and Proteins (ed. E. Clementi, G. Corongiu, M.H. Sarma and R.H. Sarma), Adenine Press, Guilderland, NY, p 49 (1985).
53. E. Clementi, in "Supercomputer Simulations in Chemistry," Symposium, Montreal, Canada, Aug. 25-27, 1985.
54. P. Lengyel, Ann. Rev. Biochem., **51**, 251 (1982).
55. S. Pestka and S. Baron, Meth. Enzymol., **78**, 3 (1981).
56. W.P. Levy, M. Rubinstein, J. Shively, U. Del Valle, C.-Y. Lai, J. Moschera, L. Brink, L. Gerber, S. Stein and S. Pestka, Proc. Natl., Acad. Sci., U.S., **78**, 6186 (1981).
57. D.V. Goeddel, D.W. Leung, T.J. Dull, M. Gross, R.M. Lawn, R. McCandliss, P.H. Seeburg, A. Ullrich, E. Yelverton and P.W. Gray, Nature **290**, 20 (1981).
58. C. Weissmann, S. Nagata, W. Boll, M. Fountoulakis, A. Fujisawa, J.-I. Fujisawa, J. Haynes, K. Henco, M. Mantei, H. Ragg, C. Schein, J. Schmid, G. Shaw, M. Streuli, H. Taira, K. Todokaro and U. Weidle, Phil. Trans. Roy. Soc., **B 299**, 7 (1982).
59. J. Collins, Symp. Soc. Gen Microbiol., **35**, 35 (1983).
60. D.L. Miller, H.-F. Kung, T. Staehelin and S. Pestka, Meth. Enzymol., **79**, 3 (1981).

61. T.A. Bewley, H.L. Levine and R. Wetzel, Int. J. Peptide Proteins Res., **20**, 93 (1982).
62. R. Wetzel, H.L. Levine, D.A. Estell, S. Shire, J. Finer-Moore, R.M. Stroud and T.A. Bewley, in **Interferons** (ed. T.C. Merigan and R.M. Friedman), UCLA Symposia on Molecular and Cellular Biology, Academic Press, Vol. XXV, p. 365 (1982).
63. P. Manavalan, W.C. Johnson, Jr. and P.D. Johnston, FEBS Lett., **175**, 227 (1984).
64. R.W. Williams, J. Biol. Chem., **260**, 3937 (1985).
65. T.G. Hayes, Biochem. Biophys. Res. Commun., **95**, 872 (1980).
66. M.J.E. Sternberg and F.E. Cohen, Int. J. Biol. Macromol., **4**, 137 (1982).
67. O.B. Ptitsyn, A.V. Finkelstein and A.G. Murzin, FEBS Lett., **186**, 143 (1985).
68. G.D. Shaw, W. Boll, H. Taira, N. Mantei, P. Lengyel and C. Weissmann, Nucleic Acid Res., **11**, 555 (1983).
69. R. Wetzel, P.D. Johnston and C.W. Czarniecki, in *The Biology of the Interferon System* (ed E. De Maeyer and H. Schellekens), Elsevier, p. 101 (1983).
70. D. Valenzuela, H. Weber and C. Weissman, Nature, **313**, 698 (1985).
71. R.R. Matheson, Jr. and H.A. Scheraga, Macromolecules, **11**, 819 (1978).
72. M.O. Dayhoff, Atlas of Protein Sequence and Structure, National Biomedical Research Foundation, Silver Springs, Md., Vol. 5, p. D2 (1972).

ALGORITHMS FOR SUPERCOMPUTERS

Berni J. Alder

Lawrence Livermore National Laboratory

Better numerical procedures, improved computational power and additional physical insights have contributed significantly to progress in dealing with classical and quantum statistical mechanics problems. Past developments are discussed and future possibilities outlined.

I. INTRODUCTION

The computer, an experimental device that performs fast arithmetic, has revolutionized the field of statistical mechanics by providing a method for treating the many-body problem. Previously, because of mathematical limitations, all such problems involving many interacting particles had to be reduced to ones involving only a few degrees of freedom. Usually the many-particle system was approximated by a single particle in the averaged field of all the others. The van der Waals theory is an example of such a mean field approximation. Alternative procedures that omit the correlation between particles are called self-consistent theories, of which the Hartree-Fock scheme of quantum mechanics is an example.

To be sure the computer can not deal with an Avogadro's number ($N = 6 \times 10^{23}$) of particles, however it can simulate a sufficiently large number (typically $N = 10^3$) with the proper boundary conditions, so that bulk matter is accurately represented; corrections being order $1/N$. Phase transitions are an exception to this rule; however, for second order transitions, such as the critical point, use can be made of the renormalization theory to scale the system to larger and larger size.[1] For the quantum mechanical calculation of molecules, for the structure of nuclei and for small clusters of atoms, we only have to deal with a finite number of particles; in fact, being able to deal with finite systems represents an advantage.

The principle method of quantum chemistry, namely the variational method, is able to go beyond the one-particle approximation. However, it has been demonstrated over many years of application, that in its present form it is too slowly convergent. This is because the procedure expands the wavefunction in a complete set of one-particle functions, called orbitals, thus avoiding higher order integrals. Frequently the orbitals are chosen such that the low dimensional integrals can be carried out analytically. The price to be paid is that a huge number of expansion terms (in the millions) have to be employed, resulting in a numerical method that is not sufficiently accurate for chemical purposes, and also is extremely awkward to use on computers.

The many-body algorithms overcome these limitations and provide a much more versatile variational procedure.[2] In such a calculation the wavefunction is expanded, for example, in pair product, using any desired functional form; the resulting higher dimensional integrals are performed numerically. Even with very few terms (less than 100) higher accuracy is achieved than in the conventional variational procedures.

Furthermore, the results from the variational procedure can be employed as the importance function in another many-body algorithm that yields the exact answer, if run sufficiently long.[3] Such calculations are exact in the sense that the correct answer is always contained within the known (statistically determined) error bars. These error bars or the variances of the results decrease as the inverse square root of the length of the computer run. Thus, the accuracy with which the result can be determined depends critically on developing an efficient calculational procedure. Another enormously favorable aspect of this numerical method is that it is extremely simple and highly repetitive, thus taking full advantage of what the computer does best. The codes are at least an order of magnitude smaller than those for the conventional variational chemistry calculation and the procedure is easily adaptable to any computer architecture.

We shall in the next section describe these algorithms as they have been applied to classical and quantum systems. First, the Monte Carlo method for classical equilibrium calculations will be discussed. Then the extension of the Monte Carlo method to the description of quantum systems will be illustrated. For that purpose the separate procedures that have been developed for describing particles obeying either Boson or Fermi statistics will be described. The extension of these calculations to finite temperature requires additional considerations. We shall then discuss the molecular dynamics algorithms for classical particles from which both transport and equilibrium properties can be determined. Finally, the absence of a viable numerical procedure to deal with many-body quantum transport processes will be emphasized.

It should also be pointed out that these numerical procedures have wider applicability than is discussed here. In fact, they are increasingly used to solve all sorts of practical many-dimensional optimization problems such as that of the traveling salesman or the best layout of many electronic components on a silicon chip.[4] Interestingly enough, extension of the numerical methods outlined here are also being applied to equation of quantum chromodynamics.[5] In these lattice gauge calculations the dimensionality of the equations is so high that no analytic procedure has so far yielded results. Thus, one is forced to employ numerical procedures to test whether the theory makes correct predictions.

II. CLASSICAL MONTE CARLO

The Monte Carlo method is not a totally random process as the name implies. A completely random process would involve calculating the potential energy of the system from the potential energy of interaction between the particles whose centers are located at randomly chosen points. The Boltzmann factor, namely the negative exponential of that potential energy divided by the Boltzman constant times the temperature, can then be formed. This determines the probability of that configuration. At typical liquid densities we would overwhelmingly find that we had formed a configuration of vanishing probability, because in such a random process at relatively high density, particles are likely to be placed so close together that their interaction energy is highly repulsive. In other words, in such a random process it is highly unlikely to generate the local order of the particles in the liquid state, that order being created because particles are forced, because of their size, to keep out of each others way.

To avoid this problem, importance sampling is introduced. In this scheme, configurations are chosen according to their Boltzmann probability. That means that in any average to be performed over the configurations generated, each configuration carries equal weight.[6] Thus improbable configurations are very unlikely to be generated. The process by which this is achieved is called a Markov process, a stochastic algorithm which is of the next higher complexity than a completely random process. We shall illustrate it here for hard spheres for simplicity. The hard spheres are initially put in any allowable configuration, namely one where no spheres overlap; for example, any ordered solid configuration. A randomly chosen particle is then displaced by a random amount in a random direction. If that displaced particle does not overlap with any other sphere in the system, the configuration is accepted (the Boltzmann factor is one) and the process repeated. If the random displacement results in an overlap, the configuration is rejected (the Boltzmann factor is zero) and the particle is placed back where it was and that old configuration is counted again in any averaging process. The reason for recounting is that the original configuration is more probable since a displacement from it was rejected. In more technical terms recounting is required because of microscopic reversibility.

In the above process, memory of the initial configuration is rapidly wiped out. Any desired average taken over the subsequently generated configurations. This simple algorithm was discovered even before the advent of electronic computers[7] and has basically not been improved. It is, on the one hand, as efficient as one can get, since each configuration counts toward the average equally. One might think that the random moves could be more cleverly selected so that one would move through phase space more efficiently. In some sense that is true. However, in trying to be clever, more computing has to be done. This generally overcomes the increased efficiency.[8] Molecular dynamics provide a more efficient way to move through phase space by following the equations of motion of the particles, with the importance function, so to speak, built in. As shall be discussed later, this method is comparable in efficiency to the Monte Carlo method.

The Monte Carlo method has been applied to almost every conceivable molecule and physical phenomenon, from argon to DNA and from melting to bilayer formation. There are two principle limitations to the method. One is that the intermolecular potential must be known. In the next section we show that this can be avoided by solving the Schrödinger equation. Then the only input are the masses and charges of the species and their Coulomb intermolecular potential. The other possible limitation is that the Markov process has to reach all of phase space. This may be difficult if the initial state is connected to the state of lowest free energy by a very improbable occurrence. Even worse is the case in which there is no way to reach the thermodynamically stable state from the initial state. An example is an initially assumed crystal structure in a given unit cell for a given number of particles. Under these circumstances, no transformation to another, possibly more stable, crystal structure can take place. However, a recent algorithmic development[9] allows the system, in principle, to find its most stable form by mimicking nature. This involves performing a constant pressure rather than constant volume simulation. At constant pressure the unit cell can change volume as well as shape. Some complex crystal structure changes with pressure have been successfully predicted by this approach.

III. QUANTUM MONTE CARLO

To solve the Schrödinger equation stochastically involves several additionally processes. The first one is a branching process or a birth and death rate process. This is because the very thing we want to solve for, namely the wave function, determines the underlying probability with which phase space has to be sampled. Concretely, the wave function squared replaces the Boltzmann factor of classical statistical mechanics. Hence, a guess is made for this wave function, called a trial wave function, which serves as the importance function.[10] This trial wave function is usually obtained from a variational calculation and with its use some thousand snapshots of the system are prepared. The kinetic energy term in the Schrödinger equation is analogous to the random walk process of solving the diffusion equation (Ficks' law in 3 N dimensions with a diffusion coefficient of $\hbar^2/2m$) Using this correspondence the random displacement of each point in each of the one thousand snapshots is carried out. For any displacement the energy is calculated and compared to the trial wave function energy to determine whether it results in too large or too small a probability at that point in phase space. If the probability is too large the configuration is killed; if too small, a number of replicas of that snapshot are made according to the difference to the guessed trial energy. Thus, the number of snapshots changes from the initial one thousand as the calculation progresses and the correct trial energy is adjusted till the number of snapshots has reached a constant average value over some number of Monte Carlo moves. At that point the birth and death rate cancel. This indicates that on the average, the trial wave function has been corrected to the true wave function and the correct energy eigenvalue has been found.

From the computational point of view, the process is not much more complex than the classical one. However, in quantum chemistry calculations very high precision is required, typically better than one part in 10^6. This is because the chemical bond energy is a small part of the total energy. Thus high accuracy in the trial wave function is required, since the variance, though still inversely proportional to the square root of the length of the run, has a coefficient which depends on the difference between the trial energy and the true energy.[11] However, high accuracy in the trial wave function generally requires a function with many terms which, slows down the calculation, since it has to be sampled every move. There is thus a high premium for a concise, accurate trial wave function, the search for which is still in progress.

The search can be helped by an adaptive Monte Carlo process. In such a procedure, the trial wave function is continuously improved using the output improved from the Monte Carlo process. This self-learning process could lead in principle to arbitrarily accurate results, depending on how often the process is repeated. Unfortunately, the process is difficult to carry out in practice because of the highly dimensional nature of the wave function and the statistical noise in the output wave function.

Another possible way to circumvent the high accuracy requirement in the total energy is to employ a differential Monte Carlo scheme, wherein only differences in energy are calculated.[12] In such a scheme, the difference in the energy between, say, two different configurations of the atoms, are calculated by starting each calculation with the same random number. In this way both the statistical fluctuations are reduced and identical configurations in the two different arrangements eliminated. So far, neither of these schemes to increase precision has been very successful.

The most serious numerical problem in quantum statistical mechanics involves Fermi statistics. The scheme so far discussed works well for Bosons at zero temperature, because the wave function is then everywhere positive and hence may be interpreted as a probability density. Unfortunately, the only physically accessible realization of such systems is helium four. For all other practical systems, such as ones containing electrons, the wave function has positive and negative regions, separated by nodes where the wave function goes through zero. This is because of the Pauli exclusion principle, which requires antisymmetry with respect to exchange of particles. In order to avoid the problem of having to deal probabilistically with negative functions, the fixed node approximation was introduced.[13] The fixed nodes were those provided by the trial wave function. Under this approximation one still deals with a wave function of one sign only; the Monte Carlo random walk is killed if a walk tries to cross the borders determined by the nodes. One can show that such a constraint, in a variational sense, leads to an upper bound to the energy. That bound is close to the correct answer as can be shown numerically. The placement of the nodes is not very important to the calculation of the energy, since the wave function is small there. The principle contribution to the energy comes from the region where the wave function is large.

To get the exact answer, however, the nodes must be released and this results in a numerically unstable process or transient estimate.[14] In the release process, once a random walk crosses a node, that walk is no longer eliminated but instead assigned a negative weight. Thus positive and negative populations are generated and each approaches its nodeless ground state. This manifests itself in an exponentially growing birth process for both populations, the desired antisymmetric wave function being the difference between these two exploding populations. That difference is, however, an increasing smaller fraction of these two populations as the Monte Carlo run proceeds and is thus easily overwhelmed by the statistical noise. The trick is to project out the difference before the noise becomes too large. That depends on having an accurate fixed node solution to start from. For low charged systems, the process has been successfully implemented. However, for high Z elements, the correct energy is projected out only with difficulty or large variance. This is because the exponential growth process has a higher rate (the difference between the fixed node energy and the nodeless Boson energy is larger).

Alternative algorithms which avoid this numerical instability have yet to be fully explored. The most promising is the so called Hubbard-Stratonovich transformation, which rigorously transforms a many-interacting Fermion particle system to a one-day calculation of the Hartree-Fock type by introducing a random field in the calculation.[15] It should however, be pointed out that even the present quantum Monte Carlo scheme has led to impressive results for low Z molecules and bulk system. For chemical calculations the energies are at least as accurate as the most elaborate variational calculations in comparable computer time. For bulk hydrogen starting with protons and electrons, the pressure versus volume curve agrees perfectly with experiments and allows reliable predictions beyond the experimental range.[16] Furthermore, for hydrogen detailed predictions of other properties have been made that have yet to be experimentally explored. For example, it is feasible to investigate the properties of clusters of lithium atoms, letting both the nuclei and electrons sample their accessible configurations, and predict the properties of the surface as well as the changed nature of a hydrogen molecule absorbed on that surface. One of the difficulties to overcome in that calculation is that the lithium nuclei diffuse at a much slower rate than the electrons, namely by the ratio of their masses. This multiple time scale problem can lead to difficulties in sampling the phase space accessible to the lithium nuclei. A possible way out of this difficulty is to combine a classical Monte Carlo calculation for the nuclei with a quantum mechanical one for the electrons.

The discussion on quantum systems has so far been confined to ground state calculations or zero temperature calculations for bulk systems. Excited states which are the lowest energy of a given symmetry class can readily be calculated by the appropriate choice of the trial wave functions.[17] Alternatively, excited states can be calculated by choosing a trial wave function which is orthogonal to the exact lower energy eigenstates.

Finite temperature calculations for Boson systems have recently been perfected. For the first time one can calculate the Boson condensation of an interacting helium gas.[18] The algorithm is an improved implementation of the analogy of the Feynman path method with the classical interacting polymer rings system. The polymer rings are made as small as possible by replacing the usual semi-classical density matrix representing the polymer links by a more accurate (at low temperature) pair-trial density matrix. To move this collection of smaller polymer rings more efficiently through all accessible configurations, the polymer segments are not moved one at a time but rather sections of several links are cut out and reconnected. That reconnection is more efficiently started at the middle of the cut out section by a process involving importance sampling, which forces reconnection at both ends. Furthermore, this procedure facilitates the sampling of permutation space required for Boson statistics. This involves possibility of different polymers cross linking, connecting one end of a polymer to the end of another while the corresponding segment does the opposite reconnection. Finite temperature Fermion algorithms (to deal, for example, with plasmas) are in the formative stage and the equivalent process to the fixed node approximation is in the process of being implemented.

IV. CLASSICAL MOLECULAR DYNAMICS

The simultaneous solution of the coupled Newtonian equations of motion by which both equilibrium and transport properties of many-body systems can be determined was developed shortly after the Monte Carlo method, immediately after the availability of the first electronic computers.[19] The method presents no great problem from the numerical point of view and proves to be competitive to the Monte Carlo method for equilibrium properties. To sample collective motions, such as occur in crystal transformations, the molecular dynamics method, once generalized to the constant pressure approach for equilibrium properties, appears both intuitively and actually to be more efficient. Again, the numerical methods developed early on, have not been improved upon in any significant way.

An essential difference between the Monte Carlo and molecular dynamics algorithms is that the former is easily adaptable to vector and parallel computer architecture while the latter is not because that calculation is necessarily sequential. It is possible to vectorize that part of the molecular dynamic calculation that involves calculating all the new forces on each particle after a time step has been taken, provided the interaction is long range, since then each particle interacts with all the other particles in the system. For short range forces even that vector is only as long as the number of particles within the range of the potential. The only way parallelism can be taken advantage of is to run several systems simultaneously to increase statistics.

Since it is highly desirable for some applications, as will be mentioned, to run very large systems of particles (10^6 particles) with short range forces for very long times (10^{-6} sec), a good case can be made for developing a special purpose machine to that end. Such machines that do the dynamics well have already been built[20] and it is desirable to incorporate that special hardware into a general purpose supercomputer so as to make the analysis of the efficiently generated trajectories, also efficient. It would then be possible to take fuller advantage of what has been the principle discovery of molecular dynamics, namely the surprising validity of hydrodynamics on a nearly microscopic scale.[21]

The idea is to study the initial phases of a hydrodynamic phenomenon on a microscopic scale by solving the Liouville equation (the simultaneous Newtonian equations) for a given initial condition. As the phenomena evolve in time and spread spacewise, it is then possible to analyze the results by averaging over spacial regions or grid regions. Each region represents the average properties of all the particles within that region, such as the average velocity, leading to determination of the velocity field. As the phenomena reach the boundary of the molecular dynamic system and the grid regions become large enough to be accurately represented by a continuum, it is possible to continue the calculation by using the molecular dynamically generated grid conditions as input to a Navier-Stokes continuum hydrodynamic calculation. The advantage of this procedure is that one can follow the evolution of the system from the microscopic frequency and wave length scale all the way to the macroscopic ones without invoking any approximations. Particularly, one can run the molecular dynamic calculation to the point where the hydrodynamic calculation is well represented by the Navier-Stokes equations, that is, where non-linearities associated with initially high gradients have become negligible. This appears to be a promising approach to study hydrodynamic instabilities, including turbulence. For example, the vortex shedding phenomena leading to a van Karmanwake has been successfully generated on a microscopic scale by the molecular dynamics method.[22]

In some physical situations, the event to be investigated occur only rarely, such as the jumping of a vacancy, the exchange of a water molecule in the neighborhood of an ion, or in chemical reactions. Special algorithms can be designed to study such occurrences, provided there is no coupling between the fact and slow degrees of freedom.[23] The particle participating in the rare event can then be considered to be in equilibrium with all the other particles. Thus it is possible by various artificial means to constrain the particle along the reaction path, using an external field, for example. The particle's potential of interaction with the rest of the system is calculated, that is, the adiabatic potential surface for the reaction or probability of the particle being at various positions along the reaction path. The reaction rate can be determined by calculating in addition the conversion probability. Given some constrained position along the reacting path, usually the point at the top of the barrier, one can calculate the conversion probability by removing the constraint and running the molecular dynamics trajectory forward and backwards in time. This determines whether the two resulting end configurations represent a successful reaction. The process is repeated for a number of initial constrained configurations to get an average conversion probability. By placing the particle in the rare event position and determining its probability to be there, we have avoided wasting all the computer time waiting for that event to occur.

V. QUANTUM MOLECULAR DYNAMICS

No viable algorithm exists to carry out many-body quantum molecular dynamics calculation to study, for example, the unusual superfluid transport properties of liquid helium four. One possible numerical approach illustrates the difficulties. By the so-called Madelung transformation[24] the Schrödinger equation can be rewritten in terms of the amplitude and phase of the wave function resulting in a coupled set of Navier-Stokes like hydrodynamic equations. In other words, instead of having to solve a coupled set of Newtonian equations, one for each particle, as in classical mechanics we now have to solve a very much more complicated coupled set of equations. It is difficult enough to solve a single Navier-Stokes equation in three dimensions. We now have to solve a set of such equations for a fluid mixture of N components, where each of the N particles can be considered to have a different color. To determine how these colors mix from some initial state is a formidable numerical problem.

Some information about the time evolution of a many-body system, namely the short time or slow relaxation aspects, can be obtained from equilibrium quantum Monte Carlo calculations. These equilibrium calculations solve the Schrödinger equation in the imaginary time and hence it is possible to evaluate the various correlation functions which determine the transport coefficients in imaginary time and extrapolate them to real time. Such extrapolation can be done empirically via

Pade approximates, for example.[25] Rigorously all derivatives of the real time correlation function at the time origin can be determined (at $t = 0$, where the real and imaginary time axis coincide). However, it is difficult to evaluate more than a few of these numerically.[26] From them a functional form can be made up that predicts the entire dependence on time, but we can reliably predict only the slow time dependence.

The computational difficulty of the quantum dynamics problem can best be demonstrated in the single particle case. This is the most frequently studied situation. For example, in the simplest chemical reaction, a deuterium atom exchanges with a hydrogen atom of a hydrogen molecule. Even this motion of a single atom on a potential surface in three dimensions can only be solved with great difficulty. For a two dimensional example, that of the Lorentz gas, where a quantum particle moves through a random maze, the calculation takes 10^4 times longer than the corresponding classical calculation. Part of the numerical difficulties is that the time dependent Schrödinger equation is conventionally solved on a grid, which makes the calculation even in low dimensions extremely time consuming.[27]

For steady state calculations, as in the above chemical rate calculation, quantum Monte Carlo methods ought to be a more efficient way to calculate the rate then the numerical methods used so far. Using quantum Monte Carlo method, we have so far only examined equilibrium properties (by setting the rate of change of the wave function with respect to time equal to zero), but clearly one could also obtain steady state properties. The method is a generalization of the familiar scheme whereby one obtains the inversion rate of ammonia by calculating the splitting of the degeneracy of the lowest vibrational energy level in a double minimum potential well, a purely equilibrium calculation.[28] The equivalence of the imaginary and real time solution of the Schrödinger equation has recently been demonstrated for the linear version of the deuterium exchange with hydrogen molecules.[29] Steady state rate calculations of radioactive decay have also been performed using Monte Carl methods.[30] The advantages of the Monte Carlo approach to steady state rate calculations are that it is easily extendable to higher dimensions and that via the proper importance function the rare event aspects of the calculation can be overcome. This is another of many examples where new numerical procedures can make much more realistic calculations feasible.

REFERENCES

1. R.H. Swendson, *Topics in Current Physics* (Ed. T.W. Burkhardt and J.M.J. van Leeuwen, Springer, Berlin, 1981) Vol. 30.
2. W.L. McMillan, Phys. Rev. **A138**, 442 (1965).
3. M.H. Kalos, D. Levesque, and L. Verlet, Phys. Rev. **A9**, 2178 (1974).
4. S. Kilpatrick, C.D. Gelatt, and M.P. Vecchi, Science **220**, 671 (1983).
5. K.G. Wilson, Phys. Rev. **D10**, 2445 (1974).
6. N. Metropolis, A.W. Rosenbluth, M.N. Rosenbluth, A.H. Teller, and E. Teller, J. Chem. Phys. **21**, 1087 (1953).
7. B.J. Alder, S.P. Frankel, and V.A. Lewinson, J. Chem. Phys. **23**, 417 (1955).
8. M. Rao and B.J. Berne, J. Chem. Phys. **71**, 129 (1979).
9. M. Parrinello and A. Rahman, Phys. Rev. Lett. **45**, 1196 (1980).
10. K.E. Schmidt and M.H. Kalos, *Monte Carlo Methods in Statistical Physics* (Ed. K. Binder, Springer, New York, 1984).

11. B.J. Alder, D.M. Ceperley, and E.R. Pollock, *Accounts of Chem. Research* (1985).
12. B.H. Wells, Chem. Phys. Lett. **115**, 89 (1985).
13. P.J. Reynolds, D.M. Ceperley, B.J. Alder, and W.A. Lester, J. Chem. Phys. **77**, 5593 (1982).
14. C.M. Ceperley and B.J. Alder, J. Chem. Phys. **81**, 5833 (1984).
15. G. Sugiyama and S.E. Koonin, Ann. Phys. (1985).
16. C.M. Ceperley and B.J. Alder, Science (1986).
17. P.J. Reynolds, M. Dupuis, and W.A. Lester, J. Chem. Phys. **82**, 1983 (1985).
18. E.L. Pollock and D.M. Ceperley, Phys. Rev. **B30**, 2555 (1984).
19. B.J. Alder and T.E. Wainwright, J. Phys. Soc. Japan, **26**, 267 (1969).
20. A.F. Bakker, "Design and Implementation of the Delft Molecular Dynamics Processor," Delft University of Technology Thesis (1983); H.J. Hilhorst, A.F. Bakker, C. Bruin, A. Compagnier, and A. Hoogland, J. Stat. Phys. **34**, 987 (1984).
21. B.J. Alder, W.E. Alley, and E.L. Pollock, Ber. der Bunsengesellschaft fur Phys. Chem. **85**, 944 (1981).
22. E. Meiburg, Phys. Fluids (1985).
23. C.H. Bennett, *Algorithms for Chemical Computation* (Ed. R.E. Christofferson, Am. Chem. Soc., Washington, D.C., 1977).
24. E. Madelung, Z. Physik **40**, 322 (1926).
25. W.H. Miller, S.D. Schwartz, and S.W. Trump, J. Chem. Phys. **79**, 4889 (1983).
26. H. Mori, Progr. Theoret. Phys. **34**, 399 (1965).
27. M.D. Feit and J.A. Fleck, J. Comp. Phys. **47**, 412 (1982).
28. D.H. Thouless, Proc. Phys. Soc. **86**, 893 (1965).
29. M.D. Feit and B.J. Alder, NATO Summer School, Varenna (1985).
30. J.W. Negele, J. Stat. Phys. (1986).

PARALLEL COMPUTING FROM FERMION SYSTEMS TO HYDRODYNAMICS: WATER AS AN EXAMPLE

M. Wojcik, G. Corongiu, J. Detrich, M. M. Mansour, E. Clementi, and G. C. Lie

IBM Corporation
Dept. 48B/MS 428
Neighborhood Road
Kingston, New York 12401

ABSTRACT

With the use of attached parallel processors as described by Clementi in this conference, we have carried out a program where starting from the Schrödinger equation, we can obtain useful information for quite complicated real macroscopic systems. As an example, we have employed configuration interaction method to calculate the potential function for the water-water interaction, which was then used in Monte Carlo and molecular dynamics simulations to study static and dynamic properties of liquid water. Various hydrodynamic constants, such as viscosity, thermal conductivity, and sound speed were obtained from Green-Kubo relations and the density and current correlations of the simulated water. Effects due to many-body interactions and intramolecular vibrations are reported. Correlation functions resulting from solving the fluctuating hydrodynamic equations for a dilute gas subjected to a temperature gradient are also presented.

I. INTRODUCTION

One of the main scientific goals is to explain the properties and the dynamics of macroscopic objects from the interactions and/or arrangements of their constituent atoms. With the advent of the quantum mechanics, this objective had *in principle* been realized in the late 1920s. What remains to be done is, as P. A. M. Dirac pointed out long ago, to find ways to solve the resulting complicated mathematical equations.

We have witnessed an explosion of numerical computations with the appearance of the electronic computers in the 1950s. Employing today's supercomputers, quantum chemistry has progressed to a stage that molecular systems with up to about 20 electrons can be routinely calculated to very high accuracy. In this brief review we will discuss mainly the work on liquid water carried out in our laboratory over the past ten years. We shall show that with the help of supercomputers, for example of the type currently being developed in our laboratory, we are able to obtain structural and dynamical properties, as well as hydrodynamic correlation functions, from *first principles* for a system as complicated and important as water.

The energy of an N-molecule macroscopic system, relative to infinitely separated N individual molecules, can be written as

$$U(1,2,...) = \sum_{i>j} U_2(i,j) + \sum_{i>j>k} U_3(i,j,k) + \cdots + \sum_{i>j>k>l} U_4(i,j,k,l) + \cdots$$

Lecture Notes in Chemistry, Vol. 44
Supercomputer Simulations in Chemistry. Edited by M. Dupuis
© Springer-Verlag Berlin Heidelberg 1986

where $U_n(i_1 i_2,..., i_n)$ is the interaction energy among the $i_1, i_2...., i_n$ molecules. Section II discusses the calculations of the $U_2(i,j)$ and the properties of liquid water resulting from the use of the U_2 alone. The calculations of U_3 and U_4 and their effects on the properties of liquid water are discussed in Sec. III. The assumption of the rigidity of the water molecules used in Secs. II and III is removed in Sec. IV, and new aspects of liquid water are reviewed. Most of our studies deal with systems at equilibrium; however, as a first step toward investigating nonequilibrium phenomena, we will also present in Sec. V some of the results obtained from solving nonequilibrium Landau-Lifshitz Langevin type equations for a finite hydrodynamic system. A similar but dense system is currently being simulated from the molecular point of view.

II. TWO-BODY POTENTIAL for LIQUID WATER

Large scale configuration interaction techniques have has been employed to calculate the water-water interaction energy.[1] The basis set used is slightly better than the double-zeta quality. To reduce the number of degrees of freedom, we have constrained the geometry of the two water molecules to be frozen at the gas phase equilibrium ground state experimental value.[2] Energies for 66 points in the remaining 6-dimensional intermolecular space were calculated and the results fitted to the expression

$$U_2 = q^2 \left(1/r_{13} + 1/r_{14} + 1/r_{23} + 1/r_{24}\right) + 4q^2/r_{78}$$

$$- 2q^2 \left(1/r_{18} + 1/r_{28} + 1/r_{37} + 1/r_{47}\right) + a_1 e^{-b_1 r_{56}}$$

$$+ a_2 \left(e^{-b_2 r_{13}} + e^{-b_2 r_{14}} + e^{-b_2 r_{23}} + e^{-b_2 r_{24}}\right)$$

$$+ a_3 \left(e^{-b_3 r_{16}} + e^{-b_3 r_{26}} + e^{-b_3 r_{35}} + e^{-b_3 r_{45}}\right)$$

$$- a_4 \left(e^{-b_4 r_{16}} + e^{-b_4 r_{26}} + e^{-b_4 r_{35}} + e^{-b_4 r_{45}}\right)$$

where r_{ij} denotes the distance between atoms or negative charge centers i and j (see Fig. 1 for the numbering). The standard deviation of this fit is 0.2 Kcal/mole. We refer to the original paper for the numerical values of the fitting constants q, a_i, b_i, and the position of the negative charge center M.

The geometry of the minimum energy dimer found from the potential is given and compared with the experiments in Table 1. Both theoretical and experimental results predict the most stable dimer configuration to be a near linear hydrogen bonded structure as shown in Fig. 1. The dipole moment, μ_H, reported in Table I is the projection on the direction of the hydrogen bond. The large difference between the theoretical and the experimental dipole moments is due to the difference in the angle β, not in the wave function. As a matter of fact, the dipole moment calculated at $\beta = 58°$, 2.77 D, agrees very well with the experimental value of 2.6 D[3]. Fig. 2 compares the second virial coefficients of steam calculated from various

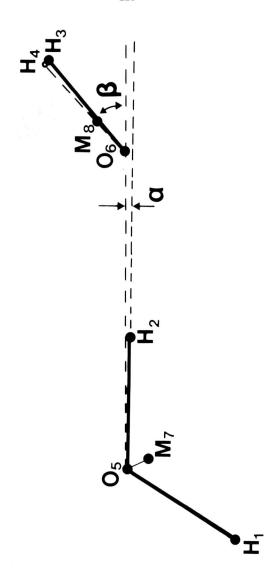

Figure 1.   Definition of water dimer geometry (illustrated with the most stable dimer configuration).

Table 1. Comparison of the most stable water dimer configuration (refer to Fig. 1 for the definition of the angles).

	Theoretical(CI)	Experimental
R_{OO} (Å)	2.87	2.98 ± 0.01
α (deg)	4	1 ± 6
β (deg)	37	58 ± 6
μ_H (debye)	3.3	2.6
E (Kcal/mole)	− 5.87	− 5.66

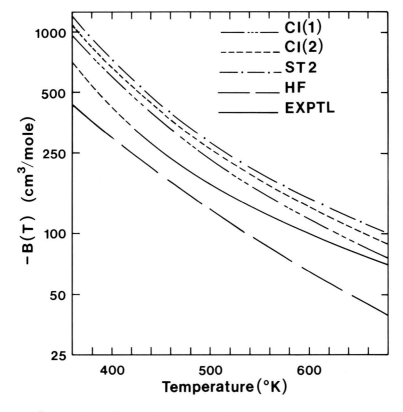

Figure 2.   Comparison of second virial coefficients for steam. CI(1) is the MCY potential; CI(2) the potential denoted by CI in Ref. 1; ST2 the semi-empirical potential taken from Ref. 38; HF the Hartree-Fock potential taken from Ref. 39; EXPTL the experimental curve taken from Ref. 40.

models with the experiments.[4] Although this potential (denoted by CI(1) in the figure) is seen to be the best in reproducing the experimental results, the deviation is however too large to be judged satisfactory. A larger number of points on the energy surface and more refined CI computations are needed to determine if this is

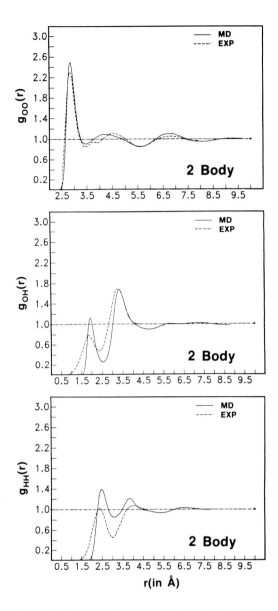

Figure 3.   Comparison of the radial distribution functions for water. MD curves
are the results of molecular dynamics simulation with two-body MCY
potential.

due to a deficiency in the potential or because of the limited number of points sampled.

The fitted potential, henceforth referred to as MCY, was used in a Monte Carlo (MC) and a molecular dynamics (MD) simulations of liquid water. Both simulations used periodic boundary conditions and minimum image to simulate an infinite system. 343 water molecules were used in a MC simulation at 298 K where long range dipolar interactions were neglected,[4] whereas the MD simulation employed 256 molecules at 296 K with Ewald summation technique to handle the long range forces.[5]

Since the expression for the MCY potential is somewhat complicated and the number of interactions to be evaluated is large, we found that the computation of forces, even with table-lookup for the exponents, requires considerable amount of computer time in the MD simulations. Thus we have turned our attention to parallel processing techniques in order to make the simulation more tractable. We have used up to 4 attached processors (FPS-164) for the parallel computation of the forces acting on any given molecule. It should be pointed out that our current LCAP-1 computer system for parallel processing (see the article by E. Clementi) is very suitable for this type of calculations, since the number of variables to be passed between the host IBM computer and each attached processor is of the order N (the number of molecules), whereas the quantities to be computed in each processor are proportional to N^2 (mutual interaction forces). Test runs indicate that up to 85% efficiency is obtained even if 10 attached processors are used in parallel. The efficiency is about 93% with 4 attached processors.

The radial distribution functions (RDF) for oxygen-oxygen, oxygen-hydrogen, and hydrogen-hydrogen obtained in the MC and MD simulations agree with each other. They are shown and compared with the so-called "experimental" curves[6] in Fig. 3. The simulated oxygen-oxygen radial distribution function, $g_{OO}(r)$, agrees quite well with the experimental data determined from x-ray scattering,[6] and shows well defined shell structures. The peak centered at ~2.9Å is due primarily to the nearest neighbors of the central molecule, and thus its position gives an approximate value for the average separation of the nearest neighbors in the liquid. The number of nearest neighbors can be estimated from the area under the first peak to be about 5, an indication that the nearly perfect tetrahedral bonding observed in ice is broken in the liquid state.

The simulated $g_{OH}(r)$ and $g_{HH}(r)$ are not in such good agreement with the experiments of Narten and Levy[6] as the case of $g_{OO}(r)$. However, it is well recognized by now, from both theoretical consideration and other experimental evidence, that the so-called "experimental" RDFs for oxygen-hydrogen and hydrogen-hydrogen are incorrect.[7] Although it is in principle possible to determine the three RDFs from three different experimental techniques or by isotope substitution, the existing experimental data are not yet accurate enough to produce such a unique set of the RDFs.[6,8,9]

For a more direct comparison with the experiment, we present in Fig. 4 the simulated x-ray and neutron scattering intensities obtained from the Fourier transform of the RDFs. Given in the figure are also the experimental x-ray[6] and neutron[10] scattering intensities. We refer to Ref. 4 for the definition of the scattering intensities plotted. Besides the general overall agreement, we note in particular that the simulation is able to reproduce the characteristic double peaks at $s \cong 2.5\text{Å}^{-1}$ observed in the x-ray scattering.

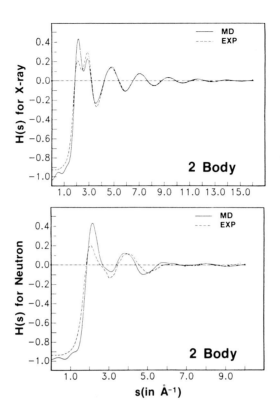

Figure 4.   Comparison of the x-ray and neutron scattering intensities. MD curves
are the results of molecular dynamics simulation with two-body MCY
potential.

The internal energy calculated after quantum corrections is -6.67 Kcal/mole,[5] a
value more than 1 Kcal/mole too high compared with the experimental value of -8.1
Kcal/mole.[11]   This discrepancy results from the neglect of the many body effects,
which will be addressed in the next section, and minor deficiencies in the two-body
potential.   The simulated heat capacity, after quantum corrections, is 15
cal/mole/deg.[5] Although this is too low compared with the experimental value of 18
cal/mole/deg, it nevertheless reproduces the characteristic high heat capacity of
liquid water.   Agreement with experiments was found if we include the
intramolecular vibrational motions to be discussed in Sec. IV.   The ratio of the
constant pressure heat capacity to the constant volume heat capacity, $\gamma = C_p/C_v$, can
be calculated from the correlation of fluctuations in pressure and temperature and
is found to be 1.07 for the MCY potential, to be compared with the experimental
value of 1.02.   We postpone the discussion of the dielectric constant calculations to
the next section.

The hydrodynamic coefficients can be obtained in the MD simulation by the use of
various Green-Kubo relations.[12]   The value we obtained for the shear viscosity is

0.64 cP, to be compared with the experimental value of 0.9 cP.[13] The bulk viscosity calculated is 0.80 cP. Experimentally this is a difficult quantity to measure accurately, and experimental values range from 1.3 cP to 2.5 cP.[13,14] The thermal conductivity obtained from the simulation is 2.9×10^{-3} cal/cm/sec/K, about twice the experimental value of 1.5×10^{-3} cal/cm/sec/K.[15] The diffusion coefficient calculated is 2.25×10^{-5} cm^2/sec, very close to the value of about 2.5×10^{-5} found experimentally.[15]

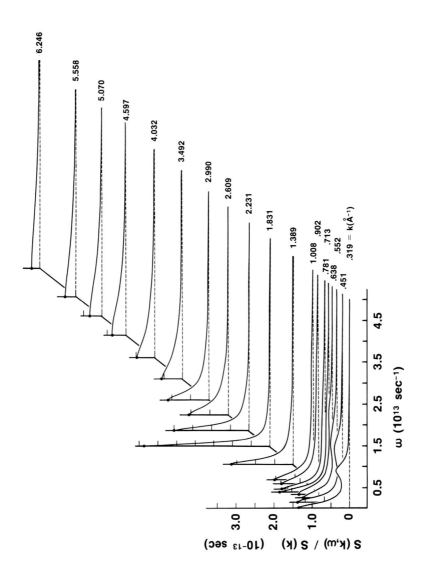

Figure 5. Normalized dynamics structure factors for liquid water obtained from molecular dynamics simulation with two-body MCY potential.

Two of the most important functions in the study of the dynamics of fluids are the density fluctuation correlation and current fluctuation correlation functions. Fig.

5 shows $S(k,\omega)$, the spatial temporal Fourier transform of the density correlation function. As the wavevector k decreases, the propagation of the sound mode begins to appear. From the peak of the $S(k,\omega)$ at the lowest k value allowed in the simulation, the *adiabatic* sound speed, c_s, is estimated to be about 2900 m/sec. This sound speed was at first thought to be the normal one (experimental ~1500 m/sec);[16] however, recent coherent inelastic neutron scattering seems to indicate that it should be identified with the high frequency sound speed found at 3310 cm/sec.[17] The isothermal compressibility calculated from k → 0 limit of the equal-time spatial Fourier transform of the density correlation function is 2.0×10^{-5} atm^{-1}, to be compared with the experimental value of 4.5×10^{-5} atm^{-1}.[18]

Fig. 6 shows the spatial temporal Fourier transform of the transverse current correlation function. It is evident from the figure that there exist shear waves in the low k region. The shear wave is estimated to propagate at 1.2×10^5 cm/sec, which has not yet been observed experimentally.

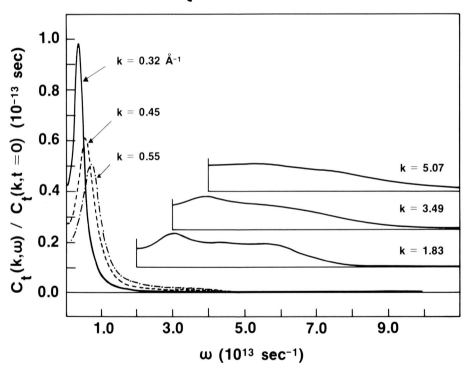

Figure 6.   Spatial-temporal Fourier transforms of the transverse current correlation functions obtained from molecular dynamics simulation with two-body MCY potential.

A property not satisfactorily reproduced in the simulation with the MCY potential is the pressure, ~8500 atm, to be compared with the experimental value of about 100 atm at the same conditions. The incorrect high pressure of the MCY model requires a short comment. Although it is a notable shortcoming, it however should not be over emphasized. Indeed, in an analysis of the MC data, we have found that high pressure comes almost exclusively from r_{OO} less than 2.85 Å. There are only 19% of the nearest neighboring pairs in that region, but they contribute more than 7500 atm to the system pressure. Alternatively stated, the MCY potential is too repulsive for compact configurations, which we recall is due to the fact that very few of them were computed in the repulsive region to obtain a good fit there.[1] This explains why despite the high pressure, the MCY potential is still capable of reproducing many properties of water. A more thoroughly study covering more points in the repulsive regions is currently under investigation in this laboratory.

III. MANY-BODY CORRECTIONS for LIQUID WATER

The first important correction to a many-body system is of course the three-body interactions. It is fortunate that the electron correlation energy is not important in the evaluation of the three-body interaction energies for water and thus Hartree-Fock calculations proved to be suffice.[19] The three-body interaction energy is calculated as

$$U_3(1,2,3) = E(1,2,3) - E(1,2) - E(1,3) - E(2,3) + 3E(1)$$

where $E(i,j,k)$ is the SCF energy for a system of three water molecules i, j, and k. By now 173 different geometrical arrangements of water trimers have been studied and the resulting energies were fitted to a polarizable bond model with the following expressions:[19]

$$\vec{\mu}_{b_i} = \alpha \vec{E}_{b_i} + \beta \vec{e}_{b_i} (e_{b_i} \bullet \vec{E}_{b_i})$$

$$\vec{E}_{b_i} = \sum_{j(\neq i)} \vec{E}_{b_i j}$$

$$U_3 = -\frac{1}{2} \sum_{b_i} \vec{\mu}_{b_i} \bullet \vec{E}_{b_i}$$

where $\vec{\mu}_{b_i}$ and $\vec{E}_{b_i}$ are, respectively, the induced dipole and electric field at the midpoint of bond b of molecule i, $\vec{E}_{b_i j}$ the electric field at site b_i due to the charges on the j-th molecule, α and β are the fitting parameters related to the perpendicular and transverse polarizabilities. We again refer to the original paper for the numerical values of the resulting parameters. The standard deviation of the fitting, 0.21 Kcal/mole, is comparable to the one for 2-body potential. We note that the basis set used here is better than that for the MCY and the counterpoise technique was employed to reduce the superposition error. This three-body potential will henceforth be referred to as the CC potential. It has been shown to improve the theoretical third virial coefficients for steam.[20]

The MCY and the CC potentials are combined and used in MC and MD simulations. While dynamical properties are currently being analyzed, we give in Figs. 7 and 8 the RDFs and scattering intensities, respectively. From the enhanced structures shown in Fig. 7, it can be concluded that the CC potential has the effect of strengthening the hydrogen bonds. One pronounced improvement of the MCY + CC potential over the two-body one is the notably improved reproduction of the height of the double peaks observed in the x-ray scattering intensity at s $\cong 2.5$Å^{-1}.

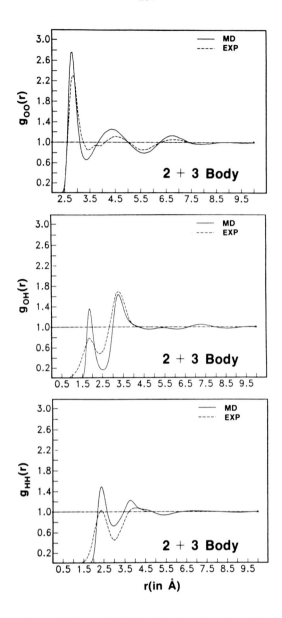

Figure 7.   Comparison of the radial distribution functions for water. MD curves are the results of molecular dynamics simulation with two- and three-body MCY + CC potential.

One of the most important properties of liquid water is the dielectric constant. This has been computed in the MD simulation using the MCY potential with a sample of 512 molecules of water and the Ewald summation to handle the long range interaction. Using the Kirkwood formula[21] and the dipole moment computed from the charge locations in the MCY water, the static dielectric constant calculated, 30, is only about 40% of the experimental value. This result may not be surprising if we recall that the dipole moment used and the Kirkwood g factor calculated may be too small due to the neglect of the many-body effects. What surprises us is a decrease in the calculated dielectric constant when the MCY + CC potential is used. We are now in the process to analyze those results. Tentatively we might suggest that the dielectric constant results from the overall interaction of a very large sample of water molecules, thus a much larger simulation system might be needed. Another possibility is that the effective dipole moment of the water molecule in the liquid state might be much larger than the value, 2.19 D, we used, due to collective induction.

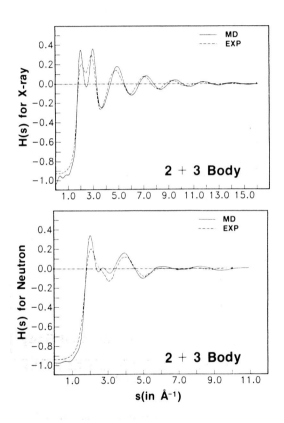

Figure 8.   Comparison of the x-ray and neutron scattering intensities. MD curves are the results of molecular dynamics simulation with two- and three-body MCY + CC potential.

The next important correction to the interaction energy of an N-body system is the four-body interaction. The computations involved for the calculations of the four-body correction to the interaction energy are much heavier than those above, since not only is the system larger, but additional subset calculations (4 triplets and 6 doublets for each tetramer) are also more. It was noted however that the next term in the expression for U_3 is the induced dipole-dipole interaction which is the dominating term in the four-body correction. Therefore it was only necessary

236

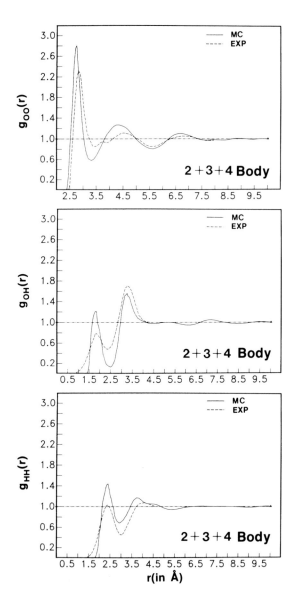

Figure 9.   Comparison of the radial distribution functions for water. MC curves are the results of Monte Carlo simulation with two-, three- and four-body MCY + CC + DCC potential.

to check if the parameters obtained previously were able to reproduce the SCF energies calculated for tetramers (defined similar to the U_3). To check on this, we have carried out several *ab initio* SCF calculations for water tetramers, using the same basis set and counterpoise technique as for the trimers.[22] The results point out that our model does in fact account correctly for the four-body interactions.[22]

The four-body potential (henceforth referred to as DCC) was then used in conjunction with the MCY and CC potentials to simulate liquid water by the Monte Carlo technique. The resulting oxygen-oxygen radial distribution function is even more structured as can be seen in Fig. 9. The heights of the double peaks in the x-ray structure function, H(s), are well reproduced as shown in Fig. 10, although the agreement with the experimental curve is worsened in the high momentum transferring region.

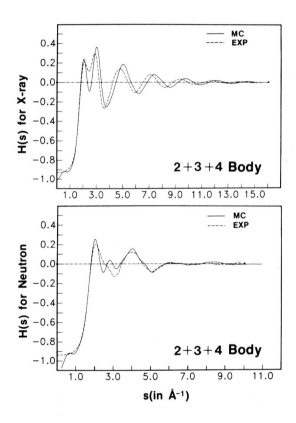

Figure 10.   Comparison of the x-ray and neutron scattering intensities. MC curves are the results of Monte Carlo simulation with two-, three- and four-body MCY + CC + DCC potential.

Fig. 11 shows the internal energy of liquid water as a function of the many-body corrections. We see a steady improvement in the calculated energy as the number of terms in the energy expression is increased. Note that the simulated internal energy with four-body included is lower than the experimental results since quantum corrections and intramolecular motions have been neglected in the above MC simulations. These problems are addressed in the next section.

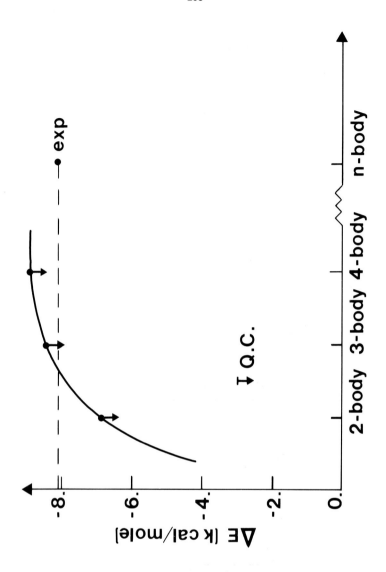

Figure 11. Total energy of liquid water as a function of many-body interaction corrections. Q.C. denotes quantum correction, which has been shown to bring the calculated classical energy higher.

IV. VIBRATING WATER MOLECULES and LIQUID WATER

All the potential models described so far assume the water molecules are rigid with geometry determined by the vibration-rotational spectra of the gas phase. There is, however, much experimental evidence that the dynamics of the water molecules in the liquid phase is different from that in the gas.[23] To account for that, it is necessary to extend all of the previous potentials to handle the vibrations of the molecules. This has been achieved, as a first step, for the MCY potential.[24]

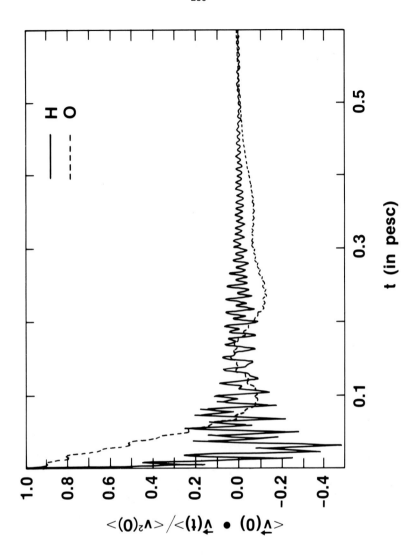

Figure 12.   Velocity autocorrelation function for liquid water obtained with the MCYL potential.

Following the same spirit that is being pursued in this laboratory, we have decided to use also an *ab initio* potential for the intramolecular motions.   The best such potential in the literature seems to be the one denoted by D-MBPT(∞) in the many-body perturbation theory calculations of Bartlett, Shavitt, and Purvis.[25]   The potential is expressed by the Taylor series expansion up to quartic terms in terms of three internal coordinates of water: changes in OH bond lengths [$\delta_i = R_i - R_e$] and HOH bond angle [$\delta_3 = R_e(\theta - \theta_e)$].   We refer to the original paper for the force constants.

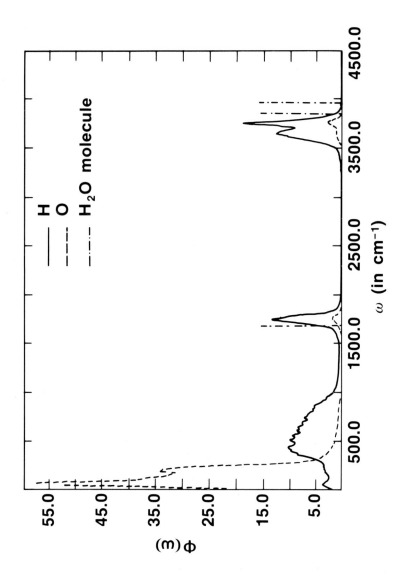

Figure 13.   Spectral density for liquid water obtained with the MCYL potential. Also shown in the figure are the harmonic vibrational frequencies for isolated water molecule.

This potential is added to the MCY potential and the combined potential is assumed to be valid for any geometry of the water molecules. Since the MCY potential contains a negative charge center M not residing on any atom, extending the potential to flexible water requires a specification of how M moves with the deformation. The one we proposed is found to be able to explain some features of the IR intensities in the liquid phase.[24]   This extended potential is henceforth designated as MCYL.

The MCYL potential was used in a molecular dynamics simulation, with 343 molecules at 301 K, of liquid water where the reaction field method was employed to handle the long range interactions. Since most properties obtained are very similar to, but slightly better than, those described previously for the MCY potential, we report here only those which are not relevant for a rigid model.

The average OH bond length in the simulated liquid is found to be 0.018Å longer than that in the gas phase, whereas the average HOH bond angle is about 1° smaller.[24] These changes compare very well with indirect experimental evidence of 0.009 ± 0.006Å and 1.7°, respectively.[26]

Fig. 12 shows the velocity autocorrelation functions for the oxygen and the hydrogen atoms. The high frequency vibrational motions of the molecules are clearly seen in the figure. These vibrations shows up as three distinct peaks centered at 1740, 3648, and 3752 cm^{-1} in the Fourier transforms of the autocorrelation functions, shown in Fig. 13. Comparing with the classical harmonic motions of the *isolated* molecules, which are also shown in Fig. 13, we find that in going from gas to liquid phase, there are an up shift of 55 cm^{-1} in the bending frequency and down shifts of 198 and 203 cm^{-1} in the stretching frequencies. These shifts are all in good agreement with the experimental IR and Raman results of 50, 167, and 266 cm^{-1}, respectively.[23]

All the motions contained in the spectral density (Fig. 13) can be treated as quantum harmonic oscillators to find the quantum corrections to the internal energy and heat capacity obtained from the classical MD simulations.[27] We have extended the analysis of our data to the MC simulation with *rigid* water and found 0.93 Kcal/mole as the quantum correction to the energy.[24] Adding this correction to the energy simulated with the MCY + CC + DCC potential yields an internal energy of -8.02 Kcal/mole for the simulated liquid water,[24] in very good agreement with the experimental value of -8.1 Kcal/mole.[11] The heat capacity calculated after quantum correction is 18 cal/mol/deg, in very good agreement with experiments.

V. NON-EQUILIBRIUM STATISTICAL MECHANICS

The statistical properties of a non-equilibrium system can be drastically different from those of an equilibrium system, mainly because external constraints can maintain the system in a state that would be highly improbable at equilibrium.[28] Consequently, the postulate of equal *a priori* probability of microstates and the hypothesis of the statistical independence of far apart subsystems are no longer applicable.[29] For the same reason, a purely microscopic approach to the study of a non-equilibrium system proves to be far more difficult than an equilibrium system .[30,31] A very promising way to tackle the problem is through the fluctuating hydrodynamic formalism, mainly because of its relative simplicity as compared to more fundamental approaches. However, it can only be justified in equilibrium situations where theory and experiments can be compared and are found to be in perfect agreement. For a non-equilibrium system, experimental data are scarce and despite some striking new results[32-37], the question of the validity of the fluctuating hydrodynamic formalism in far from equilibrium situations still remains open.

An interesting way to verify the hydrodynamic hypotheses of non-equilibrium systems is by direct molecular dynamics simulations. With the recent development of super-computers, it is now possible to simulate systems under desired non-equilibrium constraints, and not only averages,[38,39] but correlation functions as well can be measured.[40] This program has also been undertaken by our group and we hope to be able to answer some of the fundamental questions of non-equilibrium statistical mechanics, e.g., the validity of the fluctuating hydrodynamic formalism in the near future.

From the theoretical point of view, the new aspect comes chiefly from the boundary value problem which must be chosen with great care.[41,42] Since the dimension of the system we are supposed to be describing is at most of the order of a few hundred mean free paths and the correlation can be long ranged, the boundary conditions are thus expected to play a very important role. For this reason, an analytical solution of the fluctuating hydrodynamic equations proves to be much more difficult to obtain than in an infinite medium.[43] This problem is also being presently considered by our group. The hydrodynamic equations with added fluctuation terms, as first proposed by Landau and Lifshitz,[44] are solved numerically. Given the various thermodynamic parameters, such as viscosity and thermal conductivity coefficient, the program computes various correlation functions between hydrodynamic variables, which can then be compared with the results of molecular dynamics experiments. While we are still working on the molecular dynamics simulation problem, we present in the following some preliminary hydrodynamic results obtained for a dilute Boltzmann gas between two parallel walls maintained at different temperatures.

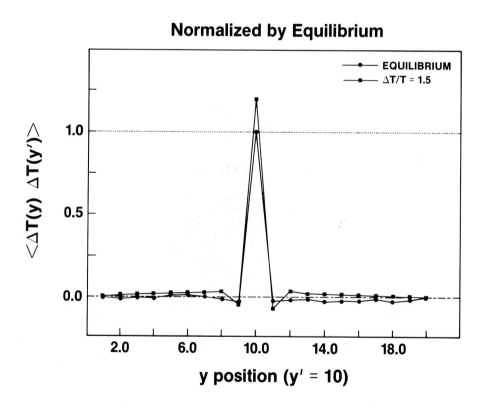

Figure 14.   Temperature-temperature correlation functions for the 10-th cell obtained from solving fluctuating hydrodynamic equations. Equilibrium means there is no temperature gradient between the walls.

The dilute gas is confined between two walls maintained at T and T + ΔT. Periodic boundary conditions are applied in the x- and z-directions with unit cell length equal to 1, whereas the walls are located at y = 0 and y = 1. The system is divided into 20 equal subcells along the y-axis. Hydrodynamic values reported are the values averaged over each subcell after steady state has been reached. Fig. 14 displays the spatial temperature-temperature correlation between the 10-th subcell and the rest. It can be seen from the figure that when the system is driven out of equilibrium, the *local* temperature fluctuation grows rapidly. But despite this, the spatial correlation remains very small. This is not the case for the velocity-temperature correlation, as can be seen in Fig. 15. The spatial correlations here grow very quickly and encompass the entire system. In the absence of a gravitational field, the system remains in a state of mechanical rest and the heat is transferred from the hot wall to the cold one only through conduction. As we increase the temperature gradient, the system is expected to approach more and more closely to a transition point, reflecting the fact that a small gravitational field would be sufficient to break the stability of the system leading thus to a non-equilibrium transition from zero convection to finite convection, like the Benard instability in liquids.

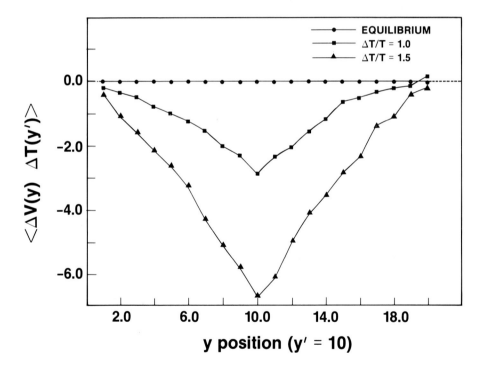

Figure 15.   Velocity-temperature correlation functions for the 10-th cell obtained from solving fluctuating hydrodynamic equations. Equilibrium means there is no temperature gradient between the walls.

VI. CONCLUSIONS

Using water as an example, it has been demonstrated that with supercomputers under development here, we can simulate macroscopic systems, and thus obtain macroscopic properties traditionally being measured through laboratory experiments, starting from aggregates of electrons and nuclei. Plugging these macroscopic constants into the hydrodynamics equations, we can then solve problems of daily life scales, thus making a practical connection between the microscopic and the macroscopic worlds. Other objectives which we are pursuing include understanding phase transitions, leading to the construction of phase diagrams, and simulations of non-equilibrium systems and hydrodynamic fluid flows from molecular point of view.

REFERENCES

1. O. Matsuoka, E. Clementi, and M. Yoshimine, *J. Chem. Phys.* **64**, 1351 (1976).
2. W.S. Benedict, N. Gailar, and E.K. Plyler, *J. Phys. Chem.* **24**, 1139 (1956).
3. T.R. Dyke, K.M. Mack, and J.S. Muenter, *J. Chem. Phys.* **66**, 498 (1977).
4. G.C. Lie, E. Clementi, and M. Yoshimine, *J. Chem. Phys.* **64**, 2314 (1976).
5. M. Wojcik, *IBM Technical Report KGN-28* (1985).
6. A.H. Narten and H.A. Levy, *J. Chem. Phys.* **55**, 2263 (1971).
7. G.C. Lie, *IBM Technical Report KGN-52* (1986).
8. G. Palinkas, E. Kalman, and P. Kovacs, *Mol. Phys.* **34**, 525 (1977).
9. A.K. Soper and R.N. Silver, *Phys. Rev. Lett.* **49**, 471 (1982).
10. A.H. Narten, *J. Chem. Phys.* **56**, 5681 (1972).
11. N.E. Dosey, *Properties of Ordinary Water Substance* (Reihold, New York, 1940).
12. J.P. Boon and S. Yip, *Molecular Dynamics* (McGraw-Hill, New York, 1980).
13. C.M. Davis and J. Jarzynski, in *Water: A Comprehensive Treatise*, Vol. 1, edited by F. Franks (Plenum Press, New York, 1972).
14. J. Rouch, C.C. Lai, and S.H. Chen, *J. Chem. Phys.* **66**, 5031 (1977).
15. G.S. Kell, in *Water: A Comprehensive Treatise*, Vol. 1, edited by F. Franks (Plenum Press, New York, 1972).
16. R.W. Impey, P.A. Madden, and I.R. McDonald, *Mol. Phys.* **46**, 513 (1982).
17. J. Teixeira, M.C. Bellissent-Funel, S.H. Chen, and B. Dorner, *Phys. Rev. Lett.* **54**, 2681 (1985).
18. G.S. Kell and E. Whalley, *Philos. Trans. R. Soc. London* **A258**, 565 (1965).
19. E. Clementi and G. Corongiu, *Int. J. Quantum Chem.: Quantum Bio. Sym.* **10**, 31 (1983).
20. G.C. Lie, G. Corongiu, and E. Clementi, *J. Phys. Chem.* **89**, 4131 (1985).
21. S.W. de Leeuw, J.W. Perram, and E.R. Smith, *Proc. Phys. Soc. Lond.* **A373**, 27 (1980); *ibid.* **A373**, 57 (1980).
22. J. Detrich, G. Corongiu, and E. Clementi, *Chem. Phys. Lett.* **112**, 426 (1984).
23. D. Eisenberg and W. Kauzmann, *The Structure and Properties of Water* (Oxford University, New York, 1969).
24. G.C. Lie and E. Clementi *IBM Technical Report KGN-36* (1985); to appear in *Phys. Rev. A* (1986).
25. R.J. Bartlett, I. Shavitt, and G.D. Purvis, *J. Chem. Phys.* **71**, 281 (1979); Professor Shavitt has pointed out in the conference that he has obtained an even better potential.
26. W.E. Thiessen and A.H. Narten, *J. Chem. Phys.* **77**, 2656 (1982).
27. P.H. Beren, D.H.J. Mackay, G.M. White, and K.R. Wilson, *J. Chem. Phys.* **79**, 2375 (1983).
28. G. Nicolis and I. Prigogine, *Self Organization in Nonequilibrium System,* (Wiley, New York, 1977).
29. For review, see A.M. Tremblay, in *Recent Development in Nonequilibrium Thermodynamics* J. Casas-Vasquez, D. Jou, and G. Lebon ed. (Springer Verlag, Berlin, 1984).
30. T.R. Kirkpatrick, E.G.D. Cohen, and J.R. Dorfman, *Phys. Rev. Lett.* **44**, 472 (1980).
31. T.R. Kirkpatrick, E.G.D. Cohen, and J.R. Dorfman, *Phys. Rev.* **A26**, 995 (1982).
32. D. Ronis, I. Procaccia, and I. Oppenhein, *Phys. Rev.* **A19**, 1324 (1979).
33. A.M. Tremblay, M. Arai, and E.D. Siggia, *Phys. Rev.* **A23**, 1451 (1981).
34. G. Van Der Zwan, D. Bedaus, and P. Mazur, *Physica* **107A**, 491 (1981).
35. T.R. Kirkpatrick, E.G.D. Cohen, and J.R. Dorfman, *Phys. Rev.* **A26**, 950 (1982).

36. D. Beysens, Y. Garrabos, and G. Zalczer, *Phys. Rev. Lett.* **45**, 403 (1980).
37. R. Penney, H. Kiefte, and J.M. Clouter, *Bull. Can. Ass. Physicists* **39**, BB8 (1983).
38. A. Tenebaum, G. Ciccotti, and R. Gallico, *Phys. Rev.* **A25**, 2778 (1982).
39. C. Trozzi and G. Ciccotti, *Phys. Rev.* **A29**, 916 (1983).
40. M. Mareschal and E. Kestemont, *Phys. Rev.* **A30**, 1158 (1984).
41. P.G. Wolynes, *Phys. Rev.* **A13**, 1235 (1976).
42. G. Satten and D. Ronis, *Phys. Rev.* **A26**, 940 (1982).
43. M.M. Mansour, J.W. Turner, and G. Nicolis, *to be published.*
44. L.D. Landau and E.M. Lifshitz, *Fluid Mechanics,* 3rd Edition (Pergamon Press, New York, 1959).
45. F.H. Stillinger and A. Rahman, *J. Chem. Phys.* **60**, 1545 (1974).
46. H. Kistenmacher, G.C. Lie, H. Popkie, and E. Clementi, *J. Chem. Phys.* **61**, 546 (1974).
47. G.S. Kell, G.E. Malaurin, and E. Whalley, *J. Chem. Phys.* **48**, 3805 (1968).

SUPERCOMPUTER SIMULATIONS IN SOLID STATE CHEMISTRY

C. R. A. Catlow

Department of Chemistry
University of Keele,
Keele,
STAFFS,
ST5 5BG, U.K.

ABSTRACT

We describe those areas of solid state chemistry in which computer simulation studies using supercomputers have proved of value. Following a general survey of the field we focus on three recent examples which concern the simulation of ion transport in superionic $\beta'' Al_2O_3$, the study of sorption in zeolites, and the properties of $BaTiO_3$ ceramics.

1. INTRODUCTION

Computer simulation studies are now having a major impact on solid state chemistry and physics. Much of the recent progress has depended on the availability of supercomputers which have allowed us to examine materials and problems of greater complexity and to include interatomic potentials of greater sophistication which extend the range of solids that can be reliably modelled. In this chapter we will discuss some of the recent advances in this rapidly developing field which now encompasses the study of materials as diverse as zeolites, ceramics and superionic conductors and which is establishing increasingly strong links both with experimental work and with other theoretical techniques e.g. quantum mechanical calculations.

2. SCOPE AND METHODOLOGY

As there are several recent reviews of the field[1,2,3], our account here will be brief. The basis of simulation methods in condensed matter research is the use of effective potentials to describe the interatomic forces, from which we then aim, in the case of solids, to predict the following properties.

(a) *Structure*, including both cell-dimensions and unit-cell coordinates. Structure calculations, in addition to their role in validation of interatomic potential models can be used to assist the refinement of complex crystal structures (for which perhaps only approximate information is available from powder data) and to predict, for example, structures at very high pressures - a topic of great contemporary geophysical interest.

(b)   *Elastic, Dielectric and Lattice Dynamical Properties*, which we have grouped together as they all rely on the calculation of dynamical matrix, i.e. the second derivative of the internal energy with respect to particle coordinates. Again, calculation of these quantities is commonly used to investigate the quality of the interatomic potentials. In addition, such calculations have a predictive role in the study of complex materials and materials under conditions (e.g. high pressure) where experimental study is difficult.

(c)   *Defect Properties*, which are known to control transport, thermodynamic and mechanical properties of solids, but whose detailed characterization by experimental methods is often difficult. The calculation of the energies and more recently the entropies,[4] of defect formation, aggregation and migration has been a highly successful field of simulation studies which has been important in establishing the reliability of the simulation procedures. We note a recent development in which the field has been extended to study the properties of sorbed molecules in porous solids which may be modelled using the same techniques as those employed in classical defect simulations.

(d)   *Detailed Ion Transport Mechanisms*, the study of which may be affected by the same techniques as those discussed immediately above, but which often require the use of dynamical simulation techniques in contrast to the static simulations which are normally used in studying defect properties. We shall see in section (3.3) how molecular dynamics studies have been of particular value in the study of ion transport in 'superionic' solids, i.e. solids with exceptionally high ionic conductivities.

(e)   *Surface Structure and Defect Properties*. This important field which has been developed principally by Mackrodt,[5] Tasker[6] and co-workers allows predictions to be made of surface structures of ionic materials and more importantly defect and impurity states at surfaces. The techniques have recently been extended to the study of other types of interface especially grain-boundaries,[7] which are known to play an important role in many materials properties.

2.1 Methodology

We are concerned with two classes of calculation, i.e. static-lattice and dynamical simulations. The former which involve no explicit treatment of thermal motions are based on energy minimization procedures. They may be subdivided into two categories, i.e. *perfect lattice* and *defect lattice* simulations. In the former case, a unit cell must be specified, the lattice energy of which is minimized with respect to atomic coordinates and cell dimensions to give the equilibrium structure. From the dynamical matrix we may, as noted, then calculate elastic, dielectric and lattice dynamical properties of the crystal. The theory on which such calculations are based is well established and is described in detail elsewhere.[8] Two features to which special attention should be drawn concern first the treatment of the Coulomb summation for which the Ewald procedure[9] is employed; this method overcomes the convergence difficulties which occur when the sum is handled in real space, by a transformation into reciprocal space. Secondly, the use of efficient energy minimization procedures is important. Newton-Raphson procedures which make use of 2nd in addition to 1st derivatives with respect to the energy are found to be most effective.[10]

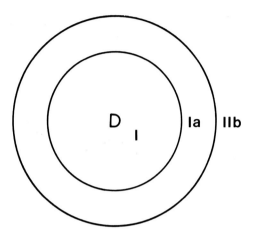

Figure 1.   Two region strategy for defect calculations.

Defect simulations use essentially the same simulation and minimization procedures as employed in the perfect lattice calculations. An essential component of such calculations is, however, the treatment of lattice relaxation around the defect. This is handled by a well-known procedure, commonly referred to as the 'two-region' strategy, which is illustrated diagrammatically in Fig. (1). A region of the crystal around the defect (region I), containing typically 100-500 atoms is relaxed to equilibrium; a full atomistic treatment is used in this region, as is necessary due to the strength of the defect forces. In contrast, the more distant, weak field regions may be described by a pseudo-continuum approach. In ionic crystals a favored method is that due originally to Mott and Littleton:[11] it involves calculation of the polarization per unit-cell which may then be divided up into atomistic displacements. Details of the mathematical development of the theory are available in references (12), (13) and (14). It is, however, well established that provided good interatomic potentials are available for the simulation of region I, and provided the later is large enough, then accurate defect energies are calculated by these procedures.

For many solid state properties, static simulation are adequate. But for very high temperature solids and for lower temperature materials which show high ionic conductivities, it may be necessary to use *dynamical simulation procedures* The molecular dynamics technique is discussed elsewhere in this work (see e.g. the chapter of Lie). In applications to solids, the following features deserve special attention.

(i)   The omission of any representation of electronic polarizability (which is normal in M.D. studies due to the computational cost of including polarizability), may have especially serious consequences as polarization may be an important phenomenon especially in defective solids.

(ii)   The short time-scale of the simulations (only a few picoseconds of real time is simulated) may mean that an insufficient number of diffusion events is observed. When used to study atomic transport, the method has, however, proved very powerful in simulating superionic solids; an example will be described in section (3).

To date the majority of *Surface Simulations* have employed static techniques. Surface structures are calculated by energy minimization techniques; and surface defects are modelled using a modified 'two-region' strategy employing a hemispherical region I. For details we refer to references (5)-(6).

All these classes of simulation require specification of *interatomic potentials*. In work on insulating solids these have, in general, been particularly simple, and have included the following features:

(i)   The assignment of charges to atoms. These are commonly integral charges, even for materials such as transition metal oxides and silicates where the bonding is known to have appreciative covalence. We have argued elsewhere,[15] however, that the use of formal ionic charges in a potential does not necessarily require a 'fully ionic' electron density distribution.

(ii)   The representation of short-range forces by simple analytical functions usually of a two-body, central-force nature. A commonly used form is the Buckingham potential:

$$V(r) = Ae^{-r/\rho} - Cr^{-6} \tag{1}$$

although Morse and Lennard-Jones potentials have been employed. Numerical potentials (i.e. tabulations of v(r)) may also be used. Recently Sanders et al.[16,17] showed that in modelling framework structural silicates, it was necessary to include an additional 'bonding-bending' term of the form

$$E(\theta) = K(\theta - \theta_o)^2 \tag{2}$$

where θ is the 0-Si-0 bond angle and θ_0 the tetrahedral angle. The inclusion of these terms helps to constrain the SiO_4 groups to remain tetrahedral; their use has greatly improved our ability to model framework structured silicales.

(iii)   *Electronic Polarizability*. Inclusion of the effects of polarizability is, as noted, especially important in defect calculations since defects polarize the surrounding lattice. In simulations of ionic materials, greatest success has been enjoyed by the shell model originally developed by Dick Overhauser.[18] The model is a crude, but effective, mechanical description of polarization: the ion is described in terms of a mass-less shell, representing the polarizable valence shell electrons, and a core which represents the nucleus and the core electrons; core and shell are coupled by a harmonic spring, and development of a dipole moment is described in terms of the displacement of the shell relative to core. Short range forces are taken as acting between shells - a central feature of the model as it means that there is a resulting coupling between polarization and short range repulsion, the omission of which led to serious inadequacies in simpler point-dipole models.

Despite its simplicity, no major advance on the shell model treatment of polarizability has yet been developed; and in the majority of cases the model seems to incorporate the essential physics of polarization in ionic crystal.

Having developed potential models, it is necessary to parameterize them. The simplest approach uses empirical procedures,[19] and adjusts variable parameters, usually via least-squares fitting routines, in order to reproduce as accurately as possible observed crystal properties including crystal structures, elastic and dielectric constants and phonon dispersion curves where these are available. The weakness of the approach is that for many important materials the necessary experimental data are not available. And there is a more fundamental problem as empirical procedures necessarily yield information only about interatomic potentials at perfect lattice spacings, whereas in studies of defects very different internuclear separations may occur. For this reason, there is a great incentive to develop reliable non-empirical procedures. Mackrodt and co-workers[20,21] have shown that a good measure of success can be achieved by simple electron-gas procedures. The same group have also used Hartree-Fock methods on 'super-molecules' to study pair potentials. The future is undoubtedly in this direction, and this is indeed a field where supercomputer calculations will clearly play a major role.

The development of potential models for metals is far more difficult than for ionic and semi-ionic solids. In many metals, especially transition metal oxides, the pair potential approximation is less justifiable and there are, of course, volume dependent terms due to the conduction band electrons. Nevertheless useful models have been developed and are discussed by Taylor.[22,23]

3. APPLICATIONS

A large number of successful simulation studies of solids have been reported on the last few years. Predications of the structures and properties of transition metal oxides and silicates are reported by Cormack, Parker and co-workers.[24,25,26,27] Studies of defect energies are now routine and straightforward; and a good recent review of the state-of-the-art on this field is produced by Mackrodt.[2] Defect simulations have also been used to examine defect aggregation in non-stoichiometric and heavily doped materials; earlier work is described in reference (28) and (29), while interesting applications to extended, planar defect formation in non-stoichiometric ReO_3 structured transition metal oxides are discussed in references (30) and (31).

The field is therefore well established, and in the remainder of this article we will try to illustrate its contemporary scope and diversity by describing some recent applications: first to suberionic conduction, secondly, to zeolite chemistry and thirdly, to the electronic properties of $BaTiO_3$.

3.1 Ion Transport in $\beta'' Al_2O_3$

$\beta'' - Al_2O_3$ is an important layer structured superionic. As illustrated in Fig. (2), condition planes containing the mobile Na^+ ions are sandwiched between spinel structured alumina layers which also contain Mg^{2+} (the formula of the stoichiometric compound be $Na_2OMgO5Al_2O_3$). The compound is closely related to the celebrated $\beta - Al_2O_3$, in which, however, the spinel blocks are thicker and in which there is no Mg^{2+}.

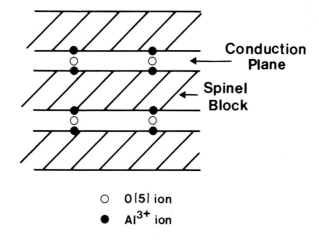

○   O (5) ion

●   Al³⁺ ion

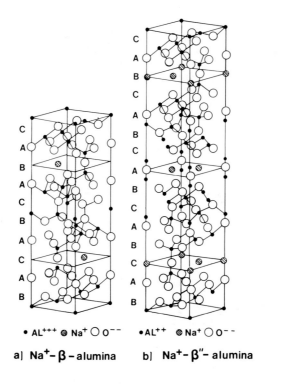

●AL⁺⁺⁺ ⊘Na⁺ ○O⁻⁻   •AL⁺⁺ ⊘Na⁺ ○O⁻⁻

a) Na⁺-β-alumina   b) Na⁺-β″-alumina

Figure 2.   Structure of $\beta'' - Al_2O_3$.

As normally prepared, both β and β″Al_2O_3 are non-stoichiometric. In the latter case the non-stoichiometry arises from the replacement of Mg^{2+} by Al^{3+} with the consequent creation of Na^+ ion vacancies in the conduction plane. And among the many unanswered questions concerning this material one of great importance concerns the role of non-stoichiometry in controlling the conductivity of the material. Other questions concern the variation of the conductivity with temperature, as there is good evidence from experimental studies that there is a change in the ion migration mechanism at about 500 K.[32] And in general the nature of the dynamics of the mobile Na^+ ion is poorly characterized.

To investigate these problems, Wolf[33] and Wolf et al.[34] performed an illuminating dynamical simulation study. They used a simulation box in which the triply primitive hexagonal unit cell of the crystal (containing 90 ions) had been quadrupled normal to the C-axis. The interatomic potentials were taken from those of the appropriate binary oxides; the time steps used in the simulation were 5×10^{-14} sec or 4×10^{-14} sec, depending on the temperature which was varied from 300 K to 700 K. The simulations were performed on the CRAY-IS computer at the University of London Computer Centre, U.K.

The first problem to be investigated concerned the effect of non-stoichiometry on the conductivity. In non-stoichiometric β $- Al_2O_3$, one in six of the Na^+ sites are vacant; the closest approach to this composition which could be achieved with the box size which could be used in our calculations was one in which one in eight of the Na^+ ions are missing. Simulations were therefore performed on the latter composition and on the stoichiometric material.

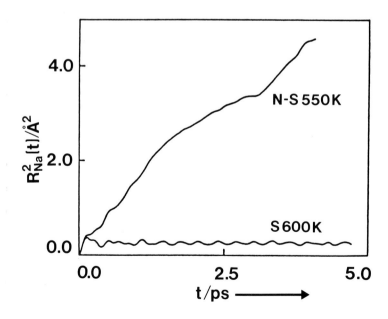

Figure 3.   Plot of mean square displacement of sodium ions vs. time in non-stoichiometric β″ $- Al_2O_3$ at various temperatures.

The contrast between the behavior of the two systems is very marked. Plots of the mean square displacement of Na^+ ions vs. time (the slope of which are proportional to the diffusion coefficient) are illustrated for the two compositions at 550 K in Fig. (3). They show that in the stoichiometric material, the Na^+ diffusion coefficient is small, whereas in the non-stoichiometric solid it is high. Table (1) gives the calculated conductivities (obtained from the diffusion coefficient via the Nernst-Einstein relationship assuming a Na^+ vacancy migration mechanism) for the non-stoichiometric material at three temperatures. Agreement between theory and experiment is certainly acceptable at the two higher temperatures. The less satisfactory result for 300 K is, however, explicable as diffraction data[35] indicate the presence of a vacancy super-lattice at this temperature. Our simulation box is too small to permit the formation of such supercell. Our calculated conductivity will therefore be too high as the supercell would be expected to reduce the Na^+ conductivity by locking the Na^+ vacancies in to an ordered array.

Table 1. Calculated and Experimental Conductivities of Non-Stoichiometric $Na^+ - \beta'' - $ Alumina

T/K	Calculated $D/cm^2\ s^{-1}$	Calculated $\sigma/\Omega^{-1}\ cm^{-1}$	Observed $\sigma/\Omega^{-1}\ cm^{-1}$
300	1.41 x 10^{-5}	1.53	0.014 - 0.160
550	2.24 x 10^{-5}	1.23	0.80
700	5.78 x 10^{-5}	2.49	1.24

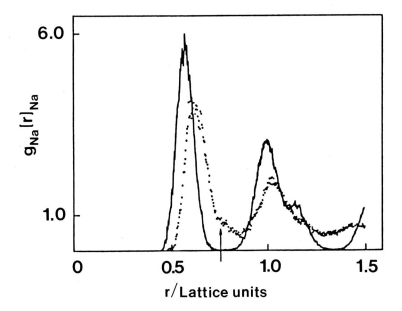

Figure 4.   Radial distribution function for stoichiometric and non stoichiometric $\beta'' - Al_2O_3$ (dotted line) at 300 K.

Information on the structural consequences of the non-stoichiometry is provided by the $Na^+ \cdots Na^+$ radial distribution functions, which are illustrated in Fig. (4) for both stoichiometric and non-stoichiometric solids at 300 K. The greater diffuseness of the r.d.f. for the later is indicative of the greater disorder in the conduction plane of the non-stoichiometric compound.

We also note that the first peak in the r.d.f. of the non-stoichiometric material has a shoulder at high r, which is indicated by the arrow shown in the diagram. As argued by Wolf[33] and Wolf et al.[34] this feature can be explained in terms of relaxation of the n.n. Na^+ ions towards a Na^+ vacancy, and its occurrence indicates that there are well defined vacancies in the conduction plane. Migration would therefore be expected to take place via a conventional vacancy hopping mechanism. In contrast, at 700 K, the shoulder has disappeared suggesting that vacancies are present at lattice sites for too short a period for the occurrence of appreciate relaxation of the surrounding lattice, indicating a much more 'continuous' type of migration mechanism.

Further evidence for this fascinating change in the nature of the Na^+ ion dynamics is provided by study of the moment ratio, $P\alpha(t)$, defined as:

$$P\alpha(t) = \frac{3 < r\alpha(t)^4 >}{5\xi < r\alpha(t)^2 >^2} \tag{3}$$

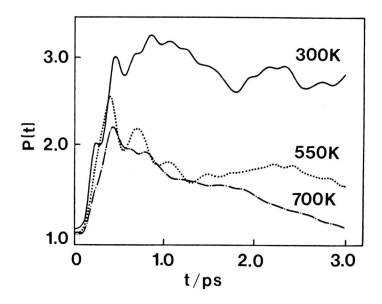

Figure 5.   Plot of the monomer ratio, P(t), vs time in non-stoichiometric $\beta'' - Al_2O_3$ at various temperatures.

It has been shown[36,37] that this ration will tend to unity at large t when a continuous diffusion mechanism is operative. In contrast, strong deviations from unity indicate a hopping mechanism. Fig. (5) shows $P(t)$ vs. t plots for three temperatures. At higher temperatures $P(t)$ approaches unity from which, however, it deviates strongly at the lower temperature.

References (33) and (34) provide far greater detail of the nature of the Na^+ ion migration revealed by the simulations. It is clear that subtle effects may be predicted which it would be of great interest to investigate experimentally.

3.2 Sorption in Zeolites

Zeolites are framework structured aluminosilicates of great industrial importance. Owing to their porosity, organic molecules can diffuse into them and undergo a variety of reactions, in many of which protonation is the first step (Zeolites invariably contain protons and other cations to balance the reduction in positive charge due to replacement of Si^{4+} by Al^{3+} in the aluminosilicate framework). Moreover, the geometry of the pores may control the nature of the product and for this reason zeolites are often referred to as shape selective catalysts. A compilation of recent papers on zeolites is given in Reference (38).

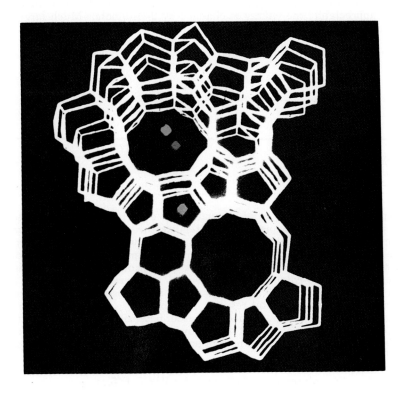

Figure 6.   Sites for **sorption** of Kr in silicate: site(1) is the lowest energy position, site(2) the next lowest.

If we are to understand the details of how the architecture of the zeolite framework controls the nature of the reactions of sorbed molecules, it is clearly necessary to obtain detailed information on the sites at which sorption occurs. Recent work has shown that it may be possible in certain cases to derive such information from diffraction techniques.[39,40] There is clearly, however, an incentive to devise reliable theoretical methods for predicting sorption sites and energies. To this end, we recently undertook a detailed study of a relatively simple problem: we investigated the sorption of krypton into silicalite, a 'pentasil' structured zeolite (see Fig. 6) - so called as it contains 5 and 10 rings. Silicalite and the closely related ZSM-5 zeolite have high Si/Al ratios (>200) and have attracted considerable attention as they catalyze the methanol-gasoline interconversion.

Our simulation study of the sorption of Kr in silicalite first established that presently available potentials could reproduce the structure of this zeolite. We found that an energy minimized structure in good agreement with experiment could be generated using the potentials which incorporate bond bending terms developed by Sanders et al..[16] We then examined a wide range of sites for sorption of the Kr atoms. The potentials used were derived and are discussed by Hope[41] and Hope et al..[42] Three sites were clearly identified. Their positions are shown in Fig (6). The sorption energies are given in table (2). We should stress that these calculations included effects of framework relaxation around the sorbed atom, which was indeed handled by treating the atom as a 'defect' around which the structure was relaxed using the techniques summarised in section (2). The results of including these effects are appreciable; the energy is changed by 2-4KJ/mole. Much larger effects could be anticipated for large complex molecules.

Table 2. Experimental and Calculated Energy Values For Krypton Energy Sites

Sites	$-U_{CALC}(kJmol^{-1})$	ΔU_{CALC}	$-U_{EXPT}(kJ/mole)$	ΔU_{EXPT}
1	20.9		12.8	
		2.0		0.7
2	18.9		12.0	
		1.6		0.7
3	17.3		11.3	

In a companion study, Hope et al.[42,43] undertook experimental sorption studies of the same problems using low pressure, low temperature volumetric techniques. It is gratifying that this work again identified three sites whose energies are compared with the calculations in table (2). The discrepancy between theory and experimental (of ≈5KJ/mole) is due to a combination of factors. Omission of thermal energies in the calculations will lead to an overestimation of the sorption energy probably by 1-2 KJ/mole. There is also some uncertainty in the 'zero' of the energy scale for the experimental results. And, of course, there will be some error due to inadequacies in the Kr... framework atom potentials. The important feature of the calculations is, however, that they show that it is possible using simulations, to identify sorption sites. Work is now in progress on the sorption of hydrocarbons in several zeolites.

3.3 The Defect Chemistry of BaTiO$_3$ and the PTCR Effect

BaTiO$_3$ is a widely used electronic ceramic material, despite which its defect structure, and the role of impurities in controlling defect and electrical properties, have been poorly understood until recently. Here we are particularly concerned with one important application of the material in thermistor devices, which rely on the dramatic 'Positive Temperature Coefficient of Resistance' (PTCR) effect observed near the Curie point. This behavior is illustrated in Fig. (7), which shows the pronounced increase in resistivity that occurs close to T_C in donor doped polycrystalline materials. Thermistor devices, which rely on the PTCR effect are widely used especially in electrical circuits where protection against 'surges' is needed.

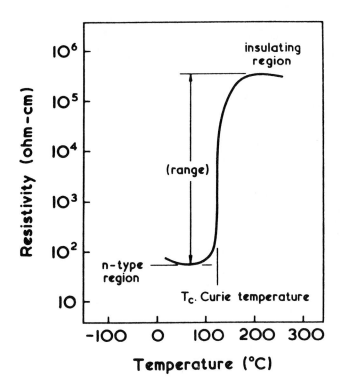

Figure 7.   The PTCR effect in $BaTiO_3$: schematic plot of resistivity profile for polycrystalline material.

For good PTCR behavior it is essential that the material be polycrystalline, and that it be donor doped. Common dopants are La^{3+} and Y^{3+} which replace Ba and are compensated by electrons at normal oxygen partial pressures. It appears, however, that for effective PTCR behavior it is essential to have a small level of *acceptors* in addition to donor impurities; a typical acceptor impurity is Mn^{3+} which replaces Ti^{4+} and is compensated by hole states at higher oxygen partial pressures.

How can such observations be rationalized? Several models have been debated in the literature. But it has been generally agreed that the key feature concerns the electrical properties of the intergranular contacts which must be either insulating or even 'p-type'. The type of model is illustrated on Fig. (8): thus it is proposed that each grain consists of n-type core surrounded by p-type or insulating skin. We can understand at least qualitatively how such a grain structure would give rise to PTCR behavior. The intergranular contacts will function as n-p-n or n-i-n junctions, which lead to the corresponding space-charge barrier and hence to high restivity above T_C. However, we might expect the effects of this barrier to be largely nullified below T_C owing to the spontaneous polarization of the material.

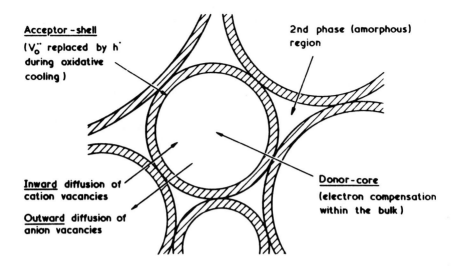

Figure 8.   Suggested structure of $BaTiO_3$.

The question remains as to how the grain structure shown in Fig. (8) arises. Insight into this problem has been provided by calculations. These were performed for impurities in sites within the bulk of the crystal and at the surface; the latter calculations were performed by E. A. Colbourn and W. C. Markrodt using the SURFIS package. The more important results are summarized in Table (3). They show that donor impurities e.g. La^{3+} have an energetic preference for sites in the bulk, whereas acceptors are commonly more stable at surface sites. The former will therefore segregate away from the surface and the latter towards the surface of the grains. Calculations on defect migration energies also suggested that there were high barriers to acceptor migration, which will also favor the retention of acceptors close to the surface of the grains.

Our results thus suggest that there are both thermodynamic and kinetic factors favoring the grain structure shown in Fig. (8) and thus provide a rationalization of the PTCR effect in terms of the defect chemistry of $BaTiO_3$. For further details of this work we refer to reference (44).

4. SUMMARY AND CONCLUSIONS

Simulation studies of solids are advancing rapidly - an advance that is facilitated by developments in both software and hardware. Future work will concentrate on the study of highly complex systems, the development of more accurate potentials (using quantum mechanical methods) and an increasing interaction with experiment. The availability of a new generation of supercomputers offers a very exciting future for the field.

Table 3. Surface defect energies for $BaTio_3$ calculated using SURFIS computer code.

Species	$E_{SURFIS}(J \times 10^{19})$	$\Delta^*(j \times 10^{19})$
	Barium (100) surface	
O_o	24.00	-2.74
V''_{Ba}	31.02	-1.91
$V_{\ddot{o}}$	28.17	-1.23
Sc_{Ba}	-42.17	-1.09
La_{Ba}	-30.86	$+0.98$
Mg_{Ba}	-6.70	-0.91
Mn_{Ba}	-4.76	-0.64
	Titanium (100) surface	
V_n''''	129.6	-3.43
$V_{\ddot{o}}$	28.20	-1.07
Ti'_{Ti}	24.00	-3.51
Sc'_{Ti}	50.26	-0.45
Mn'_{Ti}	46.69	$+0.37$
Mg''_{Ti}	86.92	-1.59
Mn''_{Ti}	90.43	-1.30
Mn_{Ti}	-2.26	$+0.70$
Al'_{Ti}	46.85	$+0.22$

$^*\Delta_{bulk}$ refers to difference $(E_{SURFIS} - E_{CASCADE})$; a negative value indicates that surface substitution is energetically more favorable than bulk substitution.

ACKNOWLEDGEMENTS

I would like to acknowledge the contribution to the work discussed in this chapter of M.L. Wolf, A.I.J. Hope, G.V. Lewis, E.A. Colbourn and W.C. MacKrodt.

REFERENCES

1. Catlow C.R.A. and MacKrodt W.C, *eds.*, 'Computer Simulation of Solids', Lecture Notes in Physics, **vol 166**, (Springer-Berlin), (1982).

2. MacKrodt W.C, in Transport in Non-Stoichiometric Compounds (*eds*. G. Petot-Ervas, H.J. Matzke and C. Monty), North-Holland (1984).
3. Catlow C.R.A., Ann. Rev. Mat. Sci. - in press.
4. Harding J.H., Phys. Rev. B, **32**, 6861 (1985).
5. Colbourn E.A., MacKrodt W.C. and Tasker P.W., Physica **B131**, 41 (1985).
6. Tasker P.W., Surf. Sci, **87**, 315 (1979).
7. Duffy D. and Tasker P.W., Phil. Mag. **A47**, 817 (1983).
8. Catlow C.R.A. and Norgett M.J. UKAEA report, AERE-M2936 (1976).
9. Tosi M., Solid State Physics (*eds*. F. Seitz and S. Turnbull) vol **16**, p1.
10. Norgett M.J. and Fletcher R., J. Phys. C.**3**, 163 (1970).
11. Mott N. F. and Littleton M.J., Trans Faraday Soc., **34**, 485 (1938).
12. Norgett M.J., UKAEA report, AERE-R7650 (1974).
13. Catlow C.R.A., James R., MacKrodt W.C. and Stewart R.F., Phys. Rev. **B25**, 1006 (1928).
14. MacKrodt W.C. in 'Mass Transport in Solids', (*eds*. C.R.A. Catlow and W.C. MacKrodt), Plenum Press, (1983).
15. Catlow C.R.A and Stoneham A.M., J. Phys. C., **16**, 4321 (1983).
16. Sanders M.J., Leslie M. and Catlow C.R.A., J. Chem. Soc., Chem. Comm, 1271, (1984).
17. Sanders M.J., PhD Thesis, University of London (1984).
18. Dick B.G. and Overhauser A.W., Phys. Rev. **B112**, 90 (1958).
19. Catlow C.R.A., Dixon M. and MacKrodt W.C., in 'Computer Simulation of Solids' (**eds.** C.R.A. Catlow and W.C. MacKrodt), Lecture Notes in Physics, vol **166**, (Springer-Berlin), 1982.
20. MacKrodt W.C. and Stewart R.F., J. Phys. C. **12**, 431 (1979).
21. MacKrodt W.C. and Stewart R.F., J. Phys. C. **12**, 5015 (1979).
22. Taylor R. in 'Computer Simulation of Solids', (*eds*. C.R.A. Catlow and W.C. MacKrodt), Lecture Notes in Physics, vol 166 (Springer-Berlin), (1982).
23. Taylor R., Physica B., **131**, 103 (1985).
24. Catlow C.R.A., Cormack A.N. and Theobald F., Acta Cryst. **B40**, 185 (1984).
25. Parker S.C., Cormack A.N. and Catlow C.R.A., Acta Cryst. **B40**, 201 (1984).
26. Catlow C.R.A. and Parker S.C. in 'Point Defects in Minerals' (*ed*. R.N. Schock), Geophysical Monograph **31** 26 (1985).
27. Catlow C.R.A., Doherty M., Parker S.C., Price G.D. and Sanders M.J., Materials Science Forum, **7**, 118 (1985).
28. Catlow C.R.A in 'Non-Stoichiometric Compounds' (*ed*.. O.T. Sorensen) Academic Press (1981).
29. Catlow C.R.A. and James R., Proc. Roy. Soc. **A384**, 157 (1982).
30. Cormack A.N., Jones R., Tasker P.W. and Catlow C.R.A., J. Solid State Chem **44**, 174 (1982).
31. Coremack A.N., Catlow C.R.A. and Tasker P.W., Radiat. Eff. **74**, 237 (1983).
32. Farrington J.C. and Briant J.L. in 'Fast Ion Transport in Solids' (*eds* P. Vaslushta, J.N. Mundy and G.K. Shenoy) North-Holland, p 395 (1979).
33. Wolf M.L., PhD Thesis, University of London (1984).
34. Wolf M.J., Walker J.R. and Catlow C.R.A., Solid State Ionics, **13**, 33 (1984)
35. Collin G. et al. in 'Solid State Ionics 83', (*eds* M. Kleitz, B. Sapoval and D. Ravaire) (North Holland) p 311 (1983).
36. Hansen J.P. and McDonald Il, 'Theory of Simple Liquids', Academic Press (1976).
37. Rahman A., J. Chem. Phys. **65**, 4845 (1976).
38. Proceedings of the 6th International Conference on Zeolites, Reno (1983).
39. Fitch A.N., et al., J. Phys. Chem. - in press.
40. Cheetham A.K. et al., Nature - press.
41. Hope A.T.J. PhD Thesis, University of London.
42. Hope A.T.J. and Catlow C.R.A. - to be published.
43. Hope A.T.J. and Leng C.A. and Catlow C.R.A. - to be published.
44. Lewis G.V. and Catlow C.R.A., J. Amer. Ceram. Soc. **68**, 555 (1985).

MD SIMULATIONS OF THE EFFECT OF PRESSURE ON THE STRUCTURAL AND DYNAMICAL PROPERTIES OF WATER AND AQUEOUS ELECTROLYTE SOLUTIONS

Karl Heinzinger

Max-Planck-Institut Fur Chemie (Otto-Hahn Institut),
Mainz, Federal Republic of Germany

ABSTRACT

The results presented in this paper are mainly based on MD simulations of pure water at 1 bar and 22 kbar at about 70° C and of 2.2 molal NaCl solutions at room temperature and pressures of 1 bar and 10 kbar. They are compared with experimental data available in the literature. The changes in the structural properties with increasing pressure are discussed on the basis of various radial distribution functions and the orientation of the water molecules in the hydration shells of the ions. The effect of pressure on the self-diffusion coefficients has been calculated for ions, for pure and solvent water and separately for the three water subsystems - bulk water, hydration water of the cation and of the anion. Finally, the changes with pressure of the hindered translational motions and of the intramolecular vibrations of the water molecules are reported.

I. INTRODUCTION

Experimental investigations and computer simulations of the effect of pressure on the structural and kinetic properties of water can be expected to give new insight into our understanding of the structure of this liquid. The experimental investigations include neutron diffraction[1], X-ray diffraction,[2] NMR[3-5] and Raman spectroscopic studies[6-8] of water under high pressures (several kilobars). Molecular dynamics (MD) simulations have been performed at a density of 1.346 g/cm³ corresponding to that observed at the ice VI- ice VII - liquid triple point (t = 81.6° C and p = 22.0 kbar). The MD simulations reported by Stillinger and Rahman[9] and Impey, Klein and McDonald[10] used the ST2[11] and Matsuoka-Clementi-Yoshimine[12] water-water potential, respectively, whereas Jancsó, et al.[13-14] employed a modified version of the central force potential which is capable of describing the effect of pressure on the internal vibrational motions of water.[15]

A long standing question is concerned with the effect of pressure on the hydration sphere of ions. Zisman,[16] who carried out a series of measurements on the electrical conductivity of aqueous solutions under high pressure, concluded that there must be a large change in the diameters of ions due to pressure. In order to interpret the pressure dependence of conductance data, Horne suggested that the application of pressure breaks up the structure in bulk water and also the local water structure near ions.[17] According to him - if the pressure is high enough - even the innermost hydration sheath will be destroyed and a pressure-induced dehydration takes place.[18] Brummer and Gancy[19] also support the view that pressure destroys the water structures around the ions. On the other hand, Samoilov et al.[20] arrived at the conclusion, on the basis of density data on dilute aqueous solutions of NaCl and KCl at 20 and 45°C up to a pressure of 1 kbar, that the increase in pressure leads to an enhancement of the close hydration of these cations due to the breakdown of the bulk water structure. Computer simulations can contribute significantly in solving this discrepancy.

MD simulations at room temperature and a density corresponding to a pressure of about 10 kbar have been performed for a 2.2 molal NaCl solution[21,22] with the CF model for water[15] and for a calcium as well as a chloride ion surrounded by 64 water molecules[23] with the TIP4P model for water from Jorgensen et al.[24] The effect of increased temperature and pressure on the structure of a 0.55 molal LiI solution has been investigated by an MD simulation at 300 K and 500 K and constant density, where the higher temperature corresponds to a pressure of about 3 kbar[25] with the ST2 model for water.[11] Bounds has investigated the changes of the hydration shell of Ca^{++} by an temperature increase to 81° C at constant density by a simulation where again one ion is surrounded by 64 TIP4P water molecules.[23]

The experimental results of high pressure studies of aqueous solutions have extensively been reviewed in recent years.[26-28] The neutron diffraction measurements on concentrated $NiCl_2$ and LiCl solutions by Neilson[29] and the conductance and transference number data for KCl from Nekahara et al.[30] are of special relevance for a comparison with the results from computer simulations.

In this paper the results of the computer simulation of the effect of pressure on the structure and on dynamical properties of pure water and aqueous electrolyte solutions are reviewed. The simulation data are compared with experimental results available in the literature. For the details of the simulations the reader is referred to the original communications.

II. STRUCTURAL CHANGES

A) Pure Water

The change with pressure of the oxygen-oxygen, oxygen-hydrogen and hydrogen-hydrogen radial distribution functions (RDFs) $g_{xy}(r)$, and the corresponding running integration numbers, $n_{xy}(r)$, calculated from the simulation with the CF model,[13] is demonstrated in Fig. 1. Some characteristic values of the RDFs are given in Table 1.

The trends with pressure increase are qualitatively similar to those obtained from MD simulations with ST2[9] and MCY[10] potentials. The first maximum has shifted by 0.05Å (0.03 and 0.04Å for ST2 and MCY potentials, respectively) to a smaller distance as a result of compression, which indicates that the 38% density increase results in a much smaller decrease in the hydrogen bond length than would be expected on the basis of a uniform shrinkage of all intermolecular distances. In contrast to the result obtained from the MD simulation with MCY potential no significant increase has been found in the height of the $g_{OO}(r)$ first peak on density increase. It can be seen that in the high pressure (HP) liquid the second peak occurs at ≈ 1.9 times the distance at which the first peak occurs, while in the

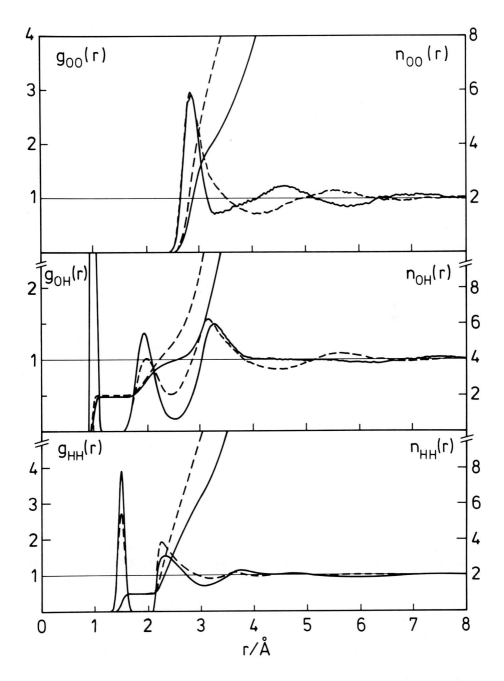

Figure 1 Oxygen-oxygen, oxygen-hydrogen and hydrogen-hydrogen radial distribution functions and running integration numbers from MD simulation of pure water at high pressure (dashed line) and normal pressure (solid line).

Table 1 Comparison of characteristic values of the radial distribution functions (intermolecular part) for normal pressure (1bar; denoted by NP) and high pressure (10kbar; HP) 2.2 molal NaCl solutions,[22] as well as for pure water at normal pressure (0.9718 g/cm^{-3}; 336 K) and high pressure (1.346 g/cm^{-3}; 350 K).[13] R_i, r_{Mi}, and r_{mi} give the distances where $g_{xy}(r) = 1$, has its ith maximum and minimum, respectively. The distances are given in Å with an uncertainty of at least $\pm$ 0.02 Å. $n_{xy}(r_{m1})$ is the running integration number at the first minimum of $g_{xy}(r)$.

x	y		R_1	r_{M1}	$g_{xy}(r_{M1})$	R_2	r_{m1}	$g_{xy}(r_{m1})$	r_{M2}	$g_{xy}(r_{M2})$	$n_{xy}(r_{m1})$
Na	0	NP	2.10	2.30	8.00	2.66	3.10	0.11	4.47	1.58	5.8
		HP	2.10	2.30	7.02	2.63	3.15	0.18	4.8	1.46	6.3
Na	H	NP	2.64	2.95	3.10	3.34	3.70	0.45	5.18	1.42	13.9
		HP	2.59	2.93	2.73	3.32	3.68	0.49	5.05	1.35	16.2
Cl	0	NP	2.95	3.18	3.37	3.60	3.87	0.71	4.62	1.24	7.7
		HP	2.91	3.16	3.10	3.8	–	–	–	–	(9.6)[a]
Cl	H	NP	1.98	2.20	2.93	2.60	2.88	0.45	3.50	1.50	6.7
		HP	1.95	2.15	2.45	2.54	2.85	0.57	3.40	1.50	7.2

Pure Water

x	y		R_1	r_{M1}	$g_{xy}(r_{M1})$	R_2	r_{m1}	$g_{xy}(r_{m1})$	r_{M2}	$g_{xy}(r_{M2})$	$n_{xy}(r_{m1})$
0	0	NP	2.64	2.85	2.93	3.13	3.28	0.71	4.63	1.23	4.2
		HP	2.60	2.80	2.95	3.49	4.05	0.71	5.48	1.14	12.2 (7.0)[a]
0	H	NP	1.80	1.92	1.40	2.05	2.53	0.19	3.23	1.51	4.0
		HP	1.95	1.98	1.01	2.02	2.40	0.51	3.15	1.56	4.6
H	H	NP	2.17	2.32	1.56	2.73	3.07	0.73	3.72	1.16	6.8
		HP	2.14	2.25	1.92	2.93	3.18	0.90	3.65	1.07	11.5

[a] These numbers result if $n_{xy}(r_{m1})$ is taken at r_{m1} of these NP cases.

normal pressure (NP) liquid the ratio of the two peak distances is ≈ 1.63, which corresponds to the open tetrahedral structure. As has been pointed out previously[9,31] the shifts of the second maximum as well as of the first minimum to larger distances indicate a more closed-packed structure for the HP liquid. The number of first-neighbour molecules obtained by integration up to the first minimum is 4.2 and 12.2 for the NP and HP water respectively. However if the integration in the case of the compressed liquid is carried out up to the first minimum in $g_{OO}(r)$ for NP water (3.28Å) a value of 7.0 results for the coordination number which is in good agreement with the value of 6.9 estimated from the volume increase of ice VII on melting.[1]

In the oxygen-hydrogen RDFs once can observe that the first intermolecular peak moves to larger distances by ≈ 0.06Å and the second one to smaller distances by ≈ 0.08Å. The first peak broadens and its height decreases significantly, the number of first-neighbour hydrogens increases a little above 2, and finally we see an increase in the correlation at larger O-H distances as a result of the compression. The shifts in the first- and second-neighbour distances indicate that the density increase is accompanied by a non-negligible distortion of linear hydrogen bonds.[1,10] This conclusion is supported by the fact that the average hydrogen bond angle increases from 23.5° in the NP water to 35.5° in the HP water.[14]

In accordance with the results obtained for the MCY potential[10] the first and second intermolecular peaks in $g_{HH}(r)$ are shifted by 0.07Å to smaller distance and the coordination number increased from 6.8 at low pressure to 11.5 at high pressure. (The effect of density increase on $g_{OH}(r)$ and $g_{HH}(r)$ has not been discussed in detail by Rahman and Stillinger for the ST2 model.[9])

After good agreement between the simulations of pure water with the three different water models for the change with pressure for the single RDFs has been found, the question arises how the results of the simulations compare with measurements. The time-of-flight neutron diffraction technique makes it possible to measure the RDFs separately. Soper and Silver have measured $g_{HH}(r)$ and 0.178 $g_{OO}(r) + 0.822 g_{OH}(r)$ at room temperature and normal pressure.[32] There is good agreement with the RDFs shown in Fig. 1. Small differences may have to be attributed to the temperature difference of 38°C[13]. The effect of temperature and pressure on the neutron diffraction of heavy water has been extensively studied by Wu et al.[1] The comparison of their data obtained at 85°C and at pressures of 0.06 and 15.6 kbar with the results from the simulations shows some differences but the general changes with pressure of the neutron weighted total structure function are in satisfactory agreement.[13]

The X-ray structure functions for high- and low-pressure water calculated from the various RDFs (Fig. 1) are shown in Fig. 2. The most remarkable feature is the disappearance of the splitting of the main peak with increasing density. This is also born out by X-ray diffraction studies of light and heavy water at pressures up to 2 kbar, at room temperature.[2] Although the compression is modest in this case, the merging of the two components of the first maximum into one peak is clearly indicated. As has been suggested previously[33] the reason for the peak splitting is the far ranging water-water structure of the central-force model and accordingly it seems reasonable to conclude that the pronounced change observed in the $g_{OO}(r)$ at higher O-O separations (≥ 4Å) with increasing density is responsible for the reduction of the double peak to a single maximum. But it must be emphasized that the MD results are in contrast to the much larger rate of contraction of the O-O first neighbor distances (0.01Å/kbar) obtained from these X-ray diffraction experiments.[2]

Another property of interest is the ensemble and time averaged geometrical arrangement of the water molecules around a central one. For this purpose a coordinate system has been defined in such a way that the plane of the central molecule coincides with the yz plane of the coordinate system, the center of mass is positioned at the origin and the oxygen atom in the positive z direction. In Fig. 3 the projections of the oxygen atom positions onto the xy plane of the coordinate system are shown separately for all nearest neighbors, the first four and the second four neighbors. It can be clearly observed that there is a decrease in the preference for the occupation of tetrahedral positions in the HP water. The neighbors five to eight in HP water are seen to be closer to the central molecule than the second neighbors in the NP water, but they do not show any preference for occupying tetrahedral positions. Accordingly, it seems reasonable to conclude that the first coordination sphere of high pressure water consists of neighbors of two different types.[14] Similar conclusions have been drawn from the results of the other two MD simulation[9,10] in which an interpenetration of the subsections of the random hydrogen bond network has been proposed.

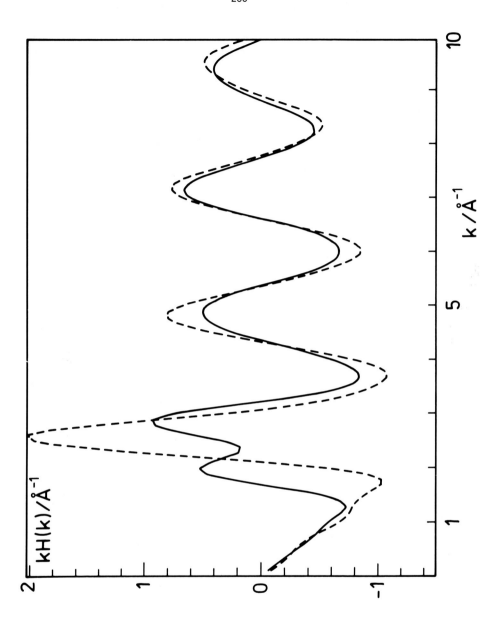

Figure 2   X-ray structure functions derived from MD simulations for low (solid line) and high (dashed line) pressure water.[13]

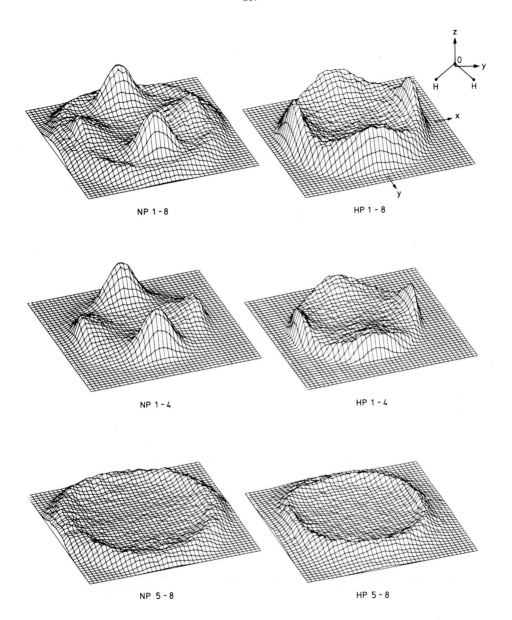

Figure 3   Three-dimensional drawings of the projections of the oxygen atom positions of the eight nearest neighbor water molecules around a central one onto the xy plane of a coordinate system as shown in the insertion calculated from the MD simulations of NP and HP water.[14]

B. Aqueous Electrolyte Solutions

The ion-oxygen and ion-hydrogen RDFs together with their running integration numbers are shown in Fig. 4 for the NP (1 bar) and HP (10 kbar) 2.2 molal NaCl solutions at room temperature.[21] Some characteristic values of the RDFs are listed in Table 1.

The position and the width of the first peak of $g_{NaO}(r)$ is unchanged while its height decreases with increasing pressure. The second peak flattens and the valley between the first and second peaks gets filled up for distances above the minimum of $g_{NaO}(r)$. The coordination number of the sodium ion, $n_{NaO}(r_{m1})$ - the value of the running integration number at the first minimum of the RDF - is 5.8 and 6.3 for the NP and HP solution, respectively. These changes are relatively small, specially keeping in mind that the HP solution has a 21% higher density than the NP solution. If the height of the first peak in the RDF is considered a measure of the degree of structure of the first hydration shell one can conclude that the increase in pressure produces a slight but still observable decrease in the hydration shell structure of Na$^+$. On the other hand it is also evident that even this high pressure does not destroy the hydration layer around the ion.

The compression of the solution leads to the broadening of the first peak in $g_{NaH}(r)$ while its height decreases and $n_{NaH}(r_{m1})$ increases by 2.3. The stronger increase of $n_{NaH}(r_{m1})$ relative to $n_{NaO}(r_{m1})$ with increasing pressure indicates the breakdown of the water structure between first and second hydration shell by compression.

It can be seen from Fig. 4 that the height of the first peak in $g_{NaO}(r)$. decreases with increasing pressure similarly to $g_{NaO}(r)$. But different from $g_{NaO}(r)$ the second peak in $g_{ClO}(r)$ disappears and the coordination number - taken at the same Cl$^-$ $-$0 distance (r = 3.87Å) - increases by two while the increase amounts to only 0.5 in the case of Na$^+$. This is an indication that in the case of the chloride ion the hydration shell is softer and thus it is easier for the water molecules to be transferred into the first hydration shell by increasing pressure.

From the positions of the first peaks in g_{ClH} (r) and $g_{ClO}(r)$ (see Table 1) it follows that linear hydrogen bonds are preferentially formed between the chloride ion and the water molecules in the first hydration shell. The increase in the number of nearest neighbor hydrogens with pressure is found to be 0.5 just as in the sodium-oxygen case. The changes observed in the Cl$^-$ $-$O and Cl$^-$ $-$ H RDFs together with those found in the cos Θ distributions (vide infra) suggest that the hydration shell structure of Cl$^-$ in the HP solution is even less developed than in the solution under normal pressure.

The changes in $g_{OO}(r)$ as well as in $g_{OH}(r)$ and $g_{HH}(r)$ with pressure in the 2.2 molal NaCl solution are similar to those observed in the MD simulation of pure water, however, their magnitude is smaller here presumably due to the lower pressure applied (10 kbar vs. 22 kbar).

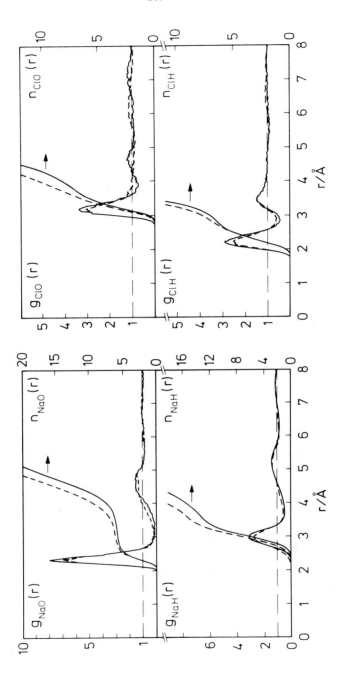

Figure 4   Ion-oxygen and ion-hydrogen radial distribution functions and running
integration numbers for high pressure (dashed line) and normal pressure
(solid line) 2.2 molal NaCl solutions.[22]

The orientation of the water molecules in the hydration shell of an ion can be characterized by the cosine of the angle between the dipole moment direction of the water molecule and the vector pointing from the oxygen atom towards the center of the ion. In Fig. 5 distributions of cos Θ are shown for the hydration shells of sodium and chloride ions in the NP and the HP solutions. The increase in pressure broadens the cos Θ distributions. This means that with increasing pressure a decrease in the preference for the trigonal orientation in the case of Na$^+$ and for linear hydrogen bond formation in the case of Cl$^-$ can be observed.

A second computer simulation in which the effect of pressure on the hydration shells of ions has been investigated is reported by Bounds.[23] He performed MD simulations where a Ca^{++} and a Cl$^-$ were surrounded by 64 TIP4P[24] water molecules at room temperature and also pressures of 1 bar and about 10 kbar. As far as the Ca^{++} is concerned the results are very similar to the ones found for Na$^+$: No change in the position of the first maximum (r_{M1}) in $g_{CaO}(r)$, a small decrease (hardly outside the limits of uncertainty) for r_{M1} in $g_{CaH}(r)$ and in the average value of Θ as well as a slightly larger hydration number ($+0.3$) in the high pressure case.

The effect of pressure on the hydration shell of Cl$^-$ derived from both MD simulations can be compared directly. They agree in a small decrease of r_{M1} in $g_{ClO}(r)$, and in $g_{ClH}(r)$ and of the average value of Θ with increasing pressure. The changes in the hydration number are significantly different. For the NaCl solution the hydration number is larger by two at the high pressure while Bounds found only an increase by about 0.5. The changes in $g_{ClO}(r)$ in the range $4 - 5$Å are also different in the two simulations. While the second peak in the NaCl case completely disappears at the high pressure, Bounds finds only a shift to smaller distances by about 0.5Å.[23] The reason for this discrepancy is not clear.

The neutron diffraction studies by Neilson[29] at pressures up to 1 kbar are not helpful in solving the discrepancy between the simulations. These measurements indicated some effect of pressure on the chloride-water structure in the case of a 3 molal NiCl$_2$ solution, while for a 10 molal LiCl solution no change was observed in the Cl$^-$ - water conformation. However, a detailed comparison with the present results is not warranted since the high concentration of these solutions does not allow the formation of well-defined hydration shells around the chloride ions.

The effect of elevated temperature and pressure on the ion-oxygen RDRs has been investigated for a 0.55 molal LiI solution by MD simulations at 398 K and 508K and constant density, where the higher temperature corresponds to a pressure of about 3 kbar.[25] It can be seen from Fig. 6 that the increase in temperature and pressure reduces significantly the height of the first peak in $g_{LiO}(r)$ and broadens it. At the same time the gap between first and second hydration shell begins to get filled up. In this way, the number of water molecules (six) in the first hydration shell of Li$^+$ remains constant as can be seen from $n_{LiO}(r)$ at r = 4.5Å which slowly disappears with increasing distance. In the case of the iodide ion the first hydration shell is, even at normal temperature and pressure, not well pronounced (lower part of Fig. 6). Consequently not much change can be observed. Only a slight smearing out of the first peak results from the temperature and pressure increase.

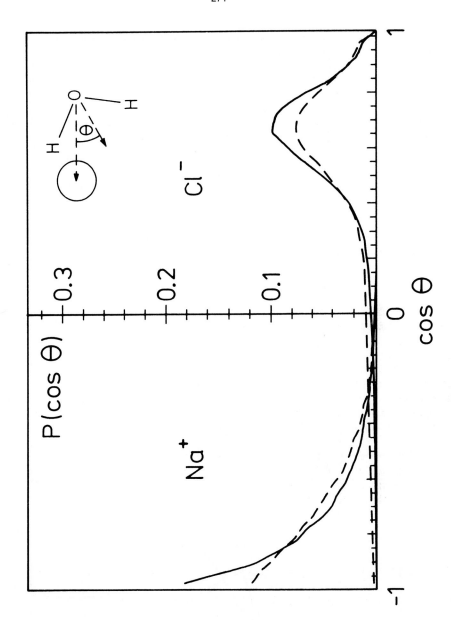

Figure 5   Distribution of cos (θ) in the first hydration shells of the ions from MD simulations of 2.2 molal NaCl solutions at normal pressure (solid line) and high pressure (dashed line). θ is defined in the insertion. The distributions are normalized and given in arbitrary units.

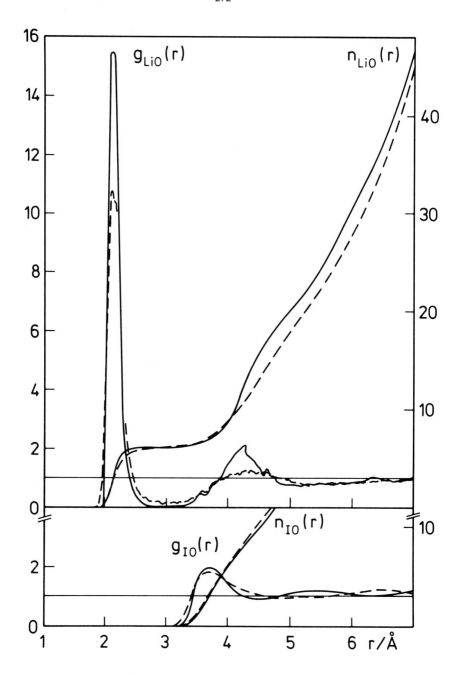

Figure 6 Ion-oxygen radial distribution functions and running integration numbers for a 0.55 molal LiI solution at 508 K (dashed) and 308 K (full) with the same density of 1.05 g/cm³. At the high temperature the pressure is about 3 kbar.[25]

Bounds has investigated the effect of temperature and pressure increase on the hydration shell of Ca^{++} by performing simulations at 286K and 354K and constant density. He did not find any significant difference in the structural properties calculated from both simulations.[23] The difference between the large effect on the hydration shell of Li^+ (Fig. 6) and the negligible one in the case of Ca^{++} must have to be attributed to the strongly different temperature increase (200° and 70° in the case of Li^+ and $Ca,^{++}$ respectively). This is in keeping with the results by Impey, Madden and McDonald.[34] They performed simulations where a Li^+ was surrounded by 124 water molecules at 278K, 297K and 368K and a constant pressure of 1 bar. The temperature increase resulted in a decrease of the first maximum in $g_{LiO}(r)$ by only about 10%.

III. CHANGES IN THE DYNAMICAL PROPERTIES

(A) Self-Diffusion Coefficients

The self-diffusion coefficients can be determined from the simulation through the velocity autocorrelation functions with the help of the Green-Kubo relation

$$D = \lim_{t \to \infty} \frac{1}{3} \int_0^t \, <\vec{v}(0)\circ\vec{v}(t)> \, dt \, .$$

The averages have been calculated according to

$$< \vec{v}(0) \circ \vec{v}(t)> \, = \, \frac{1}{N_T N} \sum_{i=1}^{N_T} \sum_{j=1}^{N} \, \vec{v}_j(t_i) \circ \vec{v}_j(t_i + t),$$

where N denotes the number of particles, N_T the number of time averages and $v_j(t)$ the velocity of particle j at time t.

Table 2 Self-diffusion coefficients of the ions (D_{Na^+}, D_{Cl^-}) and solvent water (D_W) from MD simulations of 2.2 molal NaCl solutions at normal pressure (NP) and high pressure (HP) and from experiments at room temperature in units of 10^{-5} cm²/s.[22]

			D_W	D_{Na}^+	D_{Cl}^-
MD	NP	(299 K)	1.60 (1.83)[a]	0.82	1.08
	HP	(303 K)	1.51	0.55	1.25
Exp.	NP	(298 K)	1.92[36]	1.12[37]	1.60[37]
				1.30[38]	1.85[38]

[a] extrapolated to 303 K according to Ref. (35).

The velocity autocorrelation functions for the sodium ions, chloride ions and solvent water at normal pressure (1 bar) and high pressure (10 kbar) calculated from MD simulations of 2.2 molal NaCl solutions at room temperature are shown in Fig. 7.[22]

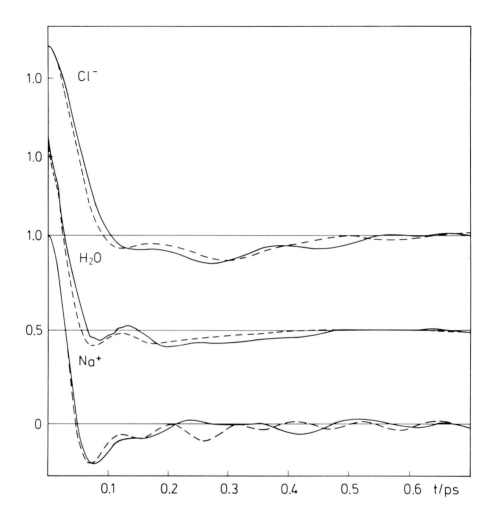

Figure 7   Normalized velocity autocorrelation functions for the Na^+, Cl^- and the oxygens from MD simulations of normal pressure (solid line) and high pressure (dashed line) 2.2 molal NaCl solutions.[22]

The pressure dependence found for solvent water is very similar to the one for pure water,[13] in agreement with the similarity in the pressure induced changes of the structural properties as mentioned above. It can be seen from Fig. 7 that in all three cases the velocity autocorrelation function decays faster at the higher pressure.

The resulting self-diffusion coefficients are given in Table 2 together with experimental results at normal pressure. Measurements of the pressure dependence of the self-diffusion coefficients of the ions in aqueous solutions are not known.

The comparison of the self-diffusion coefficients obtained from the MD simulations with the experimental results shows that the simulation leads to smaller values in all cases. In the normal pressure solution the self-diffusion coefficient of the sodium ion D_{Na^+} is found to be significantly smaller than that of the chloride ion D_{Cl^-} in good agreement with the experimental results.[37,38] This can be rationalized as in case of the LiI solution[39] by assuming that the Na^+ is moving through the solution with its hydration shell attached while the Cl^- diffuses essentially without its hydration shells. Considering the uncertainties in the experimental results for D_{Na^+} and D_{Cl^-} as indicated by the values reported from different laboratories, the differences between simulation and experiments seem to be small enough to justify the conclusion that the changes in the self-diffusion coefficients for water and ions with increasing pressure as calculated from the simulations are reliable - at least qualitatively.

Table 2 shows that the D_{Na^+} and D_{Cl^-} show an opposite pressure dependence: While D_{Na^+} decreases, D_{Cl^-} increases with increasing pressure. This result is in qualitative agreement with the conductance and transference number measurements in aqueous KCl solutions at room temperature and pressures up to 2 kbar by Nakahara et al.[30] These authors found that with increasing pressure the mobility of Cl^- increases while the one for K^+ decreases. Obviously the direct effect of the density increase is not the only factor to be considered in understanding the effect of pressure on the dynamic properties of ions in aqueous electrolyte solutions. The opposite behavior of cations and anions is not yet understood completely. A possible explanation might be that the decrease in the preference for linear hydrogen bond formation with the Cl^- (Fig. 5) makes the anion more mobile at the high pressure. Because of the strong interactions of the Na^+ with its hydration shell water molecules the change in P (cos Θ) with pressure may not significantly influence D_{Na^+} and the direct effect of density increase becomes effective.

It can be seen from Table 2 that the 21% density increase is accompanied by about 20% decrease in the self-diffusion coefficient of solvent water D_W similar to the pure water case, where at 77° C a 38% density increase yielded about 35% decrease in the self-diffusion coefficient.[13] Wilbur et al.[4] have measured the self-diffusion coefficients in liquid heavy water in the temperature range 10 − 200° C and the pressure range 1 bar to 9 kbar. The experimental data suggest $\approx$ 40% decrease in D_{D_2O} at 77° C for a 20% density increase. Thus the MD results obtained with the CF potential are more consistent with the experiment than that obtained with the ST2 potential in which case only a 10% decrease in D_{D_2O} has been predicted for a 35% density increase.[9]

The self-diffusion coefficients of the water molecules in the three water subsystems - bulk water, hydration water of Na^+ and of Cl^- - have been obtained separately from the velocity autocorrelation functions. For all three subsystems the change of the self-diffusion coefficient with pressure is the same and amounts to about -20% at 10 kbar.[22]

(B) Hindered Translational Motions

The spectral densities $f(\omega)$ of the hindered translational motions have been calculated by Fourier transformation from the normalized velocity autocorrelation functions for pure water, for the ions and separately for the water molecules of the three water subsystems - bulk water, hydration of Na^+ and Cl^- at normal and high pressure.

In Fig. 8 $f(\omega)$ is shown for pure water at 1 bar and 22 kbar and about 70° C.[13] the main distinct peak at $50cm^{-1}$ is usually assigned to the O-O-O flexing motion and the broad peak around $195cm^{-1}$ to O-O stretching motions (see, e.g., Ref. 40 and references cited therein). The strong reduction and shift to higher frequencies of the peak around $50cm^{-1}$ with rising density might be accounted for by the significant distortion of O-O-O angles. the HP case also shows about twice the intensity of the NP case in the region between 300 and $400cm^{-1}$. It should be noted that the effect of pressure on the frequency region corresponding to the hydrogen bond stretching and bending motions is similar to that of Li^+ , obtained from the MD simulation of a 2.2 molal LiI.[39]

The effect of pressure on the hindered translational motions of the water molecules in bulk water and the hydration water of Cl^- (calculated from the simulation of the 2.2 molal NaCl solution at 1 bar and 10 kbar and room temperature[22]) is very similar to the pressure effect in pure water as shown in Fig. 8. The hindered translations of the water molecules in the hydration shell of Na^+ are not significantly changed by the pressure increase in agreement with only small changes in the $g_{NaO}(r)$ as depicted in Fig. 4.

The changes with pressure in the spectral densities of the hindered translations of Na^+ and Cl^- are rather small. They do not give any hint for the understanding of the different pressure dependencies of the self-diffusion coefficients of the two ions (Table 2).[22]

(C) Intramolecular Vibrations

The shifts in the O-H stretching and the H-O-H bending frequencies of water by pressure increase have been calculated by Fourier transformation from the normalized velocity autocorrelation functions of the hydrogens for MD simulations of pure water[13] and of 2.2 molal NaCl solutions.[22] In all cases the effects of pressure are rather small.

For pure water a redshift of $10cm^{-1}$ and a blueshift of about $15cm^{-1}$ have been found for the stretching and bending vibrations, respectively, for a pressure difference of 22 kbar at about 70° C. Raman spectral studies on liquid water have been performed by Walkrafen[6-8] around 30° C and pressure up to 12 kbar. The O-H stretching peak frequency has been found to decrease by $\approx 2cm^{-1}$ /kbar while the peak position of the H-O-H bending band remained essentially constant. The straight extrapolation of these experimental findings to higher temperatures and pressures corresponding to the MD simulations seems unwarranted, however, it can be concluded that the small but still significant decrease in the O-H peak of the Fourier transform is in reasonable agreement with experiment.

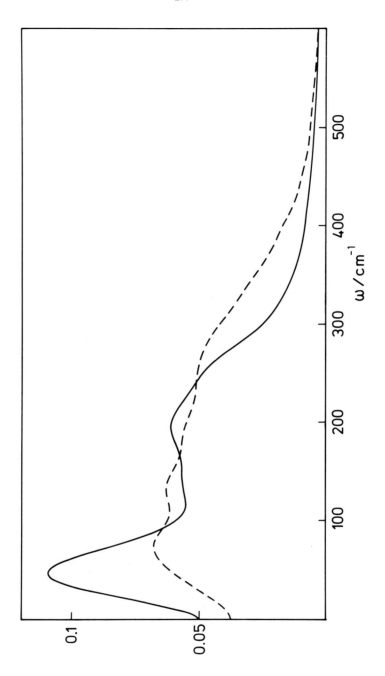

Figure 8 Spectral densities of the hindered translations of the water molecules
from MD simulations of normal pressure (solid line) and high pressure
(dashed line) pure water, in arbitrary units.[13]

From the simulations of the 2.2 molal NaCl solutions at room temperature and a pressure increase of 10 kbar the shifts have been calculated separately for bulk water, hydration water of Na^+ and of Cl^-. In this order they amount to 0, $+10$, and $+25$ cm^{-1} and $+10$, and $+10$, 0, and $+10$cm^{-1} for stretching and bending vibrations, respectively. IR spectral investigations by Inoue et al.[41] showed for a 3 molal NaCl solution at 25° C and 4 kbar a redshift of 10cm^{-1} for the maximum of the combination band $2v_1 + v_3$ which would correspond to a smaller change of the O-H stretching frequency in the fundamental region. Because of the small effects a statement on agreement or disagreement between simulation and experiment seems not to be justified.

IV. CONCLUSIONS

At a pressure of 22 kbar and a temperature of 77° C the hydrogen bond network in liquid water is strongly distorted, resulting in an increase of the number of nearest neighbours from 4.2 at 1 bar to 7. The self-diffusion coefficient decreases by about 35% with the density increase of 38%. The changes in the properties of the solvent water in a 2.2 molal NaCl solution at room temperature and 10 kbar are very similar to the changes in pure water if the smaller pressure increase is taken into account. The structure of the hydration shells of Na^+ and Cl^- are slightly decreased with increasing pressure. The self-diffusion coefficient for all three water subsystems - bulk water, hydration water of Na^+ and of Cl^- - are lower by about 20% at the high pressure (10 kbar correspond to a density increase of 21%). The self-diffusion coefficient of Na^+ decreases and of Cl^- increases with increasing pressure by about 20%. The effect of pressure on the intramolecular frequencies of water is very small.

REFERENCES

1. A.Y. Wu, E. Whalley, and G. Dolling, Mol. Phys. **47**, 603 (1982).
2. G.A. Gaballa and G.W. Neilson, Mol. Phys. **50**, 97 (1983).
3. J. Jonas, T. DeFries, and D.J. Wilbur, J. Chem. Phys. **65**, 582 (1976).
4. D.J. Wilbur, T. DeFries, and J. Jonas, J. Chem. Phys. **65**, 1783 (1976).
5. T. DeFries and J. Jonas, J. Chem. Phys. **66**, 896 (1977).
6. G.E. Walrafen, J. Sol. Chem. 2, 159 (1973).
7. G.E. Walrafen, in: L'Eau et les Systemes Biologiques. **No. 246** p. 223 (CNRS, Paris 1976).
8. G.E. Walrafen and M. Abebe, J. Chem. Phys. **68**, 4694 (1978).
9. F.H. Stillinger and A. Rahman, J. Chem. Phys. **61**, 4973 (1974).
10. R.W. Impey, M.L. Klein, and I.R. McDonald, J. Chem. Phys. **74**, 647 (1981).
11. F.H. Stillinger and A. Rahman, J. Chem. Phys. **60**, 1545 (1974).
12. O. Matsuoka, E. Clementi, and M. Yoshimine, J. Chem. Phys. **64**, 1351 (1976).
13. G. Jancsó, P. Bopp, and K. Heinzinger, Chem. Phys. **85**, 377 (1984).
14. G. Pálinkás, P. Bopp, G. Jancsó, and K. Heinzinger, Z. Naturforsch. **39a**, 179 (1984).
15. P. Bopp, G. Jancsó, and K. Heinzinger, Chem. Phys. Lett. **98,** 129 (1983).
16. W.A. Zisman, Phys. Rev. **39**, 151 (1932).
17. R.A. Horne, Nature **200**, 418 (1963).
18. R.A. Horne, in: Advances in high pressure research, **vol. 2,** ed. R.S. Bradley (Academic Press, New York, 1969) p. 169.
19. S.B. Brummer and A.B. Gancy, in: Water and Aqueous solutions, ed. R.A. Horne (Wiley-Interscience, New York, 1972) p. 745.
20. N.A. Nevolina, O. Ya. Samoilov, and A. L. Seifer, Zh. Strukt. Khim. **10**, 208 (1969).

21. G. Jancsó, K. Heinzinger, and T. Radnai, Chem. Phys. Lett. **110**, 196 (1984).
22. G. Jancsó, K. Heinzinger, and P. Bopp, to be published.
23. D.G. Bounds, Mol. Phys. **54**, 1335 (1985).
24. W.L. Jorgensen, J. Chandrasekhar, J.D. Madura, R.W. Impey, and M.L. Klein, J. Chem. Phys. **79**, 926 (1983).
25. Gy. I. Szász and K. Heinzinger, Earth Planet. Sci. Lett. **64**, 163 (1983).
26. S.D. Haman, in: Chemistry an geochemistry of solutions at high temperatures and pressures, Physics and chemistry of the earth, **vols. 13, 14**, eds. D.T. Rickard and F.W. Wickman (Pergamon, Oxford, 1981) p. 89.
27. E.U. Franck, same as above, p. 65.
28. K. Todheide, Ber. Bunsenges. Physik. Chem. **86**, 1005 (1982).
29. G.W. Neilson, Chem. Phys. Lett. **68**, 247 (1979).
30. M. Nakahara, M. Zenke, M. Ueno, and K. Shimizu, J. Chem. Phys. **83**, 280 (1985).
31. L.W. Dahl and H.C. Andersen, J. Chem. Phys. **78**, 1980 (1983).
32. A.K. Soper and R.N. Silver, Phys. Rev. Lett. **49**, 471 (1982).
33. P. Bopp, W. Dietz, and K. Heinzinger, Z. Naturforsch. **34a** 1424 (1979).
34. R.W. Impey, P.A. Madden, and J.R. McDonald, J. Phys. Chem. **87**, 5071 (1983).
35. H. Weingärtner, Z. Phys. Chem. **132**, 129 (1982).
36. K. Tanaka, JCS Faraday Trans. I **71**, 1127 (1975).
37. R.A. Robinson and R.H. Stokes, Electrolyte Solutions, Butterworths, London 1955.
38. E. Hawlicka, JCS Farday Trans. I, in press.
39. Gy. I. Szász and K. Heinzinger, J. Chem. Phys. **79**, 3467 (1983).
40. M.G. Sceats and S.A. Rice, J. Chem. Phys. **72**, 3236 (1980).
41. A. Inoue, K. Kojima, Y. Taniguchi and K. Suzuki, J. Sol. Chem. **13**, 811 (1984).

VECTOR AND PARALLEL COMPUTERS FOR QUANTUM MONTE CARLO COMPUTATIONS*[*]

P. J. Reynolds,[a] S. Alexander,[a†] D. Logan,[b†] and W. A. Lester, Jr.[a§]

[a] Materials and Molecular Research Division and
[b] Advanced Computer Architecture Laboratory
Lawrence Berkeley Laboratory
University of California
Berkeley, California 94720

ABSTRACT

Monte Carlo simulations are inherently compute-bound. Although short computations may provide order-of-magnitude estimates, long CPU times are generally required to achieve the accuracy needed for reliable comparison of Monte Carlo results with experiment or theory. The advent of supercomputers, which have made possible significantly increased computer speeds for those applications which are amenable to vector or parallel processing, thus offers promise for Monte Carlo applications. In fact, Monte Carlo codes are often highly parallel, and offer multiple avenues for both parallelization and vectorization.

We explore the gains to be obtained with supercomputers for the quantum Monte Carlo (QMC) method. The QMC algorithm treated here is used in quantum mechanical molecular calculations, to obtain solutions to the Schrödinger equation. This approach has recently been shown to achieve high accuracy in electronic structure computations. QMC is here demonstrated to fully take advantage of parallel and vector processor systems. Levels of parallelism are discussed, and an overview of parallel computer architectures, as well as present vector supercomputers is given. We also discuss how one adapts QMC to these machines. Performance ratios (versus scalar operation) for a number of supercomputer systems are given.

I. INTRODUCTION

Many-body problems in physics are often treated by a Monte Carlo approach.[1] The Monte Carlo method is statistical in nature; based on the generation of *random* numbers or *coin tosses*, it derives its name from a city famous for the random numbers embodied in its games of chance. Easy as it is to imagine using the Monte Carlo method for treating inherently statistical models, or even for numerical integration,[2] it is less obvious how to solve many-body problems. Nevertheless, many such problems are readily treated by Monte Carlo. Of particular interest to chemistry are the quantum mechanical Monte Carlo methods.[3] These have recently been applied

---

[*]   This work was supported by the director, Office of Energy Research, Office of Basic Energy Sciences, Chemical Sciences Division of the U.S. Department of Energy under contract number DE-AC03-76SF00098.

[†]   Present address: Quantum Theory Project, University of Florida, Gainesville, FL 32611

[†]   Permanent address: IBM Research Laboratory, Kingston, NY 12401

[§]   Also, Department of Chemistry, University of California, Berkeley, CA 94720

successfully to a number of molecular problems.[4-9] where they have achieved very high accuracy compared to experiment and exact results (where available). In most cases, 90-100% of the correlation energy has been obtained. Such quantum Monte Carlo (QMC) approaches stochastically *solve* the Schrödinger equation. Thus QMC provides an alternative to the conventional techniques of quantum chemistry.

Monte Carlo algorithms are almost entirely compute-bound, doing almost no I/O and requiring minimal memory. Since Monte Carlo is a statistical procedure, high precision (i.e., knowledge of many decimal places) can require long computer runs. This is because statistical uncertainty decreases slowly with computational time, going as $c(time)^{-1/2}$, where c is a constant. Because high precision is required in chemistry, where most properties are obtained as differences of almost equal large numbers, long computation times are often necessary.

One direction in addressing the problem of high precision is strictly algorithmic. An algorithm which has a significantly smaller value for the constant c will run faster by the ratio of these constants. Importance sampling[2,3] falls into this category of algorithmic approach. Other algorithmic approaches, like differential Monte Carlo methods[10] increase precision by computing energy differences directly. A complementary direction in obtaining higher precision is machine oriented. In Monte Carlo applications, increases in computational speed translate into increases in precision. Vector and parallel processors promise vast enhancements in speed for those codes able to take advantage of these architectures. Most Monte Carlo codes, and the molecular QMC code in particular, can use these architectures extremely efficiently.

In Section II we highlight some features of QMC theory and the corresponding algorithm, exploring levels of parallelism. In Sect. III we present an overview of *vector* processors, vectorizing QMC for these machines, and the enhancements in speed thus obtained. Sect. IV gives a discussion of *parallel* processing, including the various types of architectures which are categorized under this heading. Further, we discuss the gains that QMC can obtain through the use of parallel machines in terms of efficiency in using all processors. In conclusion, Sect. V gives a brief comparison of the quantum Monte Carlo speed enhancements achieved on supercomputers with those achieved by *ab initio* methods.

II. QUANTUM MONTE CARLO

For the purpose of this discussion, our goal will be the solution of the time-independent Schrödinger equation $H\Psi = E\Psi$ for the energy E. Expectation values, $<A> = <\Psi|A\Psi>$, can also be obtained, however these details[11] obscure the main features of the approach, and so here we focus on E. It is readily apparent that the ground state of the time-independent Schrödinger equation is the steady-state solution to the equation.[5-6]

$$\frac{\partial \Psi}{\partial t} = D\nabla^2\Psi + [E - V(\underline{R})]\Psi(\underline{R},t) \tag{1}$$

Here $\underline{R}$ is the vector of coordinates of all N particles in the system. If the particles being treated quantum mechanically are all electrons, ∇^2 is simply the 3-N dimensional Laplacian and $D = \hbar^2/2m_e$. Note that Eq. 1 is simply a diffusion equation combined with a first-order rate process. The function $\psi(\underline{R},t)$ has the meaning of the density of diffusing particles, which may increase or decrease locally according to the rate term $[E - V(\underline{R})]\Psi(\underline{R},t)$. such a diffusion process is readily simulated on a computer, and such simulations have been performed for simple molecules as long as a decade ago.[12]

A key issue in treating molecular systems is the proper handling of the Fermi nature of electrons. The requirement that the wave function Ψ be antisymmetric with respect to particle exchange leads to a wave function which, for more than two electrons, must have negative as well as positive values. In this situation, Eq. 1 may no longer be simulated as a diffusion process, since the density of diffusers $\Psi(\underline{R},t)$ would no longer be everywhere positive. A number of methods have been proposed to deal with fermions;[13-14] the most stable algorithmically, as well as precise statistically, entails a small variational approximation known as the fixed-node approximation.[13] The antisymmetric constraint on $(\Psi\underline{R})$ is built in by specifying in advance the nodes of the wave function to be those of a trial wave function $\Psi_T(\underline{R})$. A random walk simulation is performed separately in each volume element bounded by these nodes. Since the sign of Ψ does not change within a volume element, the problem with Ψ being a density is removed. (Excited-state Fermi calculations may be done similarly, using such built-in nodal constraints.)[15] The function $\Phi_T(\underline{R})$ can further serve as a guiding function for importance sampling, during the Monte Carlo random walk. Such a guided walk preferentially samples regions of configuration space where the true wave function is expected to be large, and avoids divergent rate coefficients, such as that which occurs in Eq. 1 when an electron approaches either another electron or a nucleus.

The result of introducing importance sampling and the fixed-node approximation into Eq. 1, is to leave it as essentially a diffusion equation with branching -- though now with an overall drift, and a modified branching term. The resulting equation is solved stochastically by allowing a set of diffusers ("random walkers") to evolve in time until a steady-state asymptotic distribution, is obtained.[5] Properties such as the energy are measured as the walkers proceed in this asymptotic distribution, and are thus averages over that distribution. The time evolution is achieved by using a Green function $G(\underline{R}', \tau)$ which evolves a walker at time t at coordinate $\underline{R}$ to time $t+\tau$ at coordinate $\underline{R}'$. The algorithm which performs this time evolution by Monte Carlo is quite simple. It is summarized here.

1. <u>Nested loops over initial conditions.</u> In most cases, a number of different initial conditions are of interest. For example, in calculating a potential-energy surface, various separate calculations at different nuclear geometries must be performed. In other calculations, a number of different trial functions may be used, for example to ascertain basis-set dependence. In still other instances a number of calculations with different time-steps τ are required for extrapolation to $\tau \to 0$, if the short-time approximation[5-6] is used. One may also want to perform a set of runs differing only in the initial *seed* for the random number generator. Such runs would not suffer from the serial correlation present in the block averages (see below). It is furthermore conceivable (and common) to need to loop over multiple initial conditions. For example, one may need to loop over geometries, at each geometry perform a loop over τ, and perhaps, at each τ perform a loop over random seeds. Each of these loops is entirely independent of one another.

For a given set of initial conditions, one generates an ensemble of N_c (typically 100-500) spatial "configurations" of the N-electron system. These coordinates may be chosen randomly, or for greater efficiency in reaching the asymptotic distribution, they may be drawn from the distribution $|\Psi_T(\underline{R})|^2$.

2. Loop over blocks. A block is an almost statistically independent "sampling unit." Each block is a complete Monte Carlo "run," which provides estimates of the properties being sampled. Generally, the total microscopic sampling time (or "target time") in a block is taken as approximately one atomic unit or longer, in order to minimize the statistical (serial) correlation between blocks. This target time defines the length of a block. Calculation of a block entails:

   3. Loop over configurations in the ensemble. For a *particular* configuration, one "measures" the properties of interest, such as the energy. (When using a trial function, the energy measured is the local energy. In the case of Eq. 1, the potential energy is measured.) Perform the random walk by:

      4. Loop over the electrons. Each electron will diffuse and drift as prescribed by the Green function. One must check if a node has been crossed in this process. In the fixed-node approximation, if the answer is "yes," eliminate the configuration. If no,
      Continue the loop over electrons.

   After all electrons have been moved, one advances the time in the current configuration by τ. The branching factor, or multiplicity M, which is also given by the Green function, is computed. This factor comes from the rate part of the differential equation (see e.g. Eq. 1). M copies of the current configuration are placed in the ensemble in place of the starting configuration. Averages for this configuration are weighted by M.
   Continue the loop over configurations.

At this point, all surviving configurations in the ensemble have reached the target time. The current block is finished. Store the current averages. "Renormalize" the ensemble back to N_c to avoid the otherwise inevitable overflow, or total death, of configurations in the ensemble.[3]
Continue the loop over blocks

The pre-determined number of blocks have now been computed. Averages and standard deviations over the blocks are computed. This provides a Monte Carlo estimate of the mean and standard errors of the quantities of interest for a particular initial condition.
Continue the loops over initial conditions.

In this outline, we have not included many of the details of the calculation, in order to focus more on the levels of parallelism and nesting. We *have* mentioned the key loops--many of which are totally independent of one another. This allows for a variety of options in creating efficient vector and parallel algorithms. The nominal communication overhead is particularly attractive, as it is beneficial in virtually every parallel architecture. We note, however, that the choice of loops vectorized can have an appreciable impact on the amount of memory required. Replicating the whole program over initial conditions is only practical if the original code uses very little memory--as is the case for QMC.

III. VECTOR PROCESSORS

Scalar computers perform sequential arithmetic operations on individual data elements. These are single-instruction, single-data (SISD) architecture machines. In contrast, vector processors are designed to perform identical arithmetic operations simultaneously on multiple data elements (single-instruction, multiple-data or SIMD). Thus, in order that vectorization can be performed, there needs to be one operation to be performed on many data elements at a given moment. (However, only certain operations on these elements are allowed.) Generally the compiler, together with some user directives, is responsible for finding such operations. What the compiler can recognize, even with user directives, is critically dependent on a program's overall structure. Thus structure as much as such conventional factors as cycle time, compiler efficiency, and memory access time, has a major effect on speed. A useful measure of the adaptability of a specific program to a vector machine is the ratio of its scalar run time to its vector run time on the same machine. For inherently scalar programs this number will be close to unity. Programs which make full use of a vector processor's capabilities will have much larger ratios. To examine the adaptability of QMC to vector processors, we performed test runs on a two-pipe and a four-pipe CDC Cyber 205, and on a Cray 1S and one processor of a Cray XMP.

The CDC Cyber 205,[16] consists of a fast scalar processor in addition to a memory-to-memory vector processor. Vector arithmetic instructions are executed in two phases. General-purpose functional units (pipes) are initially filled or emptied during the start-up phase. The time consumed during this period is independent of the vector length, and typically takes on the order of 50 machine cycles. During the stream phase, CPU time consumed is directly proportional to the vector length, and inversely proportional to the number of pipes. On a two-pipe machine, for example, a vector addition uses the first pipe to add the odd pairs of operands, while the second pipe simultaneously adds the even pairs. This sort of operation yields two results per cycle. Similarly a four-pipe machine produces four results every cycle. Up to a maximum of four pipes are available on the Cyber 205. It should be noted, however, that not all vector operations benefit from increasing the number of pipes. Furthermore, for a fixed number of pipes, certain operations are performed faster than others. Thus the ultimate speed attainable on the Cyber 205 is dependent not only on the degree to which the code can be vectorized, but also on the type of vector operations being performed. Further, although floating-point operations are normally done in full precision (64 bits), a facility has been provided for the user to code in half-precision (32 bits). In most cases this doubles both the speed and the amount of memory available.[17] Another feature of the Cyber 205 is the large, diverse instruction set available to the FORTRAN user. This enables one to convert many standard FORTRAN constructs directly into vector instructions. Thus vectorization of an algorithm does not depend strictly on the cleverness of the compiler.

The Crays,[18] like the Cyber 205, combine fast scalar capability with vector processing. Unlike the Cyber 205, however, Crays are not memory-to-memory machines. Instead, vectors are loaded from memory into one of 8 vector registers, each holding 64 words of 64 bits each. From these registers the numbers are sent to one of 12 specialized functional units where arithmetic operations are performed. Operations involving separate functional units can proceed concurrently. All result vectors are returned to the vector registers. Since each operand does not have to be fetched or each result stored in memory, operations are performed much faster. The vector registers act as fast cache memory. As a result Cray vector instructions are characterized by relatively short start-up times, ranging from 2 to 14 clock cycles. Clock cycles range from 12.5 nanoseconds on the Cray 1S to 9.5 ns on the Cray XMP -- in contrast to 20 ns for the Cyber 205. On the other hand, Cray machines have a relatively small number of instructions available to the FORTRAN user, which can hinder vectorization of some codes.

Direct quantitative comparison of the Cyber 205 and the Crays is difficult because their different capabilities give rise to different optimal coding techniques. Nevertheless, some general conclusions may be drawn. In applications involving short vectors, the Crays seem to be superior due to their shorter start-up times. Their faster clock cycles will also give them an advantage in many applications. However, for long loops, and in cases where explicit vector instructions are necessary, the Cyber 205 (especially with 4 pipes) may be more desirable, due to its higher processing rate per machine cycle, and its large, diverse set of vector FORTRAN instructions.

The QMC algorithm, as described in Sect. II, is well suited to scalar machines. However, this structure prevents it from being vectorized efficiently. Typically only the inner loops of a program can be vectorized. The longer a loop the more efficient the process.[19] The innermost loops, however, are over the electrons and over the basis functions. The number of electrons, and even the number of basis functions, is relatively small for the systems currently being considered. For example, vectorizing these loops for CH_2 yields[20] a scalar/vector ratio very close to unity on both the Cray 1 and the Cyber 205. A similar result is obtained for the carbon atom.[21] This algorithm is clearly not taking advantage of the vector architecture. Even writing explicit vector code (for the Cyber)[20] and using it only for vectors longer than an optimized length $C^* \approx 16$ (optimized to minimize the effect of the vector start-up time) leads to a scalar/vector ratio only in the range of 1-1.5, depending on the number of basis functions. Much longer vectors are clearly needed for these machines to show their abilities.

A more appropriate arrangement of the code -- that will allow the compiler to create a long vector -- is to make the loop over configurations innermost. This involves storing a linear array the size of the ensemble for every quantity in the original loop over configurations. Since each configuration is independent of the others, and since the number of configurations is usually quite large, this restructured algorithm should be much more efficient. Certain operations in the inner loop pose something of a problem, however, since they are performed on some configurations but not others. These can nevertheless be vectorized by using special mask operations available on all the machines we considered. The restructured algorithm does, however, require considerably more memory than the original form. Nevertheless, the minimal memory requirements of the scalar algorithm are such that the vector code still requires only a small memory allocation.

We note that even the scalar VAX time obtained with the restructured algorithm shows a speed-up in execution of about 40%. This type of tradeoff between speed and memory is common. When automatic vectorization is invoked on the Crays and Cybers, CPU time drops by a factor of roughly three. The current generation of compilers, however, have only a limited capability to recognize vectorizable code. To achieve faster execution it was necessary to explicitly hand-vectorize portions of the program. In Tables 1 and 2 we present the results achieved.

Table 1. Comparative run times for the restructured algorithm with an ensemble size of 100. Single precision is used on all machines.

Machine	scalar seconds	vector seconds
VAX 11/780	1620.0	--
Cyber 205 (2 pipe)	82.4	16.7
Cyber 205 (4 pipe)	81.7	13.9
Cray 1S	70.5	14.5
Cray XMP[a]	49.8	9.9

[a] All calculations were done using only one processor.

Table 2. Comparative run times for the restructured algorithm with an ensemble size of 500. Single precision is used on all machines. Other than the ensemble size, all parameters are the same as in Table 1. Thus the scalar time rises by a factor of 5, but the vector times rise less rapidly.

Machine	scalar seconds	vector seconds
VAX 11/780	8100.0	--
Cyber 205 (2 pipe)	413.6	69.2
Cyber 205 (4 pipe)	415.9	43.6
Cray XMP[a]	246.7	44.4

[a] All calculations were done using only one processor.

Scalar/vector ratios range from 5-6 for a vector length of 100, and from 6-10 for a vector length of 500. (In actual QMC calculations the ensemble size is generally in the range of 100 to 500.) Thus overall, for a vector length of 500, we achieve a factor of 117-186 over a VAX 11/780 running the same code, and a factor of 192-304 over the same VAX running the original algorithm. It is also important to note that in comparing speeds with the VAX we are comparing single precision on the VAX (32-bits--this is all that QMC requires in most cases) with single precision on the supercomputers (64 bits). For comparisons with equal numbers of significant figures, one must use single precision on the VAX and half precision on the supercomputers (where available), or double precision on the VAX and single precision on the supercomputers. In such a comparison (for a vector length of 500) we expect to achieve a speed-up over the VAX (for the same code) of roughly 370 on the Cray XMP and about the same on the four-pipe Cyber 205. This is a significant increase, especially when compared with the naive vectorization[20] of the scalar algorithm, which only led to factors of 20 to 30. Expressed in millions of floating-point operations per second (MFLOPS), we obtain an average rate of roughly 70 MFLOPS on the Cray XMP and on the four-pipe Cyber 205 at a vector length of 500. This is to be compared to the 14 Livermore kernels, which range from 3-150 MFLOPS on the Cray XMP, and average either 50 or 58 MFLOPS depending on the use of compiler directives.[22a] The harmonic mean, which is a better indicator of the actual speed of a code running sections going at different rates, gives an average rate of roughly 14 MFLOPS for the Livermore kernels on the Cray XMP.[22b] On the other hand, in half-precision we would expect our code to achieve an average rate of about 140 MFLOPS. We note especially that these rates are for the code *overall*, not just for selected parts of it.

IV. PARALLEL PROCESSING

A more ambitious approach than the SIMD vector machines are the fully parallel MIMD (multiple-instruction, multiple-data) structures, in which parallelism is achieved at the *processor* level, rather than with the functional units. The two approaches, however, should not be viewed as mutually exclusive. Ideally, one can envision a system that encompasses both strategies, i.e., a multiprocessor system whose component processors are capable of vector processing. Such an approach allows for more rapidly solving those problems that are vector decomposable, as well as solving a multitude of problems that are not.[23] QMC (and other Monte Carlo) can be decomposed simultaneously into vector and parallel parts, gaining from both types of architecture.

A number of new computational issues arise when approaching a parallel processing system. The first, of course, is in understanding the level of parallelism that a problem manifests. Such parallelism may exist, for example, at the "fine-grained" instruction level. Pipelining techniques of the vector supercomputers currently avail themselves of this form of parallelism. More ambitious approaches have been concerned with the construction of multiple-instruction pipes[24] and with data-flow architectures.[25] At the opposite extreme, the highest level of parallelism may be the segmentation of a problem into a group of concurrent cooperative subtasks, or even replication of the entire program with, for example, differing initial conditions, constraints, etc. (see Sect. II).

Closely related to the understanding of a program's parallelism is the question of how best to express this in a programming language. Is it adequate to modify an existing language with appropriate constructs, as done for vectorization, or must a new language be defined? Further, to what relative degrees shall the programmer or the compiler be responsible for the program decomposition? To a first approximation, it is generally conceded that the finer the grain size, the more emphasis must be placed upon the "intelligence" of the compiler. The higher levels will require greater participation of the programmer with perhaps interactive compilers.

Even with some understanding of the level of parallelism that a problem exhibits, and of how to map it onto a system of cooperative processors, there remains the critical issue of how best to implement processor coordination. This is related to the general issue of communication. Two broad approaches have been most extensively analysed in this regard: tightly- and loosely-coupled systems. Tightly-coupled systems may be thought of as a collection of processors that share a global memory. Important for such systems is the selection of a processor/memory interconnection network.[26] This network must ensure that the potentially very high interprocessor communication bandwidth is not degraded by memory contention conflicts and slow arbitration schemes. Further, in the event that the processors retain an additional private memory or data cache, it is vitally important that local data be correct in a global sense, i.e., data updates or modifications must be distributed across private memory boundaries. This latter criterion is the essence of what is known as the cache coherency problem.[27] The loosely-coupled alternative approach encompasses multiprocessor systems that operate as high-speed local-area networks. Here the problem of resource contention and validity is replaced by the concern with how interprocessor communication may be most effectively implemented, and the degree of system-wide connectivity that the applications merit. In the event that complete pairwise connectivity is required, bus structures may be employed. Higher communication band-width, however, requires more elaborate and costly interconnection networks.[26] Alternatively, in some applications static geometries of processors with solely nearest-neighbor communication abilities are sufficient. These generally are special-purpose machines with only a few applications. A number of such systems have been designed with topologies such as grids, rings, trees, pyramids, hypercubes, and other exotic structures.[26,28] In examples such as these, the architecture is designed to embody as closely as possible the "model of computation" required by the application. Every problem has some *natural connectivity*.[28] Having established the connection scheme, there still remains the specification of communication protocols and communication method (e.g., packet switching, memory circuit switching, etc.).

However the issue of processor coordination is resolved--whether in favor of loosely- or tightly-coupled architectural approaches--there still remain other issues. For example, should an application require intensive data throughput (>10 Mbyte/sec), it may be necessary to design specialized input and output processors, an additional interconnection network to distribute data, and sophisticated mass storage subsystems. Other system-level issues to be dealt with are job dispatching, fault tolerance, the ability to control synchronization and data broadcasting among processors, and of vital importance, the ability to monitor system performance while experimenting with problem decomposition.

At present there exists no consensus as to what approach is best. While many systems have been suggested, relatively few have actually been constructed and tested. Moreover, those that have, of necessity have been prototypes with small numbers of processors ($\approx 2\text{-}64$) and usually with limited memory and I/O capability. Thus the benchmarking of real, large-scale problems has consisted primarily of extrapolation of the results obtained on these prototypes, into the region of the large numbers of processors ($\approx 100\text{-}1000$) envisioned for future systems. These considerations led to the design and construction of a parallel processing system called MIDAS[29] at the Advanced Computer Architecture Laboratory at the Lawrence Berkeley Laboratory. MIDAS (Modular Interactive Data Analysis Systems) has combined many of the advantages of both loosely- and tightly-coupled architectures (high communication speeds without resource contention), and dealt explicitly with I/O intensive problems.

Since Monte Carlo problems are so ideally suited to multiprocessor systems, we have adapted QMC to MIDAS (and MIDAS to QMC).[30] The most intuitive and usually most efficient means of distributing a Monte Carlo computation is such that each processor is responsible for performing an independent statistical sample. An ensemble average replaces a time average in those Monte Carlo applications where equilibrium time averaging is performed. Actually, one has an ensemble average of time-averaged quantities. Increasing the number of processors substitutes more elements in the ensemble for parts of the time average. This, however, also raises the overhead of equilibrating all members of the ensemble. Some Monte Carlo applications involve only ensemble averaging, and these are most readily modified for parallel execution. The decomposition of QMC chosen for MIDAS was to give each processor initially an equal fraction of the total ensemble. This choice, rather than e.g. breaking up the calculation over blocks or initial conditions, was dictated by the limited memory size of the individual processors (128 Kwords).

In Monte Carlo, the necessity of inter-processor communication is primarily dependent on whether the sampling conditions or rules change as a function of previous sample estimates. In the event that such conditions remain constant, communication may be unnecessary (other than that required for averaging the final estimates of each processor). If, on the other hand, the set of estimates from all processors needs to be assessed for the purpose of defining a new sampling condition, periodic communication is required. Such a synchronization step represents a "critical section" of the problem. The computation may be thought of as a "fork and join" process, wherein processors fork to arrive at their estimates, followed by a join operation to define the new sampling condition. The join represents the critical section that must be performed prior to another fork. This type of critical section was explicitly realized in the QMC calculation: Each processor calculated an estimated energy by allowing its configurations to perform random walks for a fixed number of time steps. When the last processor completed this operation, all local estimates of the energy were combined to update the trial energy for the next sample period.

The MIDAS structure can be configured in a number of ways. The structure chosen to implement QMC was that of a master/slave topology. In this configuration, a single processor acts as the master and performs the join operation of updating the trial energy. Thus it was responsible for polling the slaves to ascertain whether they had completed their samples. In the event that all were completed, it read the current energy value, recalculated the trial energy, wrote this value into each slave processor's memory, and initiated (forked) the next sample period. At a lower level, the master was also responsible for detecting abnormal conditions or failures in each of the slave processors (underflow, overflow, memory parity errors, etc.). In the event that such conditions were uncorrectable, the master deallocated the processor in question, while increasing the sub-ensemble populations in the remaining processors such that the total ensemble remained constant. If any processor's population of configurations died entirely during a sample period (as happens more frequently with larger time steps) the master was responsible for downloading a new set of configurations to that processor prior to initiating a new sample period. This process of augmenting a processor's population was made by randomly selecting surviving configurations from neighbor processors.

Modifications to the original serial QMC code were fairly minor.[30] A subroutine was added that performed all of the master functions described above. It is called in the master once all initial conditions have been set, and the sampling portion of the code downloaded and initiated in each slave processor. This routine, consisting of several hundred lines of FORTRAN, executes system routines[31] which read and write selected variables within slave processors' memories, which poll processor status conditions, and which start execution at selected program addresses. Modification of the QMC code that was executed in each slave was minimal. It consisted of setting software flags for various conditions at the end of a sample period, and thereafter calling a system routine that suspended execution at a particular instruction address. This latter address and condition was recognized by the master as the termination of a normal sample period.

We now address how well QMC performs on a parallel system. Since the efficiency of a multiprocessor calculation may be defined as the fraction of time that an average processor is employed in performing useful work, 100% efficiency corresponds to a P-fold increase in speed over a uniprocessor, for a P-processor machine. Because of the random nature of the QMC birth/death process, the time for a given processor to complete (before the next "join") is itself randomly distributed. Hence, in this implementation, a processor upon completion of its calculation, is forced to remain idle until the last processor has finished. Only at that point does the master compute the new trial energy and reinitiate the slave processors. Thus the efficiency can be approximated as t_{ave}/t_{max}, where t_{ave} where t_{max} is the average time at which the processors finish and t_{max} is the average time at which the last processor finishes. With MIDAS configured with 8 slave processors, calculations were performed for the saddle-point energy of H_3, and the ground-state energies of N and of N_2. These calculations ran at approximately 95%, 85%, and 80% efficiencies respectively. Decreasing the number of processors increased the efficiencies to a small extent, but, of course, at the expense of decreasing the overall computational rate. The decrease in efficiency with the number of processors is attributable to t_{max} increasing with P, as one samples further under the tail of the distribution of finishing times. The decrease in efficiency with the larger molecules may be attributed to sampling from a broader distribution.

Although the actual calculations were performed as described above, the parallel algorithm can be readily modified to improve overall performance. The master can be programmed to perform dynamic load balancing as the sampling process proceeds. Any processor completing its work can be given additional configurations, extracted from processors which still have unfinished ones waiting to run. It should also be noted that the need for even the minimal interprocessor communication, and its resultant inefficiency, could have been entirely eliminated had each slave processor had sufficient, memory to hold the entire ensemble. In this case the loop over blocks or any (or all) of the loops over initial conditions (see Sect. II) could have been decomposed over the processors. Thus each processor could have run at 100% efficiency. This demonstrates clearly how the actual strategy for parallelizing depends on external factors, such as memory constraints.

The number of available processors also dictates the strategy for parallelizing an application. Consider the availability of an unlimited number of processors. Then an ideal parallelization of QMC would follow the nested nature of the algorithm. At the top level, independent processors would run differing sets of initial conditions--e.g. nuclear geometries. Each such processor, however, would itself be a parallel processor, running multiple (identical) codes, differing e.g. in the time-step size. Each of these processors would be a multi-processor also. These next-lower-level processors could run statistically independent samples by using different random number generators, or (if this is deemed unnecessary) by using differing seeds for the same generator. These processors could be further subdivided, with separate processors handling parts of the ensemble -- as was actually implemented on MIDAS and described above. All the above steps can be run at virtually 100% efficiency, leading to increases in execution speed of P, the number of processors. In a truly massively parallel architecture, if there are more processors available than can be used in this description, one can further parallelize the random-walk algorithm, for example, parallelizing the computation of the Coulomb potential. At this stage, efficiency would begin to drop, although speed would continue to increase, but no longer as rapidly. At a certain point saturation will set in, and speed will no longer rise even linearly with the number of processors. For QMC however, this stage apparently will not be reached until P is quite large indeed.

V. COMPARISON WITH CONVENTIONAL AB INITIO METHODS.

As we have seen, Monte Carlo is a computationally intensive method, but one that requires relatively little memory. The independence of the configurations, the near independence of the blocks, and the possibility of parallelizing over initial conditions, makes QMC particularly well suited to vector and parallel machines. Furthermore, the relatively small size of most Monte Carlo codes (roughly 1500 lines of FORTRAN in the present case) allows the user to easily optimize the algorithm to a specific machine. In this section we discuss whether these factors allow QMC to make more efficient use of current supercomputers and upcoming parallel computers, than more traditional *ab initio* methods (e.g., Hartree-Fock, multiconfiguration Hartree-Fock, configuration interaction) -- as the latter also benefit from the new computer architectures.

Scalar/vector ratios of from 10 to 50 have been quoted by Rappe[32] for selected portions of SCF and MCSCF codes on a two-pipe Cyber 205. The corresponding rates are from 26-100 MFLOPS. (This implies that at least in part his high ratios result from a scalar execution rate of only approximately 2 MFLOPS, in contrast to the more usual scalar rate of 5 MFLOPS.) It should also be noted that only 14% of his CPU time is spent in the sections going at the highest rates.[32] In programs containing a mixture of vectorizable and nonvectorizable code, the nonvectorizable part dominates the calculation.[33] The question of how much vectorization helps his SCF codes overall is not addressed. Sanders and Guest[34] discuss rates for somewhat larger program sections than Rappe. Their rates on a Cray 1S range from 10 MFLOPS for the Hartree-Fock section to 120 MFLOPS for the CI part. However the *overall* benefit remains an open question here too. (This question has, however, been addressed recently for *ab initio* calculations performed on an FPS 164 array processor[35] where relative to a VAX an overall enhancement of 10-12 was achieved.)

For QMC, virtually the *entire* code vectorizes. In fact, using a formula from Ahlrichs *et al*[36] we estimate that over 90% of the code must vectorize to achieve the rates attained. Further, QMC can be decomposed in a number of ways into virtually identical, non-communicating parts suitable for parallel processing. Although *ab initio* codes can be decomposed over some initial conditions as well, the increase in memory required for such program replication makes this degree of parallelism impractical in most cases. Analysis of recent work in optimizing *ab initio* codes for parallel processors shows efficiencies ranging from 50-95%, depending on the type of code, and the number of processors.[37] We note, however, that these efficiencies are generally *not* linear in P, and begin to saturate (i.e. deviate from linearity) for P quite small (on the order of 4-10 processors). Thus, overall it appears that QMC can more easily and efficiently take advantage of the new computer architectures than conventional *ab initio* approaches.

ACKNOWLEDGEMENTS.

We thank the Control Data Corporation and Cray Research Inc. for grants of computer time on their machines, and for assistance in carrying out parts of these calculations. Helpful comments on the manuscript by R. N. Barnett and B. L. Hammond are also gratefully acknowledged.

REFERENCES.

1. See, e.g., *Monte Carlo Methods in Statistical Physics*, K. Binder, ed. (Springer - Verlag, Berlin, 1979).
2. J.M. Hammersley and D.C. Handscomb, *Monte Carlo Methods*, (Chapman and Hall, London, 1964).
3. M.H. Kalos, Phys. Rev. **128**, 1791 (1962); J. Comp. Phys. **1**, 257 (1967); M.H. Kalos, D. Levesque, and L. Verlet, Phys. Rev. A **9**, 2178 (1974); D.M. Ceperley and M.H. Kalos, in Ref. 1; D. M. Ceperley, J. Comp. Phys. **51**, 404 (1983).
4. J.B. Anderson, J. Chem. Phys. **73**, 3897 (1980); F. Mentch and J.B. Anderson, J. Chem. Phys. **74**, 6307 (1981).
5. P.J. Reynolds, D.M. Ceperley, B.J. Alder, and W.A. Lester, Jr., J. Chem. Phys. **77**, 5593 (1982).
6. J.W. Moskowitz, K.E. Schmidt, M.A. Lee, and M.H. Kalos, J. Chem. Phys. **77**, 349 (1982).
7. P.J. Reynolds, R.N. Barnett, and W.A. Lester, Jr., Int. J. Quant. Chem. Symp. **18**, 709 (1984); F. Mentch and J. Anderson, J. Chem. Phys. **80**, 2675 (1984); R.N. Barnett, P.J. Reynolds, and W.A. Lester, Jr., J. Chem. Phys., **82**, 2700 (1985).
8. P.J. Reynolds, M. Dupuis, and W.A. Lester, Jr., J. Chem. Phys. **82**, 1983 (1985).
9. R.N. Barnett, P.J. Reynolds, and W.A. Lester, Jr., "Electron Affinity of Fluorine: Quantum Monte Carlo Study" *in preparation*.
10. B. Holmer and D.M. Ceperley, *private communication*; B. Wells, P.J. Reynolds, and W.A. Lester, Jr., *unpublished*; B. Hammond, P.J. Reynolds, and W.A. Lester, Jr., *unpublished*; B.H. Wells, Chem. Phys. Lett. **115**, 89 (1985).
11. M.H. Kalos, Phys. Rev. A **2**, 250 (1970); R.N. Barnett, P.J. Reynolds, and W.A. Lester, Jr., "Molecular Properties by Quantum Monte Carlo" *in preparation*.
12. J. B. Anderson, J. Chem. Phys. **63**, 1499 (1975); **65**, 4121 (1976).
13. D.M. Ceperley and B.J. Alder, Phys. Rev. Lett. **45**, 566 (1980); D.M. Ceperley in *Recent Progress in Many-Body Theories*, edited by J.G. Zabolitzky, M. de Llano, M. Fortes, and J.W. Clark (Springer- Verlag, Berlin, 1981).
14. J. Carlson and M.H. Kalos, "Mirror Potentials and the Fermion Problem," Phys. Rev. C (September 1985).
15. R.N. Grimes, R.N. Barnett, P.J. Reynolds, and W.A. Lester, Jr., "Molecular Excited States with Fixed-Node Quantum Monte Carlo" *in preparation*.
16. M.J. Kascic Jr., *Vector Processing on the Cyber 200* (available from the Control Data Corporation)
17. See, for example, A. Dickinson, Comp. Phys. Comm. **26**, 459 (1982).
18. W.P. Petersen, Comm. of the ACM **26**, 1008 (1983)
19. See, for example, the Cyber 205 User Guide (available from Control Data Corporation), and Cray Computer Systems Technical Note - Optimization Guide (available from Cray Research).
20. P.J. Reynolds and W.A. Lester, Jr., Proceedings of the Cyber 200 Applications Seminar, NASA Conference Proceedings 2295 (1983).
21. S. Alexander, P.J. Reynolds, R.N. Barnett, and W.A. Lester, Jr., "Vectorization of Molecular Quantum Monte Carlo" *in preparation*.
22. (a) H. Bruijnes, NMFECC Buffer **9**, No. 6, 1 (1985); (b) J. Worlton, Datamation, pp 121-30, 1 September 1984.
23. G. Rodrique, E.D. Giroux, and M. Pratt, Computer **13**, 65 (1980).
24. CDC - Cyberplus announcement, based on the AFP processor designed by Control Data Corp.
25. J.B. Dennis, Proc. First Int. Conf. on Distributed Computing Systems, p. 430 (1979).

26. *Interconnection Networks for Parallel and Distributed Processing*, C-L. Wu and T-Y. Feng, eds. (IEEE Comp. Society Press, Silver Spring, 1984).
27. M. Dubois and S. Briggs, IEEE Trans Comp, C31, 1083 (1982).
28. Physics Today, May 1984.
29. For a detailed description of the MIDAS architecture see C. Maples and D. Logan, SIAM, *in press*, and references therein.
30. P.J. Reynolds, D. Logan, C. Maples, and W.A. Lester, Jr., "Parallelism in Quantum Monte Carlo: Calculation of the Binding Energy of the Nitrogen Molecule" *in preparation*.
31. D. Logan, C. Maples, D. Weaver, and W. Rathbun, Proc. 13th Int. Conf. on Parallel Processing, p. 15 (1984).
32. A.K. Rappe, J. Comp. Chem. 5, 471 (1984).
33. I.Y. Bucher and J.W. Moore, "Comparative Performance Evaluation of Two Supercomputers: CDC Cyber-205 and CRI Cray-1," Los Alamos Scientific Report LA-2629 (1981).
34. V.R. Sanders and M.F. Guest, Comp. Phys. Comm. 26, 389 (1982); Martyn F. Guest and Stephen Wilson, in Supercomputers in Chemistry, Peter Lykos and I. Shavitt eds. (American Chemical Society, Washington D.C., 1981).
35. R.A. Bair and T.H. Dunning, Jr., J. Comp. Chem. 5, 44 (1984).
36. R. Ahlrichs, H-J. Böhm, C. Ehrhardt, P. Scharf, H. Schiffer, H. Lischka, and M. Schindler, J. Comp. Chem. 6, 200 (1985).
37. E. Clementi, G. Corongiu, J. Detrich, S. Chin, and L. Domingo, Int. J. Quant. Chem. Symp. 18, 601 (1984).

SELECTED COMPUTING PROBLEMS IN HIGH ENERGY PHYSICS

Eberhard Freytag

DESY
Hamburg
Germany

1. INTRODUCTION

In June the High Energy Physicists met in Amsterdam at a conference 'Computing in High Energy Physics.' I suppose there are some trends and facts which are worthwhile to report, and I will concentrate on two main topics: compute power problems and mass data storage problems. I shall cover the first topic in a more sketchy way and the second one in more detail.

2. COMPUTING POWER PROBLEMS IN HEP

High Energy Physicists have a never ending demand for computing resources. To describe the reasons for that let me distinguish between experimental and theoretical physicists.

2.1 Theoretical Physics

For a long time the theoreticians were satisfied by a small amount of computer time. Now when they left perturbation theory and turned to Quantum Chromodynamics to find a new foundation of Elementary Particle Theory they need much more than before. QCD is supposed to be a similar step for interaction and structure of hadrons as the Maxwell Theory for macroscopic dimensions or as Quantum Electrodynamics for the atomic region.

What makes the problems is the fact that theorists are using thermodynamics and Monte Carlo methods in a 4 dimensional space. It is a problem of many degrees of freedom with nonlinear coupling. The continuous time-space is approximated by a grid of discrete steps as it is usual in finite element methods. Typical grid dimensions today are 16^4 points. Computation along a single line between two grid points is equivalent to 10^4 floating point operations, and this has to be done across the grid at least 10^4 times for collecting statistics. Each refinement of their grids by say a factor of 2 results in an increase of storage and computer time by a factor of 16. In that respect they are also concerned with matrices of reasonable size. Nobel Prize winner Ken Wilson therefore talks about Mega-Crays, and that shows you the order of magnitude they are aiming at, even if they have not reached this goal by far. But any new and bigger type of supercomputer finds their enthusiasm and they will use it - as soon as they can afford it. I am sure you will understand their attitude - not in detail but in principle.

2.2 Experimental Physics

Experimental Physicists are engaged in planning and evaluation of the HEP experiments. Planning includes design and optimization of detectors which in turn includes extensive Monte Carlo studies for exploring the phase space covered by a special design, for determining the detection probability, and for simulating events in the planned detector. In addition, supposed particle production mechanisms are tested by Monte Carlo programs and the results are compared to measured event rates. Around 1/3 of the computing time of experimenters is spent in such Monte Carlo studies.

Data evaluation is the other branch of experimenters' computing needs. Out of around .5 million possible events per second at a storage ring experiment the single one (or 2 or 3) is filtered which presumably provides relevant physics information. Filtering at this step has to be done by rather crude criteria: a minimum number of tracks is required, the total energy of the emitted particles has to exceed a given threshold etc. At this stage both hard wired logic and specialized microprocessors are used.

The events selected in that way are stored on tape today, and these raw data are processed asynchronously later on. Due to the statistical nature of the events the structure of the programs is essentially decision trees, which are hardly vectorizable. Therefore, these programs are run on general purpose processors. This off-line processing takes about 60% of the total computer time for experiments.

The requirements in compute power of National Labs for the past and the future is shown in figure 1, which is taken from a HEPAP report by J. Ballam.

As this evaluation is the end of the data chain and the budget normally is exhausted earlier, the demands of the physicists cannot be covered in total by the computer centers with their IBMs, CDCs, VAXs or whatever. Therefore the physicists tried to build their own processors - simpler and cheaper than those monsters. There is a lot of cooperation in the High Energy Physics community. Two different lines of parallel processing have been followed:

- Emulators, i.e. processors built from microprocessors which understand the instruction set of a big machine. Common developments are the 168/E, 370/E and 3081/E. The programs are compiled at the host machine and downloaded to the emulators which are grouped in farms. Thus, program development can be used both on the host and the emulators for data processing. Typical prices claimed are: $5K for one CPU, $5K for 1 MByte of memory (hardware only).

- Multi-Microprocessor farms: arrays of commercially available microprocessors. Software has to be developed specially for these complexes. A well known project of this type is the Advanced Computer Program at Fermilab which aims at 128 micros with 2 MBytes each, representing the power of about 125 VAXs at $400K total. Here, with 256K chips, the most recent price of 1 MByte is said to be $250.

Similar approaches with specially designed processors have been made by theory groups.

Computing Needs of
The National Labs

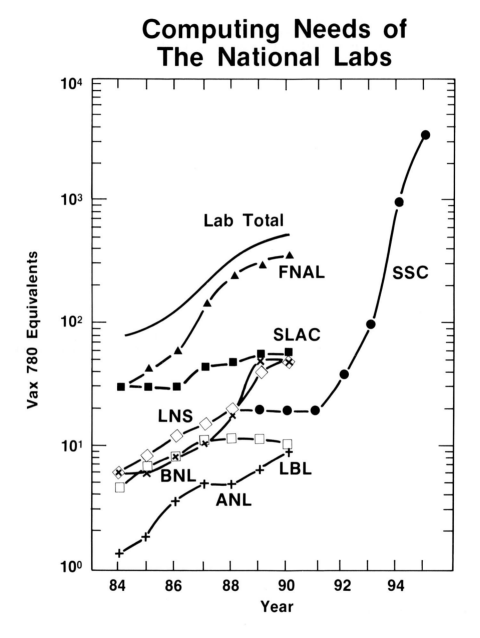

Figure 1.   Computing needs of National Labs : from report of the HEPAP
subpanel on computer needs for the next decade (April 1985).

3. MASS DATA STORAGE

3.1 Examples

The second large problem area which I shall deal with in some more detail is data storage. Where do we store all the terabytes of data?

Mass data storage is a problem in several fields of data processing. To illustrate what problems I am going to talk about, let me consider some examples of data rates and of the amount of stored data.

3.1.1 High Energy Physics

The next generation of High Energy Physics experiments at the LEP storage ring of CERN will produce something like 200-400 KBytes/sec data rate each. With 10^5 sec/day this adds up to $3*10^{10}$ Bytes/day. Assuming a useful tape volume (6250 bpi tape) of 150 MBytes, this results in 200 tapes/day, or $6*10^4$ tapes/year, i.e. 10 TBytes/year. The current tape vault at CERN holds around 250,000 tapes (around 40 TBytes) with a growth rate of about 50,000 tapes/year. Similarly at DESY there are around 70,000 tapes (10 TBytes) with a 10,000 tapes/year (1.5 TBytes/year) current growth rate.

3.1.2 Seismic Exploration

In seismic exploration echoes of a small explosion are recorded which allow a construction of geological data.[1] Sea measurements as done by Prakla-Seismos or other companies typically result in a 100 KBytes/sec data stream. Within the next years, an increase by a factor 10 is expected. Roughly 1 TByte/year has to be stored. Stored data volumes are around 100,000 to 200,000 tapes today (around 30 TBytes).

3.1.3 Weather Forecast

In 1984 about 25,000 tapes (4 TBytes) existed in the European Center for Medium Range Weather Forecast (ECMWF). The future data rate was anticipated to be 3 GBytes/day which results in about 6,000 tapes/year or 1 TBytes/year. These data will be kept for a period between 180 days and 5 years, some of them even indefinitely (Gray, Dixon and Hoffmann[2]).

3.1.4 Satellite Data

Satellites may transmit data rates up to 10 MBytes/sec and 2 GBytes/day. A single mission results in the order of 10,000 tapes or 1.5 TBytes (Helios).

3.2 MAGNETIC TAPES

3.2.1 6250 Tapes

In the past, magnetic tape has been the classical medium of mass data recording. The widespread de-facto standard of the 6250 bpi tape provides an effective data volume of about 150 MBytes/tape. At the LEP experiment data rate, a single tape is filled up in about 500 seconds or 8 minutes. Therefore, you will have to provide something like a full person for tape mounting at your experiment.

3.2.2 Higher Density Tapes

There have been attempts to introduce tapes of higher density (see the comprehensive state-of-the-art report by D. O. Williams[3]) like Videotapes (9 GBytes/volume, 750 KBytes/sec) and the Ampex super high bit rate (SHBR) digital recorder (150 GBytes/volume, 100 MBytes/sec). But because of a lack of support by the computer industry, they had no permanent success despite a vital interest of the high energy physics community.

Recently IBM has announced a new tape system (IBM 3480) which makes use of tape cassettes of smaller size than the usual reels. These cassettes are much easier to handle, the tape units are smaller than the old ones and the reliability of tape and tape unit have been increased. Also the bit density is increased: now there are 38,000 Bytes/inch, which is a factor 6 increase. However, since it is smaller, the total capacity of the cassettes remains about the same as of a 6250 reel. It is about 190 MBytes at a blocksize of 16 KBytes, compared to 160 MBytes for 6250 tape at the same blocksize. What you gain on top of the 20% data volume increase is the smaller size and the ease of handling of these cassettes. In storage size the net gain is a factor 2.

However, larger volumes of data are not easily stored and accessed with these cassettes unless:

- the tape density or the tape length is increased, increasing the total data volume, and/or

- jukeboxes become available (see below).

There is another disadvantage apart from the rather small data volume: these cassettes are not (yet) standardized, at least it will take some time until the rest of the world adopts it.

3.2.3 Jukeboxes

Jukeboxes have been talked about for some time. They are intended to take care of automatic mounting from a library of volumes. Remember that IBM once delivered something like a jukebox as its Mass Storage System MSS 3850 in which data were stored on 100 MByte volumes consisting of two cartridges. It is the automatic handling which provides and simplifies a fast access to total data volumes up to 236 GBytes. Maybe something like that will be offered also for tape cassettes. Anyway, a jukebox is a rather bulky and expensive toy. Those offered for tape reels up to now have been used in a very few places only, and even the 3850 has not been sold in large quantities.

3.3 OPTICAL DISKS

3.3.1 History

To start talking about optical disks as computer peripherals let me quote a few sentences from Datamation[4]

> "Precision Instruments Co., Palo Alto, landed a second order for its trillion-bit laser recorder/reader storage system last month and said it will market a smaller 10 million-bit version." - Datamation, July 15, 1970.

> "Lower component costs are being credited in price cuts for a read-only optical memory system announced only a year ago. A 25% drop, to 3 cents per bit for a 100 Kb model in quantities of 100, is due this month

from Optical Memory Systems, Santa Ana, Calif. Prices will get down to a penny a bit two years from now." - Datamation April 15, 1971.

"Glowing promises for optical memories in the early 70's turned out to be little more than that. Optical Memory Systems ... have vanished, apparently without having delivered a product. Precision Instrument went into Chapter 11, was revived by the Heiser Corp. with $ 26 million and a new name, Omex, but is now up for sale without ever having gone into production." - Datamation June 1, 1984.

These quotes should remind you that there have been rumors about optical disks for a long time. The rumors are louder now and the first units have been in beta test, but still I have not seen a final commercial system delivered to a customer. Therefore, anything I can tell you about optical disks as computer peripherals is taken from rumors, including advertisements.

What are currently available are the common products for the consumer market: the Compact Disk for music records and the LaserVision video disk. They are recorded once and then copied numerous times to be sold to the consumer.

From that application another market was seen by suppliers in document or image storage and retrieval systems as an alternative to microfiches or even books and journals. A compressed image of a single page needs only 50 KBytes of storage; in coded form only 5 KBytes. There are several companies offering complete office systems with optical disks.

Finally the optical disk was considered as a potential computer peripheral by companies involved in this business like CDC, Optimem (Xerox), and StorageTek. With reference to the examples of mass data given before I am not concerned with 'small disks' of < 1GByte here.

Of course, the write-once property of present offerings is not suitable for recommending optical disks as a simple substitute for magnetic disks or tapes. It aims at archiving applications, and there are many: Transactions of banks which should be impossible to change, so they can be followed by auditors, e.g., but also all the cases mentioned in the introduction, such as High Energy Physics, seismic exploration, weather data or any kind of archived data or data bases.

Different from consumer oriented products, digital data storage requires a low bit error rate (BER) of about 10^{-12} thus efficient coding and error correction schemes had to be developed. The optical disk also demands a handling by the access routines different from magnetic disks as writing is irreversible. Therefore it took considerable development time to adapt this technique for computer peripherals.

3.3.2 TECHNOLOGY

3.3.2.1 General

Optical disks and magnetic disks have the basic idea in common: both technologies employ a write/read head to make physical changes on the rotating disk medium which can be read back at a later time. Bit information is grouped together into sectors and tracks.

Recording on optical disk is done by a laser which is focused precisely on the sensitive surface of an optical platter changing its reflectivity in a small spot. For reading, a laser beam recognizes on the track parts of high and of low reflectivity. Header, tracking and clock information is provided by the unrecorded media to facilitate the storage operations. To accomplish that, the media is pregrooved by stamping from a mastering machine.

3.3.2.2 Lasers

In early developments gas lasers were used. However, currently semiconductor lasers of the GaAlAs p-n junction type are preferred due to their high efficiency, ease of modulation and compact size. They emit light of a wavelength around 800 nm and pulses up to 20 - 50 mW which sufficiently increases the local temperature at the spot of writing.

3.3.2.3 Recording Techniques

Based on this spot heating by a focused laser beam, two main techniques are currently applied for producing the reflectivity change:

1. In the ablation technique, a hole is burned into the disk medium which is made in most cases from tellurium alloys. Tellurium and its alloys are chosen because of their rather low vaporization temperature and vaporization heat.

2. In the vesicular technique, the change in reflectivity is performed by the writing laser via a blistering process: The laser beam evaporates a polymer layer thus forming a bubble in the covering ductile metal layer (Optimem). Cover metals could range up to gold and platinum with high corrosion resistance.

Both methods create non-erasable 'holes' with $<$ 1 μm diameter and around 1 μm distance. Thus, the linear recording density is similar to that of magnetic disks. The distance of tracks is about 1 - 2 μm being a factor of about 50 denser than in the magnetic disk case as there is no danger of stray fields. Thus, on a single platter of 12″ diameter, up to 1 GByte/side are recorded. Platters of 14″ diameter go up to 4 GByte/side. Sometimes the sensitive layer is sandwiched within 2 glass or plastic plates to prevent oxidation and protect from dust and scratching. The lifetime of the records is estimated to be more than 10 years.

The platter is supplied in a cassette which is easily loaded to the disk unit. Within the unit, the cassette is opened and the disk is mounted automatically.

Rough figures of recording densities are given in Table 1 for various media. They are estimates, the real number for an IBM 3380 single density is quoted as 20,000 bit/mm² e.g.

Table 1. Typical figures for high-end recording media

	linear distance	track distance	recording density
Magnetic Tape (6250)	3 μm	1000 μm	300 bit/mm²
Magnetic Tape (38K)	1.5 μm	300 μm	2,000 bit/mm²
Magnetic disk	1.5 μm	50 μm	13,000 bit/mm²
Optical disk	1 μm	2 μm	500,000 bit/mm²

3.3.2.4 Seek Mechanism

Access to a specific track is performed in two stages: A servo drive moves the arm coarsely to the region (band) of the data track to be accessed. The fine adjustment is done by tilting a small mirror which directs the light beam to the addressed track. A separate servo adjusts a focusing lens so that the focus remains at the recording surface. But unlike magnetic disks, the head keeps at a distance of the order of 1 mm from the disk compared to the 1 μm order in the magnetic case. Thus optical disk units are much less sensitive to impurities and perturbations.

3.3.2.5 Verification

To check if the data have been written correctly to the disk, three different schemes are applied currently:

1. DRDW. Direct Read During Write. Returning light of the writing laser beam is used to check the data written. In case of an error the data is rewritten on the next position.

2. DRAW. Direct Read After Write. A second laser beam is used to read the data immediately after writing. The correcting action in case of errors is the same.

3. None. Some companies rely on the inspection of the raw disks and claim that no bad spots will be tolerated. Overhead by error correction codes and checks during read/write are avoided.

3.3.3. PRODUCTS AND NUMBERS

3.3.3.1 Computer Peripherals

The optical disks which have been announced as computer peripherals are listed in the following table (Table 2). I do not refer to special developments, but only to products which should become generally available rather soon on the European market. I was told that Optimem delivered about 100 systems to customers in 1984 in the States, but I haven't found a European installation yet.

Table 2. Commercial Optical Disk Units

Company	OSI	Optimem	Thomson-Alc.	StorageTek
Type	Laserdrive 1200	(Xerox) 1000	Gigadisc GD 1001	7600
Disk diameter	12″	12″	12″	14″
one-side capacity	1 GByte	1 GByte	1 GByte	4 GByte
data rate KByte/sec	250	480	480	3000
write check	DRDW	none	none	DRAW
bit error rate	10^{-12}ECC	?	10^{-12}ECC	10^{-13}ECC
no of lasers	1	1	1	2
no of tracks	32,000	40,000	40,000	34,224
latency	62.5 ms	27 ms	27 ms	23 ms
access time	150 ms	100 ms	200 ms	62 ms
within band	?	?	5 ms	1 ms
drives/data path	?	?	8	8
data paths/contr.	?	?	1	4
disk material	glass	glass	glass	aluminum
interface	SCSI,ISI	SCSI	SCSI	IBM
SW available	?	?	no	yes
drive cost K$	12	6-14	16	130
controller cost K$	incl.	?	3.5	40
disk price $	250	400	400	180-240

Except for StorageTek, the manufacturers do not provide software or maintenance as usual for OEM manufacturers. StorageTek, on the other hand, has developed a special access method called OPSAM which is a subsystem to IBM's MVS. From this, the different application areas of optical disks can be realized: There is a so-called 'low class' for minicomputers with 1 GByte disks and low transfer rates, and a 'high class' with a 4 GByte disk and high transfer rate. They also differ in the form of the recording groove: it is spiral in the low class which is developed from consumer optical disks and has circular tracks in the high class as magnetic disks have.

OSI is a joint venture of Philips and Control Data located in Colorado. Its disk specifications are close to those of Shugart due to a cooperation agreement between those companies. An optical disk even closer to OSI is offered by Signaal, a Philips Dutch subsidiary, together with a Unibus interface. Alcatel-Thomson provides the platters to Optimem company.

3.3.3.2 Information Systems

Disks and disk units of the 'low class' with 1 GB platters are also used for delivery of information with say monthly updates. The primary market are publishers of large volumes of information. Publishers would sell the drives and the replicated optical disks to their customers as an OEM. Examples of this technology are FileNet Corporation and Reference Technology Inc.. FileNet offers its OSAR (Optical Storage And Retrieval) system with 2 GByte platters and a jukebox holding 64 cartridges. Unit purchase prices may be around $16K. The MicroVAX-II also provides a read-only optical disk for in-house information of the customer.

3.3.4 Future Developments

Setting aside for the moment that all commercial optical disks are still in the future, it is rather easy to predict the short term future of them. Presumably the companies will start with one-sided disks. However, they are prepared to go to double-sided disks soon, doubling the capacity. To make use of the other side you have to remove the disk cassette and to reinsert it into the disk unit.

In addition the automated loading procedure of the disk cassettes is well-suited for a jukebox system holding 10 - 100 platters or maybe more. This would provide access to 80 to 800 GBytes of data in the case of double-sided 4 GByte platters.

As the technology of digital optical disks is still in its infancy, increased data density and larger data volumes can be expected. Use of lasers with shorter wavelengths would support the decrease of the distances involved. Experts talk about a theoretical limit of 50 Mbit/mm^2 for optical disks as opposed to 0.5 Mbit/mm^2 for magnetic disks. Today we are a factor about 100 away from these theoretical figures.

However, all that refers to write-once optical disks. In the long term future, it will be useful to have erasable optical disks like the magnetic disks. Of course, different technical possibilities are under investigation for recording and reading. There are two main physical mechanisms for providing erasability which are considered promising today:

1. Magneto-optical recording.

   The energy of the laser is used to heat a spot of the magnetic surface above the Curie point: this part can be realigned to different vertical magnetization by a coil wrapped around the laser lens, e.g. Reading is performed by detecting the difference in the polarization plane when the polarized laser light is reflected by different magnetizations. To give you an idea: for a single layer, the change in the plane of polarization is about $0.3°$; by multilayer structures it can be increased up to $9°$. But the signal-to-noise ratio remains smaller than obtained with write-once media due to noise associated with the grain boundary diffraction. Therefore, amorphous materials may be preferable.

   Both Sony and Hitachi have announced such kind of disk using terbium-iron-cobalt alloys with a $170°$ C and a $240°$ C Curie point respectively. They will not be available before 1986.

2. Change of the crystalline state.

   - For some substances, the state of the sensitive surface can be changed from crystalline to amorphous and back again. Materials suited for this technique are tellurium and tellurium oxides, where the change has been demonstrated by Matsushita more than a million times in the lab. Energy Conversion Devices in Troy, Mich. also follows this line.

   - For other materials, two stable crystalline states exist at room temperature. Hitachi has developed a silver-zinc alloy which is silver-white in a hexagonal structure which results from sputtering the alloy to a glass substrate and tempering. By heating this substance to about $300°$ C and subsequent quenching, this is transformed to a cubic pink phase. It can be reversed by slow heating to less than $300°$ C and tempering. Possible a AgAlCu alloy also meets these requirements. Fujitsu investigated a selenium indium antimony alloy with two stable crystalline states for more than a million reversals successfully.

Among other physical effects proposed for rewritable optical disks are

- semiconductor-metal phase transitions (VO_2),

- use of colour centers in alkali-halides (KBr, KCl, ...),

- photochromic effects by photo-dimerization (Toluene sulfonate),

- photorefractive effects ($LiNbO_3$).

Thus the future development still has a wide field. It will take several years from now until a reliable re-writable optical disk is on the market. But, of course, we are waiting for such a flexible high density storage medium, too.

3.4 ECONOMICAL CONSIDERATIONS

In the running cost of a computer center, the dominating parts (apart from rental) are tapes, paper and maintenance. We are concerned here with tapes and optical disks as an alternative. There is no question that digital optical disks have several advantages, including the random access. How economical is such a technically interesting solution of our storage problems - as soon as it is available?

A 2400 ft. magnetic tape costs around $10. An IBM tape cassette with 20% increase in volume is about 50% more expensive. A 4 GByte optical disk may have a 20-fold capacity (to be cautious), and its cost is around $200 (StorageTek). Thus the price per bit is about the same as with 6250 tape. However, a 1 GByte disk having about the capacity of 5 tapes is sold for about $400 (Thomson-Alcatel) or $250 (OSI). Thus the price per bit is a factor of 8 or 5 more, respectively, than that of a tape.

One may argue, of course, that prices will go down for the new technology, whereas tape prices have stabilized. I am sure that will happen. These considerations were made mainly to emphasize the need for careful evaluation of running cost of a sweet technology - the comparison has to be repeated whenever the facts change. Today I can only stand on the offers I have.

Physical storage considerations are another part of that game. One cubic meter of a building is estimated today to be an investment of around $700 in Germany and will hold about 70 tapes. Thus on top of the tape price of $10, there is additional cost of $10 for storage. Cutting down these expenses by a factor of about 10 to 20 by using large optical platters would result in a substantial gain.

3.5 CONCLUSIONS

As physics and technology of optical disk recording are still in their infancy, a large number of varieties are possible currently and in the foreseeable future. This does not refer only to the recording substance and technique, it also refers to coding, error correction codes and cyclic redundancy check, with the result that today a disk cannot be reread on the drive of a different manufacturer.

Of course, the diverging development points to the question of standardization. The NBS decided as early as 1983 to investigate standards for digital optical disks, ISO installed a subcommittee of TC97 to look into optical disk standards, and also ECMA and Japanese bodies are looking into this field. But first draft proposals are not expected before 1986 or 1987, and they will presumably deal with small platters like 13 cm (5 1/4 inch) diameter which I have not touched at all. Thus the variety of the interesting big platters will stay with us for a few more years to come.

On the other hand, confusion will continue as progress in magnetic tape recording techniques can be well expected. This technique is under steady development and has not yet reached its limits. For a new technology it is hard to find a market sector to sell well unless it is not only competitive to well-established technologies but has additional striking advantages. My feeling is that optical disks have such an opportunity as they are:

- competitive in price
- superior in volume
- randomly accessible
- easily exchangeable.

Clearly, their present disadvantages of non-erasability, non-standards and changing development leave us with several uncertainties. At least they will provide a means to store terabytes of data conveniently and economically. However, predictions are always difficult, especially for the future.

REFERENCES

1. Butscher, W., Private communication.
2. Gray, P., Dixon, D., Hoffmann, G.-R., Large databases in a meteorological environment, SPIE Proceedings **490**, 34 (1984).
3. Williams, D.O., Trends in mass storage technology and possible impacts on high energy physics experiments, Computer Physics Communications **22**, 353 (1981).
4. Myers, E., Optical disks foreseen, Datamation, June 1, 1984.

THE SOLUTION OF THE SEISMIC ONE WAY EQUATION ON PARALLEL COMPUTERS

Werner Butscher

Prakla Seismos
Buchholzer Str. 100
D-3000 Hannover 51
W. Germany

ABSTRACT

The basic mathematical model for seismic migration calculations is the acoustic one way wave equation. Several different approaches to this partial differential equation are introduced. Special care is given to detecting inherent parallel structures in the numerical solution schemes. It is shown that alternative formulations of the problem may lead to an increase in parallelism allowing for an efficient use of tightly coupled multitasking capabilities of current vector multiprocessors.

INTRODUCTION

The purpose of reflection seismology is the investigation of unknown subsurface structures. Seismic waves are emitted by an energy source and the reflected signals are recorded by an array of receivers. The measured echo time series (traces) only give a rough picture of the subsurface structure. Seismic migration procedures are based on the governing wave equation in order to locate the origins of the reflected signals more correctly (Inverse scattering problem).

Until recently the vast majority of migration calculations were performed as two dimensional approximations to the real three dimensional world because the necessary computer power was not available. With the advent of the new generation of supercomputers there is hope to attack also three dimensional problems routinely.

THE ONE WAY WAVE EQUATION

The elastic wave equation should be used for a complete study of seismic signals. But since this equation is far too complicated, approximations have to be sought from the very beginning. Neglecting shear waves one is lead to the acoustic wave equation

$$\left(\frac{\partial^2}{\partial x^2} + \frac{\partial^2}{\partial y^2} + \frac{\partial^2}{\partial z^2} \right) P(x, y, z, t) = \frac{1}{v^2(x, y, z)} \frac{\partial^2}{\partial t^2} P(x, y, z, t) \tag{1}$$

P stands for the vertical pressure and $v(x, y, z)$ is the acoustic velocity of the medium.

The second order hyperbolic partial differential Eq. (1), however, requires boundary conditions which are not available in general. A further simplification is introduced by the 'exploding reflector' model[1] according to which reflectors are represented by sources which emit signals at time $t = 0$. These signals are forced to propagate in the upward direction only. The migration procedure now simply consists of a recursive downward continuation of the recorded pressure to a desired depth z. The 'source strength' is given by $P(x,y,z,t = o)$.

The parabolic one way equation in the space-frequency domain is given by ($i^2 = -1$; the frequency ω is a parameter)

$$\frac{\partial}{\partial z} P(x, y, z, \omega) = i \sqrt{\frac{\omega^2}{v^2} + \frac{\partial^2}{\partial x^2} + \frac{\partial^2}{\partial y^2}} \; P(x, y, z, \omega) \tag{2}$$

For actual applications the formal square root operator has to be represented by a Taylor series or continued fraction expansion and for a computer implementation the differential operators must be approximated by finite differences. A further simplification is a forced decoupling of Eq. (2)

$$\frac{\partial}{\partial z} P(x, y, z, \omega) = (0(x) + 0(y)) \; P(x, y, z, \omega) \tag{3}$$

Following Yanenko[2], a simplified solution scheme for the three dimensional partial differential Eq. (3) is obtained by applying successively the operators $0(x)$ and $0(y)$ (splitting method).

THE SOLUTION IN THE SPACE-TIME DOMAIN

Substituting $i\omega$ by $\partial/\partial t$ in Eq. (2) gives the space-time representation. A stable implicit scheme of the splitting method is obtained with the help of the Crank Nicolson scheme[3]. Within the so called 45° approximation of the square root the resulting finite difference equation for the $0(x)$ application is

$$T_1 \, \vec{P}_{t-\Delta t}^{z+\Delta z} = T_2 \, (\vec{P}_t^{z+\Delta z} - \vec{P}_t^z) + T_3 \, (\vec{P}_{t+\Delta t}^{z+\Delta z} - \vec{P}_{t-\Delta t}^z) + T_1 \vec{P}_{t+\Delta t}^z \tag{4}$$

where T_i are tridiagonal matrices and

$$\vec{P}_t^z = P_t^z(x_i) \qquad\qquad i = 1, \dots, n$$

specifies the pressure at time t and depth z for all grid points $x(i)$. A similar equation results for $0(y)$. In Eq. (4) the right hand side of the equation system is known from previous steps of the computation and the boundary conditions. Thus the main work is the setup and the solution of many tridiagonal systems for each grid point in z and t (note that time is running backward).

Although there is obviously a recursion in space and time, the setup is an inner loop problem and can be expressed by vector operations over the gridpoints x (or y). The only serious problem for vector computers is the recursive solution of the banded equation systems. However, as is well known, there are at least two ways to overcome this problem. First we can use the fact (see Fig. 1) that the partial downward continuations with $O(x)$ can be performed in parallel for all $y(i)$ (with similar arguments for $O(y)$). Thus a vectorization can be achieved by solving M decoupled systems simultaneously with vector length M. The price to pay is an increased memory requirement by a factor M. For the Cyber 205 it may be beneficial to compute with longer vectors. Following F. Url[4] the vector length can be doubled to 2M by using the trick of 'burning a candle at two ends': two eliminations can be started independently at the top and the bottom of the equation system to eliminate the off-diagonals, above and below the diagonal, respectively. The boundary region needs a special (scalar) treatment.

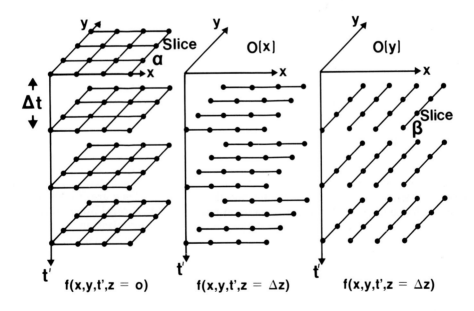

Figure 1. Downward continuation from $z=0$ to $z=\Delta z$ with the splitting method. Migrated results are slices α and β.

The other alternative is the cyclic reduction method[5] which gives the solution in a vectorized form for a single system. The price to pay in this case is an increased operation count by a factor of roughly 2.5. For memory to memory machines (Cyber 205) inconveniences arise because vector operations with non unit memory increments have to be performed. The vector length decreases as the calculation proceeds. A performance comparison between the two methods on the Cyber 205 (64 bits, two pipes) can be made with the help of table 1. One can see that already at vector length greater than four the solution across the systems beats the scalar solution of a single system. More than 16 systems at a time is faster than cyclic reduction for a single system. It is also possible to do a multiple cyclic reduction of independent systems by forming one large linear system which is composed of M decoupled block tridiagonal matrices. A comparison of multiple cyclic reduction and multiple solution across the systems shows a better performance for the multiple cyclic reduction case for M less than 60. The implementation of the 'burning at two ends' method would bring the break-even point to smaller M.

Table 1: Multiple solution of M tridiagonal systems (N=512). time for scalar solution: 1800 μs CPU times for 2 pipe Cyber 205 (64 bits)

'Across the Systems'		Multiple Cyclic Reduction	
M	T/System [μs]	M	T/System [μs]
1	5351	1	383
2	2689	2	303
3	1830	3	269
4	1362	4	254
8	708	8	226
16	407	16	212
32	257	32	205
64	196	64	201
128	150		

A recent 3D migration example from Prakla Seismos has (630, 450) gridpoints in *(x,y)*. The number of recorded echo time samples per receiver (gridpoint) is 1250. This results in a data volume of 350 million real numbers. They are stored in 32 bit mode on magnetic disks and have to be read and rewritten several hundred times in the migration process. It is obvious that an efficient I/O organization is crucial for the solution of this huge problem. About 5×10^{42} floating point operations have to be performed. With a user CPU of 20 hours, the average megaflop rate is 50 - 60 on a Cyber 205 (1 Megaword central memory). With only two disk controllers running in parallel the wall clock time is 60 hours, four controllers reduce it to approximately 40 hours.

These results show that seismic data processing not only has a very high demand in CPU power. The I/O capabilities are of equal importance.

THE SOLUTION IN THE FREQUENCE-WAVENUMBER DOMAIN

The case with constant velocity $v(x,y,z) = C$ can be treated in a straightforward way. It is known as Stolt's FK-migration[6]. Introducing the triple Fourier integral of the recorded pressure

$$\tilde{P}(k_x, k_y, z = 0, \omega) = \int \int\limits_{-\infty}^{\infty} \int P(x, y, z = 0, t) \, e^{i(\omega t + k_x x + k_y y)} \, dxdydt$$

the migrated results are given by

$$P(x, y, z, t = 0) = \int \int\limits_{-\infty}^{\infty} \int P'(k_x, k_y, z = 0, k_z) \, e^{-i(k_z z - k_x x + k_y y)} \, dk_x dk_y dk_z$$

$$\text{with } k_z = \sqrt{\frac{\omega^2}{c^2} - k_x^2 - k_y^2}$$

$$P' = \tilde{P}\left(k_x, k_y, z = 0, c\sqrt{k_z^2 + k_x^2 + k_y^2}\right) \quad \frac{ck_z}{\sqrt{k_z^2 + k_x^2 + k_y^2}}$$

This means that the migration consists essentially of six Fourier-transforms. In the discrete representation of time and space samples the numerical problem is now to perform a Fast Fourier Transform (FFT) on a vector computer efficiently. It is well known that the discrete Fourier transform of N data points x_k.

$$z_j = \sum_{k=0}^{N-1} x_k e^{2\pi ijk/N} = \sum_{k=0}^{N-1} x_k W^{kj}$$

with $i^2 = -1$ can be done in N log (N) operations. From the signal flow diagram in Fig. 2 it is easy to see that all butterflies with the same shape and coefficients w are independent. For this special example with N = 8 there are three transformation steps with vector lengths 4, 2 and 1 respectively. However, the transformed results emerge in a permuted order. The final 'unscrambling' step which is done by the bitreversal procedure can not be vectorized and may easily become the bottleneck of the FFT computations on parallel computers.

In the FK migration program of Prakla Seismos a multiple self-sorting FFT program for the more general decomposition

$$N = 2^i \; 3^j \; 4^k \; 5^l$$

has been implemented. Selfsorting means that input and final results are in natural order without an unscrambling step. The price to pay is the double amount of memory compared with the scrambled 'inplace' versions.

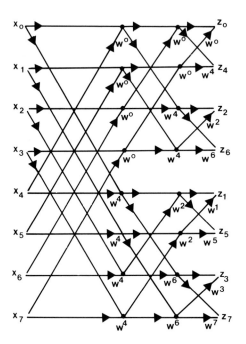

Figure 2. Signal flow diagram for N=8 (radix two case)

For most seismic applications all the transformations are independent of each other, allowing for a further degree of parallelism. On the CRAY a natural parallel implementation would be to compute 64 systems at a time with the constant vector length 64. On the Cyber 205, however, the requirement of unit memory skip increments and the necessity to generate long vectors suggests to use a different approach. The trick of storing M transforms of length N in an interleaved way such that one simulates a single transform of length M x N gives a tremendous increase in vectorlength and speed. At each transformation step the vector length is now <u>multiplied</u> by M compared to the single transform case. See the papers of Schwarztrauber[7] and Temperton[8] for an extensive discussion of all these methods. Table 2 contains the timing for the multiple FFT case with N = 1024 and different M. The CPU speed achieved with M = 100 for the real to complex case is so high that six disk controllers operating simultaneously at full speed are needed just to bring the (32 bit) data to memory to support the CPU. This again demonstrates that there is no balance of CPU and I/O power for this type of application with several hundred million of data on rotating external storage.

Table 2: Performance of multiple FFT program: N = 1024 (complex to complex)

M	T/System [μs]
10	1600
30	770
60	530
100	420

VECTOR MULTIPROCESSING - A NEW DIMENSION

In contrast to the recursive nature of the space-time migration where parallel structures appear only as vectorisable inner loops, the previous example of FK migration can be fully parallelised.

The formulation of the migration procedure in the frequency-space domain is completely equivalent in accuracy to the space-time migration. The recursion in the timesteps is no more present. All frequencies decouple and a vector multiprocessor can distribute the work per frequency over independent processors. Of course, the partial results have to be put together at certain stages of the calculation.

The synchronizable multitasking capability of the two processor CRAY-XMP has been used for a test migration in the frequency-space domain. This, of course, requires a redesign of existing programs. Multiprocessing only pays when the frequencies vary in the outermost loops of the program.

The results obtained with this approach are impressive. The two processor version is approximately 1.9 times faster (throughput) than the single processor run. This demonstrates that vector multiprocessing is a powerful tool to speed up compute intensive applications. For more details see ref. 9.

CONCLUSION

The advent of parallel, especially vector computers, allowed seismic industry to attack also three dimensional migration problems routinely. However, the computational need is much higher than offered by the present generation of computers. One should remember that the one way equation is only an approximation of the full acoustic wave equation. The solution of the elastic wave equation would give the best answer to the question of the exploration industry. To attack this problem one has to wait for computers which are orders of magnitude faster than the present generation. From the discussion of this paper it should also be obvious that much better I/O techniques than rotating devices are needed.

REFERENCES

1.   Loewenthal, D., Lu, L., Robertson, R. and Sherwood J., The wave equation applied to migration, Geophys. Prosp. **24**, 380-399 (1976).

2. Yanenko, N.N., The method of fractional steps (Springer Verlag, Berlin, Heidelberg, New York, 1971).

3. Crank J. and Nicholson P., Proc. Camb. Phil. Soc. **32**, (1947).

4. Url, F., Proceedings of the Cyber 205 user meeting Bochum (1983).

5. Hockney, R. W. and Jesshope, C. R., Parallel Computers (Adam Hilger Ltd., Bristol 1981).

6. Stolt, R., Migration by Fourier Transform, Geophysics **43**, 23-48 (1978).

7. Schwarztrauber, P.N., Vectorizing the FFT's, in: Rodrigne, G. (ed), Parallel Computations (Academic Press 1982).

8. Temperton, C., Selfsorting mixed radix Fast Fourier Transforms, Techn. Mem. No. 66, European Centre for Medium Range Weather Forecasts (1982).

9. Hsiung C., Butscher, W., paper presented at the SIAM Conference on Parallel Processing for Scientic Computing, Nov. 10-11, 1983.